RHABDOVIRUSES

Volume II

Editor

David H. L. Bishop
Professor
Department of Microbiology
University of Alabama in Birmingham
Birmingham, Alabama

CRC Press
Taylor & Francis Group
Boca Raton London New York

CRC Press is an imprint of the
Taylor & Francis Group, an **informa** business

First published 1980 by CRC Press
Taylor & Francis Group
6000 Broken Sound Parkway NW, Suite 300
Boca Raton, FL 33487-2742

Reissued 2018 by CRC Press

Library of Congress Cataloging in Publication Data
Main entry under title:

Rhabdoviruses.

Includes bibliographies and indexes.
1. Rhabdoviruses. I. Bishop, David H. L. [DNLM:
1. Rhabdoviruses. QW168.5.R4 R468]
QR415.R46 576'.6484 79-20575
ISBN 0-8493-5914-7 (v. 2)

A Library of Congress record exists under LC control number: 79020575

ISBN 13: 978-1-315-89729-5 (hbk)
ISBN 13: 978-1-351-07639-5 (ebk)

Visit the Taylor & Francis Web site at http://www.taylorandfrancis.com and the
CRC Press Web site at http://www.crcpress.com

FOREWORD

The Rhabdoviridae is a diverse family of enveloped RNA viruses (comprising 70 or more known serotypes). Member viruses infect homeothermic or poikilothermic vertebrates, invertebrates, or plants. Particular viruses are discussed in Volumes I and III in this series. Although some of the viruses are antigenically related to each other (so far only three genera of rhabdoviruses have been recognized), for many, no antigenic cross reactivities can be detected using the techniques commonly employed to categorize viruses.

A striking feature of the Rhabdoviridae family is the diversity of the host species that are susceptible to virus infection. Some rhabdoviruses are arboviruses while others have no arthropod host or vector (Volume I). Certain members of the Rhabdoviridae family cause severe diseases of man (e.g. rabies, see Volume I), domestic animals (e.g. bovine ephemeral fever virus, Volume III), fish (e.g. spring viremia of carp virus, Volume III), or plants (Volume III). Many other rhabdoviruses appear to be innocuous in their primary host (e.g. sigma virus, Volume III).

Due to their economic and disease importance, a considerable amount of research into the molecular biology of rhabdoviruses has been undertaken in many countries of the world over the past decade. While there are several important gaps in our understanding, what is known about virus structure (Volume I), the viral infection processes, assay systems, growth potential, molecular biology, and genetics (Volume II) of representative viruses are described in this series.

The process of interference of rhabdovirus replication by defective interfering virus particles has been a feature of rhabdoviruses which has received significant attention over the last few years. The subject is still under investigation and although the developments of the last few months are not covered in this series, background information on defective particle generation, replication and interference capabilities are discussed in Volume II. Other features of rhabdovirus infections which are discussed in this series are their ability to kill cells, form pseudotypes, and establish persistent infections (Volume III).

No book series on rhabdoviruses would be complete without a discussion of virus vaccines and other possible therapeutic processes. The advancements made in rabies vaccine development over the past years are described in Volume III. The development of vaccines for other rhabdoviruses will presumably be governed by the importance of the diseases they cause.

The goal of this book series has been to provide an overview of rhabdovirology as a whole (including an appraisal of current research findings), suitable for students, teachers, and research workers. To realize this goal I asked many of the research leaders in the different disciplines of rhabdovirology to contribute chapters. Only a few were not able to particpate due to prior commitments; most of those asked responded with articles which I believe do justice to what is known about the subject.

A final point, the four International Rhabdovirus Symposia that have been held every two years since 1973 have brought together rhabdovirologists from different parts of the world and from different disciplines of the field; the Symposia have been a major factor in the development of the subject and the communication that exists among the research scientists. This book series is therefore dedicated to those who had the foresight to initiate these Symposia, the participants of the meetings, and to the contributors who have given their time and energies to the compilation of this work.

David H. L. Bishop
October 1979

THE EDITOR

David H. L. Bishop, Ph.D., is Professor of Microbiology in the Medical Center of the University of Alabama in Birmingham, Alabama. He is also a Senior Scientist in the Medical School Comprehensive Cancer Center and Diabetes Research and Training Center.

Dr. Bishop was graduated from the University of Liverpool, England, with a B.Sc. (Hons.) degree in Biochemistry in 1959. He received a Ph.D. (Biochemistry) in 1962 also from the University of Liverpool. After a postdoctoral year sponsored by a Research Fellowship at the Centre Nationale de la Recherche Scientifique, Gif-sur-Yvette, France, Dr. Bishop was a Research Associate from 1963 to 1966 in the Department of Zoology, Edinburgh University, a Research Fellow from 1966 to 1969 in the Department of Microbiology, University of Illinois, Assistant Professor (1969 to 1970) then Associate Professor (1970 to 1971) at Columbia University College of Physicians and Surgeons, Department of Human Genetics and Development. Before joining the faculty as a Professor of Microbiology at the Medical Center of the University of Alabama in Birmingham in the fall of 1975, Dr. Bishop was an Associate Professor (1971 to 1975), then Professor (1975) at the Waksman Institue of Microbiology, Rutgers University.

Dr. Bishop has published more than 100 research papers over his career and has been on the Editorial Board of the *Journal of Virology,* since 1974, and *Virology* since 1979. In addition to being active in the American Society of Microbiology and the American Society of Tropical Medicine and Hygiene, Dr. Bishop is Chairman of the Bunyaviridae Study Group of the International Committee for Taxonomy of Viruses as well as a member of its Rhabdoviridae Study Group. He has also been a member of the National Cancer Institute Scientific Review Committee from 1975 to 1979.

CONTRIBUTORS

L. Andrew Ball
Associate Professor
Biophysics Laboratory and
Biochemistry Department
University of Wisconsin
Madison, Wisconsin

Amiya K. Banerjee
Associate Member
Department of Cell Biology
Roche Institute of Molecular Biology
Nutley, New Jersey

Georgine Faulkner
Doctor
Department of Biological Sciences
University of Warwick
Coventry CU4
United Kingdon

Anne Flamand
Research
Laboratoire de Genetique
Universite de Paris Sud
Centre de Orsay
France

Leslye D. Johnson
Staff Fellow
Laboratory of Tumor Virus Genetics
National Cancer Institute
Bethesda, Maryland

C. Yong Kang
Associate Professor of Microbiology
Department of Microbiology
The University of Texas Health Science Center
Southwestern Medical School
Dallas, Texas

Robert A. Lazzarini
Chief, Molecular Viology Section, LMB, NINCDS
Laboratory of Molecular Biology
NINCDS
National Institutes of Health
Bethesda, Maryland

Trudy G. Morrison
Associate Professor
Department of Microbiology
University of Massachusetts Medical School
Worcester, Massachusetts

C. R. Pringle
Member of the Medical Research Council Scientific Staff
Medical Research Council Virology Unit
Institute of Virology
Glasgow, Scotland

M. E. Reichmann
Professor of Microbiology
Department of Microbiology
University of Illinois
Urbana, Illinois

Patricia Repik
Research Associate
Department of Virus Diseases
Walter Reed Army Institute of Research
Walter Reed Army Medical Center
Washington, D.C.

John K. Rose
Assistant Research Professor
The Salk Institute
San Diego, California

William M. Schnitzlein
Research Associate
Department of Microbiology
University of Illinois
Urbana, Illinois

J. F. Szilágyi
Member of the Medical Research Council Scientific Staff
Medical Research Council Virology Unit
Institute of Virology
Glasgow, Scotland

Gail Williams Wertz
Associate Professor
Department of Bacteriology and Immunology
School of Medicine
University of North Carolina
Chapel Hill, North Carolina

TABLE OF CONTENTS

Volume I

TABLE OF CONTENTS

Volume II

TABLE OF CONTENTS

Volume III

Chapter 1

ADSORPTION, PENETRATION, UNCOATING, AND THE IN VIVO mRNA TRANSCRIPTION PROCESS

Patricia Repik*

TABLE OF CONTENTS

* Dr. Repik's affiliation at the time this chapter was written was the Department of Microbiology, University of Illinois, Urbana, Illinois.

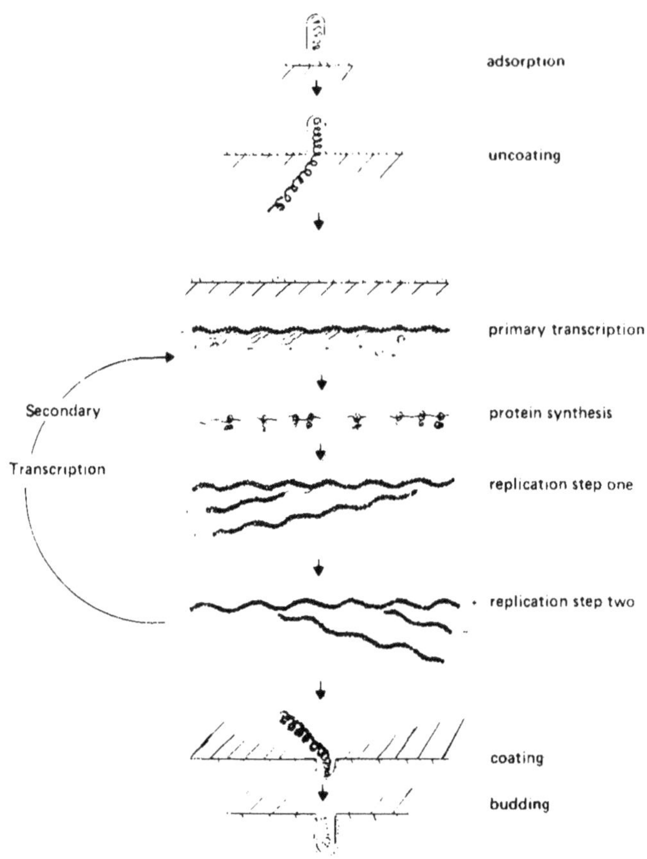

FIGURE 1. Schematic representation of a productive infection of cells by VSV-Indiana. (From Bishop, D. H. L. and Smith, M. S., *The Molecular Biology of Animal Viruses*, Vol. 1, Nayak, D. P., Ed., Marcel Dekker, New York, 1977, 176. With permission.)

I. RHABDOVIRUS REPLICATION IN CELL CULTURE

Although many rhabdoviruses have been isolated from different sources, only VSV-Indiana has been characterized well enough to allow us to draw conclusions concerning its replication cycle. Infection with rhabdoviruses can result either in a productive infection whereby the majority of progeny virions are infectious, or in an abortive infection which results in the majority of progeny virions being defective (noninfectious). The latter type of infection can be caused by the presence of defective interfering particles or by nonpermissive conditions during the replication cycle and will be discussed in greater detail in succeeding chapters. One should be aware that both infectious and defective progeny virions can be produced simultaneously within the same replication cycle.

A productive infection of cells by VSV-Indiana can be divided into several steps as shown schematically in Figure 1. These are

1. Adsorption or attachment of the virus to the cell membrane, followed by penetration into the cell and uncoating of the virion to expose the viral nucleocapsid
2. Transcription of the viral genome into mRNA species by the transcriptase enzymes present in infecting virions (primary transcription)

3. Synthesis of viral proteins (including more transcriptase enzymes)
4. Replication (probably occurring in two steps) of progeny genomes from the infecting parental genome
5. Further rounds of transcription from progeny viral-like RNA (secondary transcription), protein synthesis, and replication
6. Progeny virus assembly and release by budding

A. Adsorption

Attachment of rhabdoviruses to susceptible cell membranes involves an initial association between the viral glycoprotein spikes and some as yet undefined host cell receptor(s). The viral G protein is the sole constituent of viral glycoprotein surface projections (spikes) which stud the surface of the viral envelope. Studies have been conducted which show that the spikes are iodinated by oxidation with lactoperoxidase or chloramine T,[1,2] confirming the fact that most of the G protein is exterior to the envelope. Electron micrographs taken of cells infected with VSV and rabies viruses clearly show that the virions attach to the cell surface via their glycoprotein spikes,[3-6] with attachment noted to occur on all surfaces of the virion — the rounded end, the flat end, or on either side.

The number of adsorbed rhabdoviruses increases linearly with time during the first 30 min of infection at 37°C, and is also able to occur at 4°C.[4,6-8] Vesicular stomatitis virus adsorbs to cells inefficiently and, after adsorption, washed monolayers of infected cells desorb 20 to 40% of their virions within 30 min at 37°C.[7-9] There is also evidence that viruses differ in their efficiencies to adsorb to or desorb from different cell types. By calculating the amount of RNase resistance acquired by ^{3}H-labeled infecting VSV genomes, it was found that VSV adsorbs better to BHK cells than to mouse L cells; adsorption was least efficient on chicken (CEF) cells. By the same token, desorption was found to be less pronounced from BHK or L cells than from CEF cells.[7-10]

1. Infectivity

Since the viral glycoprotein spikes play a key role in the attachment of the virion to the host cell, anything that can alter their conformation or destroy their integrity can affect the process of adsorption and, therefore, the infectivity of the virus.

a. Reduction of Infectivity

Treatment of intact virions (Figure 2) with proteolytic enzymes removes all but a very small fragment of the G protein (fragment mol wt of approximately 7,000 daltons[11,12]) resulting in the production of spikeless virus particles (Figure 3) which retain only a small fraction of the original infectivity.[13-15] It has been found that different proteolytic enzymes have different effects on various rhabdoviruses. Using several strains of VSV, Bussereau et al.[15] studied the effects of trypsin and chymotrypsin on the infectivity, morphology, and antigenic properties. Each enzyme reduced the infectivities of VSV-Indiana and VSV-New Jersey by roughly 10,000-fold, but did not affect the infectivities of VSV-Brazil and Argentina and had minimal effect on the infectivity of Cocal (Table 1). Electron microscopy of the enzyme-treated virions and polyacrylamide gel electrophoresis of their viral polypeptides demonstrated that the enzymes removed all the glycoprotein spikes of VSV-Indiana; however, approximately one third of the spikes of VSV-Brazil were resistant to enzyme action. Similarly, it has been observed that Chandipura retains most of its infectivity after treatment with bromelain,[144] although this same virus loses all its G protein and essentially all its infectivity upon treatment with Pronase® (Table 2).

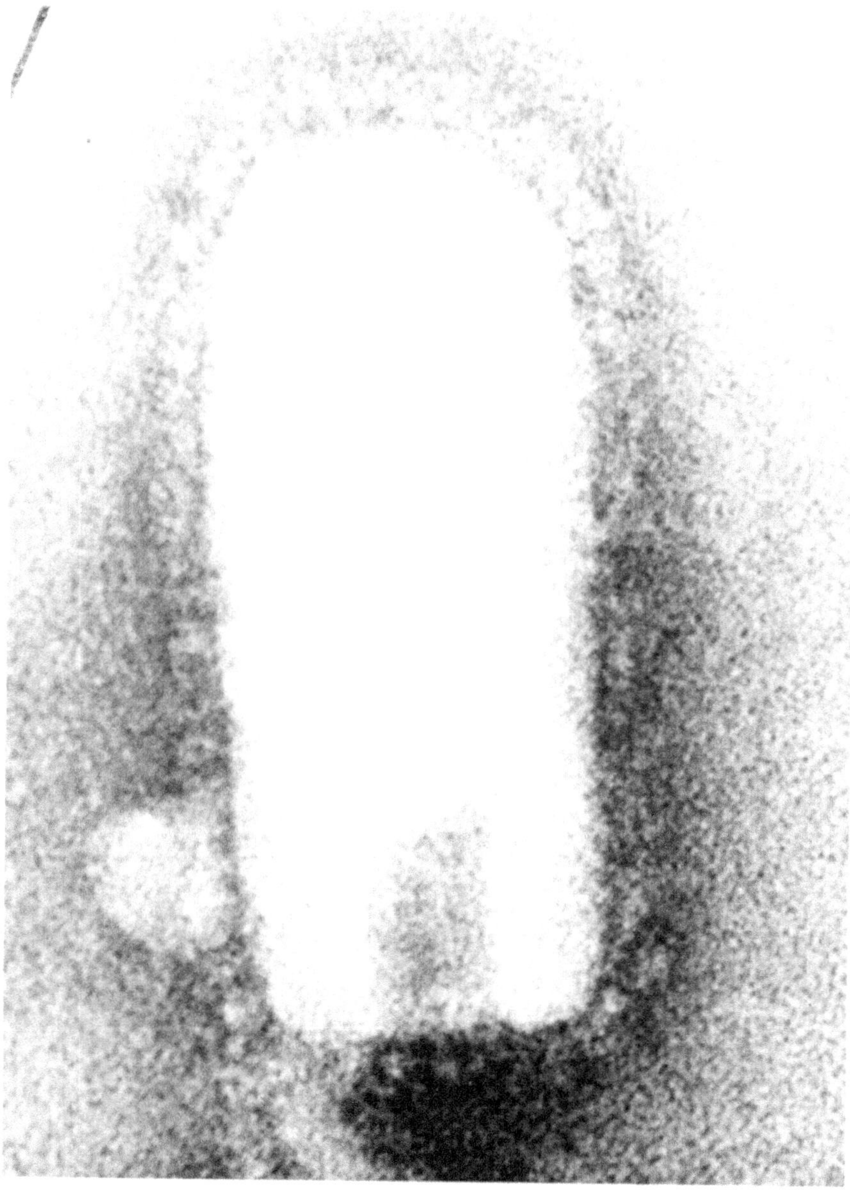

FIGURE 2. Electron micrograph of a VSV-Indiana whole virion negatively stained with sodium phosphotungstate. The particle is approximately 180 nm long, 65 nm wide, and is covered with spikes 10 nm in length. The sample was stained and photographed by Dr. R. W. Compans, University of Alabama Medical Center. Scale: 1 cm = 15 nm.

The infectivities of such spikeless preparations of rhabdoviruses can be restored to varying degrees by homologous and heterologous glycoprotein extracts (Table 2). The glycoprotein extracts contain the viral lipids and appear to be composed of mixed micellar structures (containing both lipids and protein),[16] the lipid portion forming the vesicle and the glycoproteins studding the surface (Figure 4). No infectivity could be detected among any of these extracts. As seen in Table 2, the greatest increases in infectivity resulted from mixing spikeless particles and glycoprotein extracts generated from rhabdoviruses of the VSV subgroup (VSV-Indiana, VSV-New Jersey, or Chan-

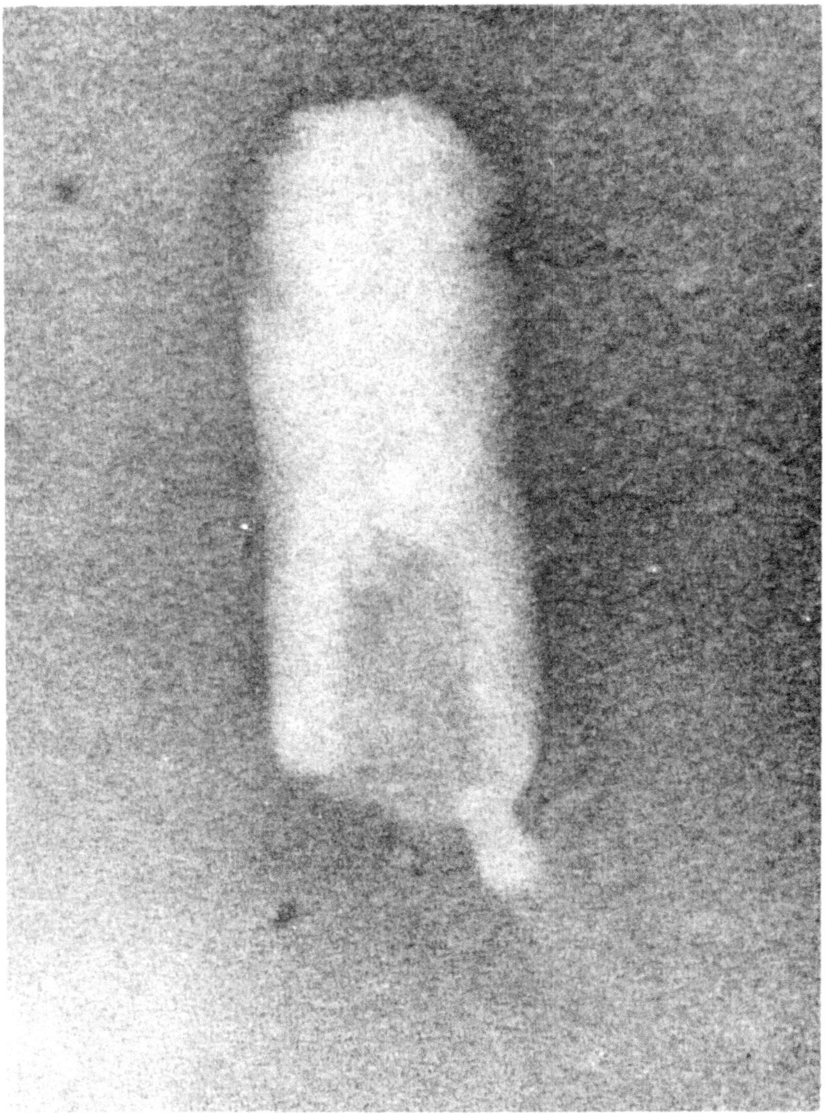

FIGURE 3. Electron micrograph of a VSV-Indiana spikeless particle. Spikeless particles were obtained by incubating purified virus samples in the presence of either bromelain or Pronase® and subsequently resolving them in a glycerol gradient as described by Bishop et al.[14] and Repik.[17] The sample was negatively stained with sodium phosphotungstate and photographed by Dr. R. W. Compans, University of Alabama Medical Center.

dipura spikeless particles plus VSV-Indiana or New Jersey glycoprotein extracts). The fact that the Chandipura glycoprotein extract appeared to be much less efficient in restoring infectivity was most likely due to the very low protein concentration in this particular extract — 0.07 mg protein per mℓ, as compared with approximately 0.3 mg protein per mℓ of most other extracts assayed. The glycoprotein extract of rabies virus strikingly increased the infectivity of VSV-Indiana spikeless particles, although it is serologically unrelated to VSV-Indiana. A substantial increase in infectivity was also observed in the mixture of VSV-Indiana glycoproteins and the spikeless particles of a DNA virus, vaccinia. Where tested, all infectivities obtained were abolished by antiserum prepared against the viruses used to generate glycoprotein extracts.

TABLE 1

Effect of Trypsin and Chymotrypsin on Five Strains of Vesicular Stomatitis Virus

	Infectivity (log ID_{50}/0.03 mℓ)		
Virus strain	Untreated	Trypsin[a] (0.1 mg/mℓ)	Chymotrypsin[a] (0.5 mg/mℓ)
New Jersey	7.9	4.5	3.3
Indiana	8.5	4.1	4.3
Brazil	8.3	8.5	8.3
Argentina	8.5	8.7	8.7
Cocal	7.5	6.1	6.1

[a] The unfractionated viruses were diluted 100-fold in 0.04 *M* phosphate, pH 7.6, and incubated at 37°C for 30 min with 0.1 mg trypsin or 0.5 mg chymotrypsin per mℓ.

From Bussereau, F., Cartwright, B., Doel, T. R., and Brown, F., *J. Gen. Virol.*, 29, 189, 1975. With permission.

"Reconstituted" particles negatively stained with sodium phosphotungstate and subjected to electron microscopy are presented in Figure 5. These particles do not contain the proper (original) arrangement of glycoprotein (compare with Figure 2) and, as such, they are *not* technically "reconstituted." Instead, it appears that the glycoprotein micellar structures randomly associate with the surface of the spikeless particles. The nature of this association is unknown. The infectivities of the reconstructed particles are probably mediated by the glycoprotein micelles which attach the complexes to surface receptors on susceptible cells. How such particles enter the cell (by phagocytosis or by fusion) is not known.

b. Neutralization of Infectivity

The glycoprotein spike of rhabdoviruses is the only viral antigen capable of binding virus neutralizing antibodies and of eliciting their formation.[18,19] Antibodies may inhibit the adsorption or penetration steps in a virus infection in various ways.[20,21] They can react directly with the virions either before or after attachment to host cells and prevent the attached virion from continuing with the normal infection sequence (penetration, etc.). They may also bring about the elution of the virion from the host cell. In some cases, the virion-antibody complexes may be taken into the host cell by phagocytosis although the subsequent steps of penetration are then usually blocked. Antibodies may also react directly with the surface of the host cell to prevent attachment of virus to specific host receptors. In any case, the end result of a virus-antibody complex is the prevention of or a reduction in the number of virions able to undergo replication in the host cell.

Antibody to G protein neutralizes the infectivity of whole virions. Mice injected with either purified VSV G protein or with antiserum specific to VSV G protein were rendered immune to subsequent challenge with infectious VSV virions.[22] Similarly, immunities to rabies and Duvenhage virus (a rabies-like virus) infections were conferred in mice injected with the homologous glycoprotein antiserum,[19,23] although some degree of cross-protection against these two virus infections was also noted.[23]

c. Enhancement of Infectivity

The infectivity of many types of viruses, rhabdoviruses included, is greatly affected

TABLE 2

Restitution of Infectivity to Spikeless Virus Particles by Homologous and Heterologous Viral Components

Glycoprotein extracts	Spikeless particle preparations							
	VSV-Ind	VSV-NJ	Chandipura	Cocal	KCV	PFR	Vaccinia	Influenza
Virus	9×10^{10}	1×10^{9}	1×10^{10}	1.4×10^{10}	7×10^{7}	9×10^{10}	1×10^{9}	5×10^{7}
SP	9×10^{3}	1×10^{4}	2×10^{4}	1.5×10^{4}	1×10^{3}	4×10^{4}	1.4×10^{5}	2×10^{2}
VSV-Ind	2×10^{6}	7×10^{5}	6×10^{6}	1×10^{6}	1×10^{2}	4×10^{5}	1.3×10^{7}	5×10^{3}
	(222)	(70)	(300)	(67)	(0)	(10)	(93)	(25)
" + As	5.7×10^{4}	2×10^{4}	1×10^{5}	7×10^{2}	—	—	4×10^{5}	1×10^{2}
VSV-NJ	4×10^{6}	2×10^{6}	2×10^{8}	2.4×10^{5}	3.5×10^{4}	ND[a]	—	—
	(444)	(200)	(10,000)	(16)	(35)	—	—	—
" + As	2×10^{4}	6×10^{4}	8×10^{5}	3×10^{4}	—	—	—	—
Chandipura	3×10^{5}	1.4×10^{5}	3×10^{5}	1.7×10^{4}	2×10^{2}	1×10^{3}	—	—
	(33)	(14)	(15)	(0)	(0)	(0)	—	—
" + As	2×10^{4}	2×10^{4}	2×10^{3}	2×10^{3}	—	—	—	—
Rabies	2×10^{6}	6×10^{4}	4×10^{5}	5.6×10^{3}	ND[a]	5×10^{3}	—	—
	(222)	(6)	(20)	(0)	—	(0)	—	—
" + As	1×10^{4}	9×10^{3}	4×10^{3}	5×10^{2}	—	—	—	—
PFR	1.5×10^{5}	—	1×10^{5}	—	ND[a]	5.5×10^{4}	—	—
	(17)	—	(5)	—	—	(1)	—	—
" + As	—	—	—	—	—	—	—	—
Piry	ND	3.6×10^{4}	4.3×10^{4}	1.4×10^{4}	5×10^{2}	—	—	—
	—	(4)	(2)	(0)	(0)	—	—	—
" + As	—	3×10^{4}	1.7×10^{4}	5.5×10^{3}	—	—	—	—
Duvenhage	5×10^{4}	4×10^{4}	4×10^{5}	—	ND[a]	ND[a]	—	—
	(6)	(4)	(20)	—	—	—	—	—
" + As	—	—	—	—	—	—	—	—
KCV	5.6×10^{5}	1×10^{5}	1.6×10^{5}	—	ND[a]	ND[a]	—	—
	(40)	(10)	(8)	—	—	—	—	—
" + As	—	—	—	—	—	—	—	—
SVCV	—	—	—	—	ND[a]	5.3×10^{5}	—	—
	—	—	—	—	—	(13)	—	—
" + As	—	—	—	—	—	—	—	—

TABLE 2 (continued)

Restitution of Infectivity to Spikeless Virus Particles by Homologous and Heterologous Viral Components

Note: Spikeless particles and glycoprotein extracts were prepared as described by Bishop et al.[14] and Repik.[17] Incubation of spikeless particles (0.05 mℓ) with glycoprotein extracts (0.10 mℓ) was carried out in the presence of 10 μg of poly-L-ornithine. Mixtures with antisera (As) were incubated 30 min at 25°C before dilution and PFU determination. After adsorption, BHK cell monolayers were washed twice with 5 mℓ of Eagle's medium containing 5% FCS to remove As before the addition of overlay medium. The results of the As-treated samples are presented as the PFU obtainable from 0.15 mℓ of the spikeless particle-glycoprotein extract mixtures. The protein concentrations of the VSV-Ind, VSV-NJ, Rabies, PFR, Piry, and SVCV glycoprotein extracts were determined by Lowry assay to be approximately 0.3 mg/mℓ. Those of the Chandipura, Duvenhage, and KCV extracts were determined to be approximately 0.06 mg/mℓ. The numbers in parentheses are ratios which indicate the increase in infectivity obtained above the spikeless particle value when glycoprotein extracts were mixed with spikeless particles. The other values represent the PFU/sample.

[a] ND, none detected.

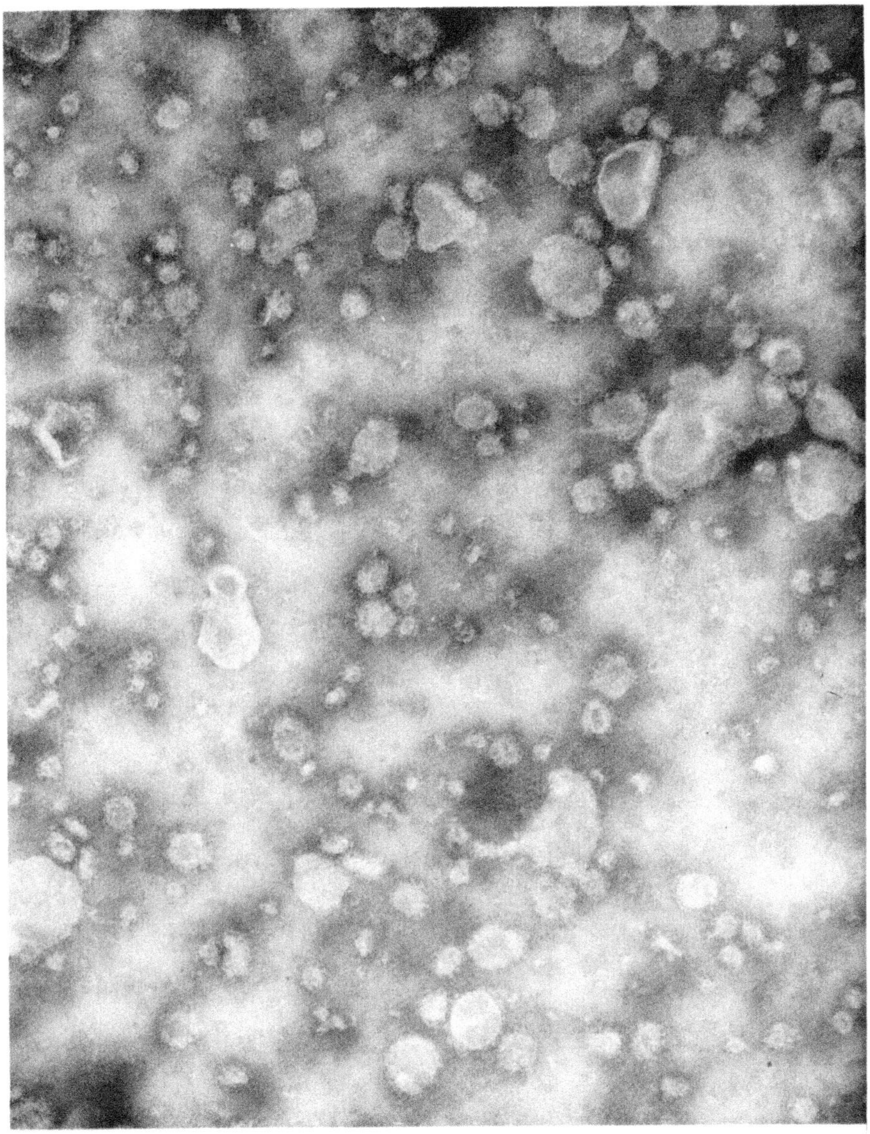

FIGURE 4. Electron micrograph of a VSV-Indiana viral glycoprotein extract. These extracts were obtained by Triton X100 extraction of virus and subsequent removal of the detergent by adsorption to Amberlite® XAD-2 resin as described by Bishop et al.[14] and Repik.[17] The sample was stained with sodium phospotungstate and photographed by Dr. R. W. Compans, University of Alabama Medical Center.

by pretreatment of host cells with polyions at certain times before infection or by the addition of polyions to the virus inoculum.[24-26] Polycations such as DEAE-dextran and protamine sulfate have been shown to markedly increase the adsorption of rabies (Table 3) and vesicular stomatitis viruses[5,7,27,28] as well as that of vesicular stomatitis subviral particles.[29] Although DEAE-dextran did not enhance the infectivity of either VSV spikeless particles or mixtures of VSV spikeless particles and glycoprotein extracts, the addition of another polycation, poly-L-ornithine, to the spikeless particle-glycoprotein mixture was effective in increasing infectivity.[14] On the other hand, polyanions such as heparin and dextran sulfate decrease the infectivity of rabies virus.[27]

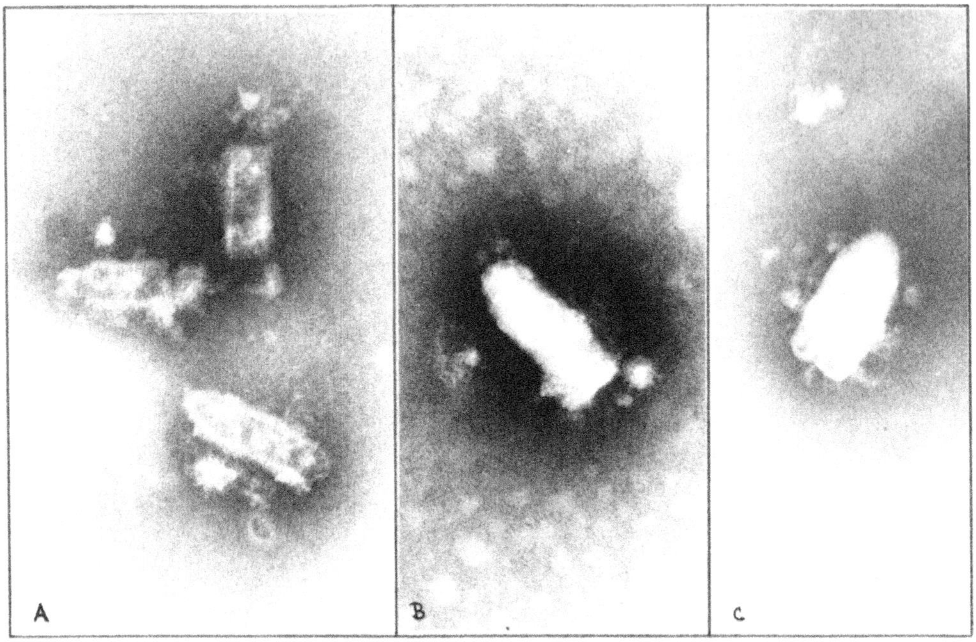

FIGURE 5. A, B, and C. Electron micrographs of reconstructed virus particles composed of VSV-New Jersey spikeless particles and VSV-New Jersey glycoprotein extract. Spikeless particles of VSV-New Jersey were incubated with the homologous glycoprotein extract in the presence of poly-L-ornithine as described by Bishop et al.[14] and Repik.[17] The mixture was dialyzed against 0.01 *M* Tris-HCl, pH 7.4, in order to remove salt. The reconstructed particles were then negatively stained with sodium phosphotungstate and photographed by Dr. R. W. Compans, University of Alabama Medical Center.

TABLE 3

Effect of DEAE dextran (DEAE) on Rabies Virus Infectivity in Various Tissue Culture Systems

	Percent FA[a]			
	24 hr		48 hr	
Cell culture	DEAE absent	DEAE[b] present	DEAE absent	DEAE[b] present
BHK 21	10	100	100	100
Nil 2	5	80	30	100
RE	10	100	30	100
RK 13	45	100	50	100
L-929	< 0.1	6	<0.1	30
$LLC\text{-}MK_2$	10	39	35	53
HDCS	2	45	5	55

[a] Percentage of cells showing presence of rabies fluorescent antigen at given times after infection.

[b] Concentration, 50μg/mℓ.

Data from Kaplan, M. M., Wiktor, T. J., Maes, R. F., Campbell, J. B., and Koprowski, H., *J. Virol.*, 1, 145, 1967. With permission.

How do polyions affect viral infectivity? The enhancing effect of polycations on virus infectivity may be attributed to charge-dependent virus-cell interactions, assuming that these viruses possess negatively charged surfaces and so would adsorb less efficiently under normal conditions to the likewise negatively charged cell surface.[24,30] Polycations would neutralize this electrostatic repulsion. The enhancement might result from the interaction of the polycation with cellular receptors, since DEAE-dextran has been shown to have a marked effect on the biochemistry of cells, and therefore, must interact with them.[31] Such an interaction might alter the spatial arrangement of cellular receptor subunits, decrease the negative charge of the cell surface, or serve as a ligand between the virus and the cell.[21] Similar interactions might also occur between the polycation and viral receptors. Polyions are also known to affect pinocytosis. However, an increase in the rate of pinocytosis was only observed after exposure of cells to anionic molecules; exposure to cationic or neutral molecules had little or no effect.[32] Attachment of polyanions to cellular or viral surfaces should increase the electrostatic repulsion and would account for the decrease obtained in virus yields. Polyanions have also been shown to abolish the virus-enhancing effect of polycations.[24] In view of what is known at the present, it seems most likely that the enhancement of viral infectivity in the presence of polycations is due to an alteration of electrostatic charges on either the virus or cell surface, or both. However, this does not preclude the possibility that other mechanisms might be involved, such as virus-virus interactions.

Structurally, which part of the viral glycoprotein mediates the adsorption of the virion to the host cell? Schloemer and Wagner,[28,33] found that treatment of VSV-Indiana with neuraminidase resulted in a 99% loss of infectivity. This was due to the fact that upon hydrolysis of their terminal sialic acid residues, the viral glycoprotein spikes were unable to attach to L cells. Infectivity was substantially restored by in vitro resialytion of the G protein using a sialyl transferase. Similar experiments were performed with rabies virus with identical results,[34] suggesting that the terminal sialic acid residues of the VSV and rabies virus glycoproteins are necessary in part for initiation of a viral infection. The sialic acid probably is not the sole component of the viral attachment site, since VSV virions treated with trypsin exhibit a 3-log greater decrease in infectivity than do the neuraminidase-treated virions.[28] Recently, however, the role of the viral sialic acid in the initiation of infection has been disputed by Cartwright and Brown,[35] who found that treatment of three VSV serotypes (Indiana, New Jersey and Brazil) with neuraminidase did not result in any significant loss of infectivity. The reason for the discrepancy of these results is unknown, since it cannot be attributed to differences in experimental techniques.

It is feasible that the glycoprotein is not the only viral component capable of interacting with the host cell, since spikeless particles of various rhabdoviruses (and of a myxovirus and a poxvirus) retain a small amount of infectivity which is not inhibited by antiserum prepared against either whole virus or purified glycoprotein (see Table 2).[14] Similarly, VSV nucleocapsids (generated by detergent treatment of whole virions and lacking the viral envelope and G protein) are also infectious.[27]

B. Entry into the Cell and Uncoating of the Virion

Once adsorbed to the cell membrane, the virions penetrate into the host cell. Two methods of entry have been observed in cell culture in vitro as schematically diagramed in Figure 6. Heine and Schnaitman[36,37] presented very convincing evidence that fusion occurs between the VSV viral envelope and the plasma membrane of L cells, followed by release of the nucleoprotein into the cytoplasm. Fusion has also been observed to some extent between the rabies viral envelope and BHK 21 cells.[6] Alternatively, it has been demonstrated by many investigators that whole virions enter the cells mainly by

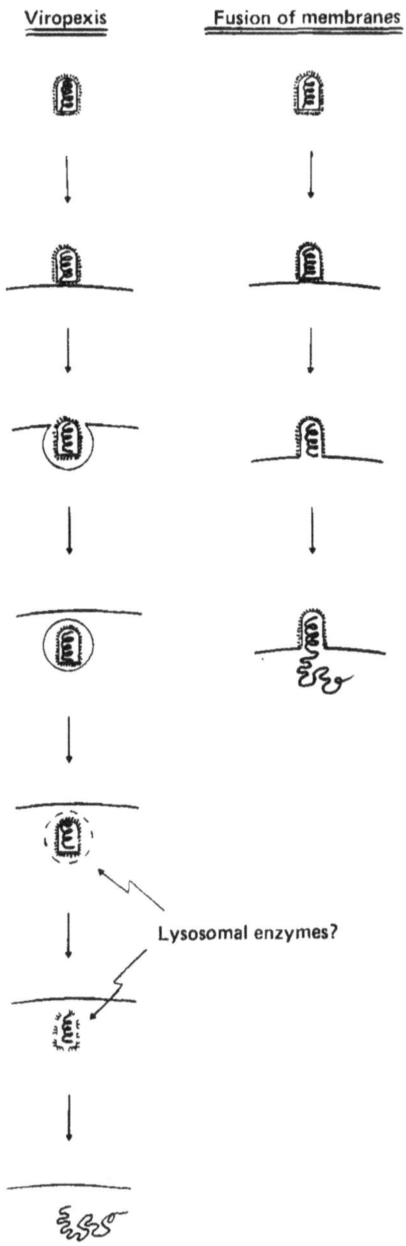

FIGURE 6. Schematic representation of two possible pathways for penetration and partial uncoating of rhabdoviruses. (This scheme is similar to that of Lonberg-Holm, K. and Phillipson, L., *Monogr. Virol.*, 9, 1, 1974. With permission.)

phagocytosis (viropexis).[3-6] Therefore, it seems likely that there is more than one way that a virus can enter the cell, although it is not known if both mechanisms lead to an equally productive infection. However, a recent study by Dahlberg[4] has shown that differences in the handling of cells and virus can partially explain the two different modes of penetration described for VSV. With the cells (L-929), virus (VSV-Indiana), and media used in his experiments, a significant level of fusion was observed only when adsorption was achieved by high speed centrifugation (method used by Heine and Schaitman[36,37]), indicating that centrifugation *per se* played a role in the incidence of fusion. Viropexis was essentially the only means of VSV entry when adsorption was allowed to proceed by incubation of cell monolayers with the virus at 37°C in the absence of centrifugation. This occurred at both high multiplicity of infection (MOI) (5×10^4 particles per cell) and at low MOI (10 to 15 particles per cell). In addition, it was found that serum had no effect on the mode of entry in either centrifuged or uncentrifuged samples. More recently, a study using outbred Swiss mice and VSV-Indiana has indicated that viropexis is probably the most common mechanism of VSV entry into susceptible cells of the central nervous system in vivo.[38]

In vitro, when tissue cultures of BHK and L cells are infected with rabies virus and VSV, respectively, virions (single particles and sometimes multiple particles) become engulfed in phagocytic vesicles. By 60 min postinfection, the vacuoles move deeper into the cytoplasm and degradation of the particles within the vacuoles is observed.[3-6] In vivo studies with VSV have revealed that generally only one virion is present in each phagocytic vacuole, and, once inside the detached vacuoles, virions are degraded and their internal nucleoprotein component is released.[38]

The removal of the membrane of enveloped viruses such as rhabdoviruses may be the most important step in their uncoating process, since the resulting nucleocapsids participate in early macromolecular synthesis without being completely disassembled.[21,29,39,40] How the viral envelope of an engulfed virion is actually removed remains a mystery. It has been suggested that the ingested virion and vacuole may fuse with lysosomes, with the result being that lysosomal enzymes uncoat the virion.[32,41] However, when penetration occurs by fusion of the viral and cellular membranes, the nucleocapsids appear to be released directly into the cellular cytoplasm, thus circumventing the need for lysosomal enzymes to uncoat the virion. This has been demonstrated by finding that the infecting virion G and M proteins (the viral envelope proteins) remain with the cellular plasma membrane when the nucleocapsid N protein enters the cytoplasm.[37]

Although the uncoating process seems to occur at 4°C for pike fry rhabdovirus,[8] it appears tha VSV-Indiana is uncoated only at higher temperatures.[42]

C. In Vivo mRNA Transcription

Infectious virions of rhabdoviruses contain single stranded RNAs with molecular weights in the range of 3.8×10^6 daltons.[43] These RNAs are considered to be "negative-stranded"; that is, they are not capable of acting as messenger RNA species, but in the infected cell their RNA must first be transcribed into viral complementary ("plus strand") mRNA before viral protein synthesis can occur.

Many groups of viruses possess virion nucleic acid polymerases. These include the arenaviruses, bunyaviruses, orthomyxoviruses, paramyxoviruses, reoviruses, rhabdoviruses, poxviruses, and RNA tumor viruses.[42] Except for the RNA tumor viruses, these virion polymerases function to synthesize mRNA which, following uncoating, is the next step to a successful infection.

The synthesis of mRNA by negative-stranded viruses in permissive cells can be distinguished operationally into two phases: primary transcription, which can be observed in cells inhibited for protein synthesis, and secondary transcription, which represents the further synthesis of mRNA in untreated cells.[7,10,44,45]

1. Messenger RNA Species

Messenger RNA species of VSV-Indiana were initially isolated from infected CHO

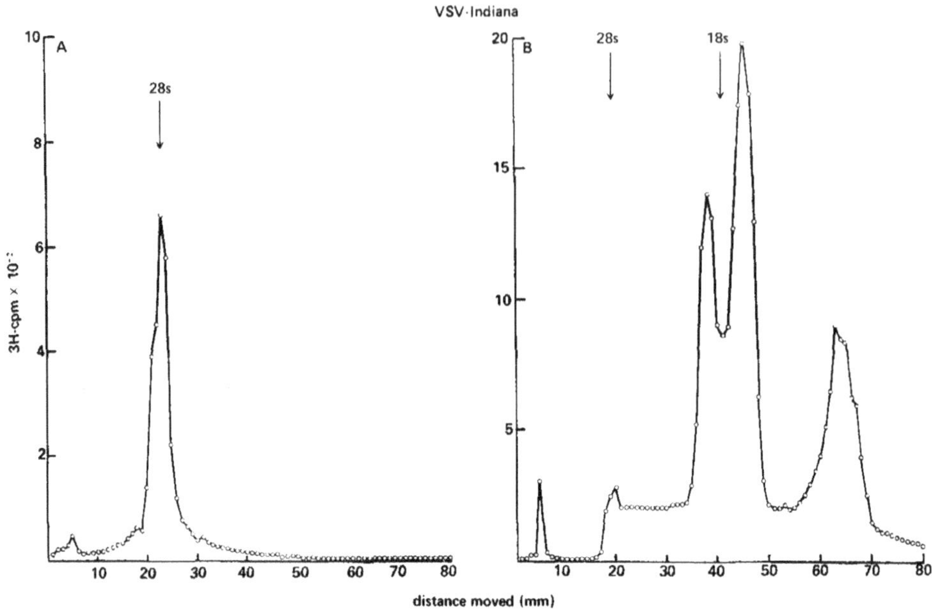

FIGURE 7. Polyacrylamide gel electrophoresis in formamide of the VSV-Indiana mRNA species synthesized in vivo in BHK cells. Each class of poly(A)-containing RNA was isolated and purified as described by Moyer et al.[50] and analyzed by electrophoresis on 4.6% polyacrylamide gels in formamide. The positions of the BHK cell ribosomal RNA markers were determined on parallel gels and are indicated by the arrows. (A) 31 S RNA and (B) 12—18 S RNAs. (Data modified and reprinted from Moyer, S. A., Grubman, M. J., Ehrenfeld, E., and Bannerjee, A. K., *Virology*, 67, 463, 1975. With permission.)

and HeLa cells and consisted of two discrete size classes, one sedimenting at 28 to 30 S and the other sedimenting at 13 to 15 S.[46,47] Upon SDS-polyacrylamide gel electrophoresis, the smaller species could be resolved into two peaks with molecular weights of 0.60 to 0.68 × 10^6 daltons (~ 17 to 19S), and 0.33 × 10^6 daltons (~ 12 to 14S). Schincariol and Howatson[48] found that infection of L cells with VSV-Indiana yielded three major mRNA species with molecular weights of 0.75, 0.59, and 0.35 × 10^6 daltons upon SDS-polyacrylamide gel electrophoresis. Further analyses by many investigators[49-55] have established that four species of mRNAs (31 S, 17 S, 14.5 S, and 12 S) are synthesized intracellularly by VSV-Indiana (Figure 7). Each of these mRNAs possesses a certain degree of secondary structure which is reflected in the variability of its molecular weight depending on the method of analysis (Table 4). These RNA species were unequivocally determined to be message species by their ability to be translated into distinct viral proteins both in vivo and in vitro. The 31 S RNA codes for the L protein,[49,56] while the 17 S and 14.5 S RNA species code for the G and N proteins, respectively.[57,58] The 12 S intracellular RNA species actually consists of two mRNA species with molecular weights of 0.278 × 10^6 and 0.25 × 10^6 daltons,[59] and are, respectively, the mRNAs of the NS and M proteins.

Studies with various other rhabdoviruses have revealed mRNA profiles similar to that of VSV Indiana, although Kern Canyon virus mRNAs appear to be unique (Figure 8; compare with Figure 7). A comparison of the mRNA species found in seven different rhabdoviruses is shown in Table 5, and while they all share similarities, distinct differences can be noted. It should also be pointed out that the mRNAs of Piry, Mokola, and KCV appear to have quite a bit of secondary structure associated with them, since these species migrate quite differently on aqueous SDS as compared with denaturing formamide polyacrylamide gel electrophoresis. Although all these mRNA spe-

TABLE 4

Molecular weight (× 10^{-6}) of VSV-Indiana mRNA species

Method of analysis	VSV-Indiana mRNA species 31 S	17 S	14.5 S	12 S	Cell type	Ref.
Sedimentation						
In sucrose gradients	2.10	0.70	0.52	0.37	—	53
In sucrose gradients with formaldehyde	2.20	0.81	0.55	—	VERO	54,55
PAGE[a]						
In SDS	—	0.68	0.60	0.35	HeLa	47
	1.75	0.75	0.59	0.35	L	48
	2.40	0.66	—	0.30	VERO	55
In 8 *M* urea	—	0.74	0.51	0.22	BHK	50
In formamide	1.85	0.95	0.64	0.37	BHK	50
In formamide-urea	1.20	0.63	0.50	0.29	BHK	52

[a] PAGE, polyacrylamide gel electrophoresis.

cies hybridize with the negative stranded viral genome and many have been shown to contain poly(A) tracts, their authenticity as message species remains contingent upon the demonstration of their ability to direct the synthesis of viral proteins.

Where and in what amounts are the viral mRNAs present in infected cells? Analysis of VSV-Indiana mRNA extracted from membrane-bound polysomes in both HeLa and BHK cells has shown that only the 17 S mRNA species is compartmentalized on such polysomes, and codes for a protein having a molecular weight of 63,000 daltons (most likely a nonglycosylated form of the viral G protein).[58,64] Since the 14.5 S and 12 S species were present in the remaining cytoplasm, it appears that the M, N and NS proteins are synthesized on free cytoplasmic polysomes. The subcellular localization of the 31 S mRNA species has yet to be determined.

The molar ratios of the VSV-Indiana mRNAs were found by Villarreal et al.[52] to be asymmetric, with both the 14.5 S and 12 S RNAs present in approximately 3- to 20-fold greater concentrations than the 31 S and 17 S RNAs respectively (Table 6). These ratios were found to be almost identical to the molar ratios calculated for their individual protein products. These results are in contrast to those reported by Stamminger and Lazzarini[65] who concluded that the intracellular VSV mRNAs may be present in equimolar concentrations.

a. Polyadenylation

All rhabdovirus mRNA species synthesized in vivo[50,51,61,64,66-68] as well as in vitro[50,69-77] contain adenosine bases in poly(A) tracts. RNA species synthesized in vivo contain heterogeneous poly(A) fragments of 64 to 150 nucleotides[50,51,66,67] (comprising 13 to 20% of the adenosines), while those synthesized in vitro contain a single length of poly(A) of approximately 200 nucleotides. The poly(A) tract is covalently linked to the 3′ terminus of the mRNAs.[72,74] In vivo, cordycepin has no effect on the poly(A) content of the viral messages from VSV-infected cells[67,68] and, similarly, cordycepin triphosphate has no effect on poly(A) synthesis in vitro.[72,73] This suggests that the adenylation process is a viral function since cordycepin does inhibit the polyadenylation of cellular mRNA.[78] In support of this view, purified virions of VSV-Indiana have been shown to contain a poly(A)-synthesizing activity,[70] and, more recently, the purified polymerase of VSV-New Jersey was able to catalyze the synthesis of poly(A) tracts which were covalently attached to the 3′ end of the mRNA products.[75] Because the

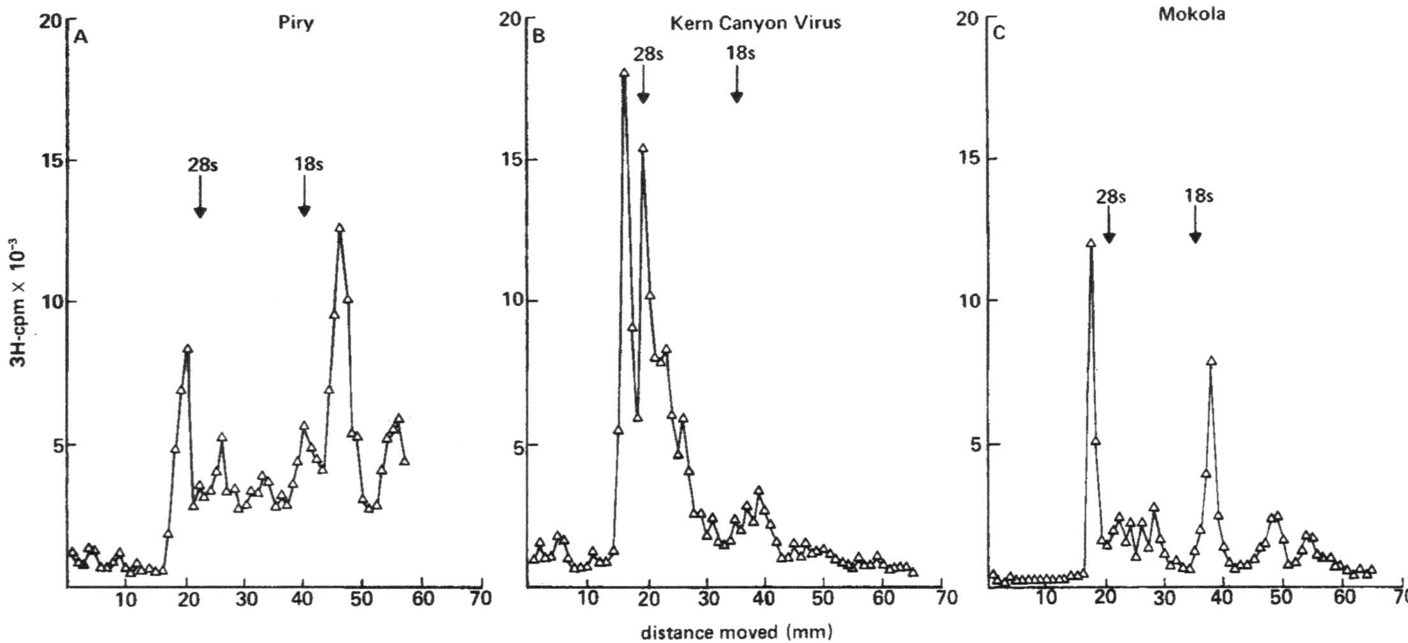

FIGURE 8. Polyacrylamide gel electrophoresis in formamide of the Piry, KCV, and Mokola mRNA species synthesized in vivo in BHK cells. Monolayers of BHK cells pretreated with Actinomycin D were infected with purified viruses at an input MOI of between 50 and 100 PFU/cell. After an adsorption period of 30 min at 4°C, the inoculum virus was removed and fresh prewarmed media containing 5 μg Actinomycin D/mℓ and 20 μCi ^{3}H-uridine/mℓ was added, and the monolayers incubated at 37°C for either 6 hr (Piry), 16 hr (KCV) or 20 hr (Mokola). mRNAs were isolated as described by Repik et al.,[112] but omitting diethylpyrocarbonate and sonication of the extracts. Aliquots of mRNAs in formamide were boiled for 1 to 3 min before electrophoresis (80 V/8 hr) on 3.5% polyacrylamide gels in formanide. ^{14}C-BHK cell ribosomal markers were coelectrophoresed with the viral mNAs in the same gel. (Unpublished data of Repik, P., and Reichmann, M. E., 1978, With permission.)

TABLE 5

A Comparison of the mRNA Species Isolated from Rhabdovirus-infected Cells

Virus	Cell type	System utilized and mRNA species isolated				Ref.
VSV-Indiana	BHK, HeLa, L, Vero	SDS-PAGE,	31 S 17 S 14.5 S 12 S	Formamide-PAGE,	28 S 20 S 17 S 11—12 S	46,47,50,55
VSV-New Jersey	BHK, HeLa, PK15	SDS-PAGE,	30 S 16—17 S 10—13 S			60,61
Piry	BHK	SDS-PAGE,	20 S 16 S	Formamide-PAGE,	30 S 26 S 18 S 16 S 12 S	145
KCV	BHK	SDS-PAGE,	33 S 26 S 21 S 17 S	Formamide-PAGE,	30 S 25—28 S 16—18 S	145
Mokola	BHK	SDS-PAGE,	20 S 16 S 14 S	Formamide-PAGE,	30 S 17 S 12 S 10 S	145
IHN	CHSE-214 (Chinook salmon)	Agarose-PAGE,	28 S 12—17 S			62
Rabies-CVS	BHK	Sedimentation in sucrose gradients	30S (possibly) 12—17 S			63

TABLE 6

Molar Ratios of VSV-Indiana mRNAs

RNA species	Gene product	RNA molar ratio	Protein molar ratio
31 S	L protein	5.6	6.9
17 S	G protein	36	34
14.5 S	N protein	100	100
12 S	M protein	107	53
	NS protein		77,48[a]

[a] The putative mRNA for NS protein does not agree in apparent size with the apparent size of the NS protein. Therefore, a corrected protein molar ratio (48), based upon a molecular weight value (32,000) corrected for mRNA size, is included for comparison with the uncorrected.

Data from Villarreal, L. P., Breindl, M., and Holland, J. J., *Biochemisty*, 15, 1663, 1976. With permission.

viral genome does not contain tracts of polyuridylic acid [poly(U)] corresponding in length to the poly(A) of the messenger species,[67-69] it appears that poly(A) sequences are added post-transcriptionally, either during or after the release of the mRNA from its viral template. The indication is that the transcriptase itself catalyzes the post-transcriptional modification of the mRNA by adding the poly(A) sequences. The molecular mechanism for the switch from transcription to poly(A) synthesis, as well as the function of poly(A) sequences in messenger activity, remains to be determined.

b. Nucleotide Sequences, Capping, and Methylation

Messenger RNAs from both eucaryotic cells and viruses have been found to be modified at the 5′ end by a "cap" — $m^7G(5')ppp(5')X$... — which serves to protect the mRNA at its terminus against attack by nucleases, and functions during the initiation of translation by enhancing the rate of mRNA binding to ribosomes.[79-81] "Capping" refers to the formation of a specific inverted 5′-5′ pyrophosphate linkage between the terminal m^7G and the penultimate nucleotide (Figure 9). The importance of the cap structure for translation in vivo was noted by Rose[82] who observed that all polysome-associated VSV mRNAs were capped, whereas uncapped molecules never associated with polysomes, presumably because they lacked the structure(s) required for ribosome recognition. Translation of the viral proteins was also found to be dependent on methylation of the viral mRNAs, since "capped" but unmethylated VSV mRNAs from infected cells were not able to be translated.[83]

Individual classes of VSV-Indiana mRNAs synthesized in vivo have been reported to contain four alternative 5′ terminal structures (Table 7). The majority of these sequences (85 to 90%) are capped and methylated and exhibit heterogeneity in their degree of methylation.[82,84,85] The relative proportions of the methylated adenosine derivatives were found by Moyer and Banerjee[84] to be dependent upon the cell type (BHK or HeLa) in which the virus was grown, but the mRNAs isolated from the same cell line contained similar methylated nucleosides. The small proportion (10 to 15%) of mRNA not directing viral protein synthesis contained uncapped 5′ termini (pppAp and pppGp) in approximately equimolar amounts.[82] Some of the modifications of the mRNA synthesized in vivo may be directed by enzymes within the virion, since two methyltransferase activities have recently been detected in purified virions of VSV-Indiana.[86] It is plausible that virion methyltransferases catalyze only part of the mRNA

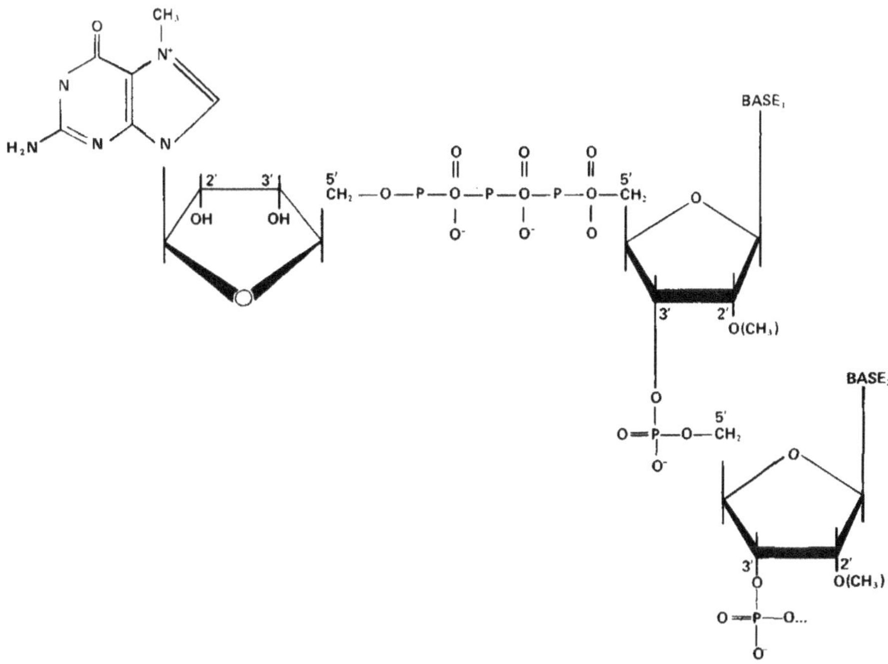

FIGURE 9. Structure of the 5′-terminal cap. (Reprinted from Shatkin, A. J., *Cell*, 9, 645, 1976. With permission.)

TABLE 7

5′ Terminal Structures Found on VSV-Indiana mRNAs[a]

Percentage of mRNA molecules	5′ Terminal structures
65—70%	$m^7G^{5'}ppp^{5'}$AmpAp . . . or $m^7G^{5'}ppp^{5'}$mAmpAp . . .
20%	$m^7G^{5'}ppp^{5'}$AmpmAmpCp . . . or $m^7G^{5'}ppp^{5'}$mAmpmAmpCp . . .
10—15%	pppAp . . . or pppGp . . .

[a] Data from Rose, J, K., *J. Biol. Chem.*, 250, 8098, 1975. With permission.

methylations demonstrated in vivo, since to date only one methylated 5′ terminal sequence has been identified in vitro, $m^7G^{5'}ppp^{5'}$AmpApCpApGp . . . [87,88] Host cell enzymes may be required to further modify the viral mRNAs.

Significant progress in determining the nucleotide sequences of VSV-Indiana mRNAs has been made by Rose[81] who recently reported the complete nucleotide sequences of ribosome recognition sites of the N, NS, and G mRNAs (14.5 S, 12 S, and 17 S mRNAs, respectively). As seen in Figure 10, the sites from the mRNAs each contain a centrally located AUG initiation codon, indicating that they are true initiation sites. The nucleotide sequences for the N and NS sites are identical to each other for 9 of the first 11 bases extending from the capped 5′ end, possibly implying some important function(s) for this region. Although this homologous region was not found in the G mRNA, the 5′ end of this mRNA does contain the sequence

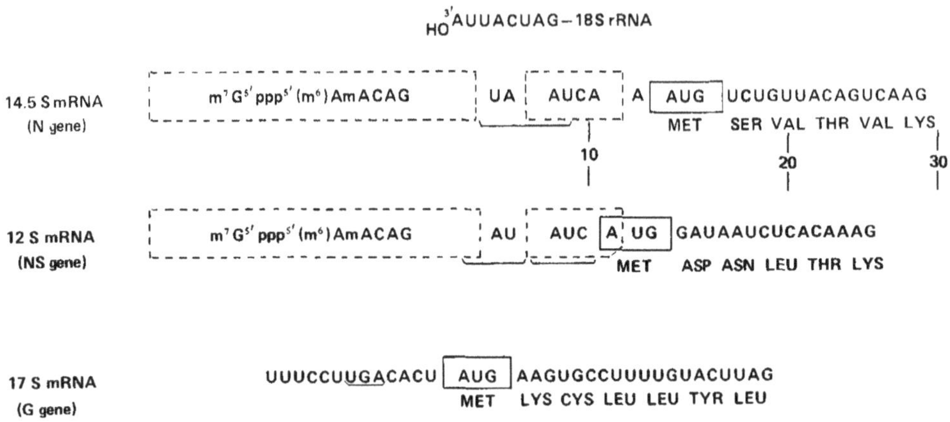

FIGURE 10. Nucleotide sequences of the ribosome binding sites from the VSV-Indiana mRNAs specifying the N, NS, and G proteins. The figure shows regions of homology (dashed boxes) between N and NS genes, and sequences complementary to the 3'-end of 18 S rRNA (brackets under sequences). (Data modified and reprinted from Rose, J. K., *Proc. Nat. Acad. Sci. U.S.A.*, 74, 3672, 1977. With permission.)

m^{7}G$^{5'}$ppp(m)AmpApCp . . . which it shares with the other VSV mRNAs. Thus it was calculated that the initiation codon in the G mRNA is at least 19 bases from the 5' capped terminus. The pyrimidine-rich region preceeding the AUG codon in this 17 S mRNA might be related to its function as membrane-associated mRNA,[58,64] while the region following the AUG codon contains four hydrophobic amino acids (leu-leu-tyr-leu) which could form part of the signal involved in the insertion of the G protein into membranes.[81,89]

The VSV mRNA species (Indiana and New Jersey serotypes) synthesized in vivo and in vitro all contain the same basic 5' terminal sequence G$^{5'}$ppp$^{5'}$ ApApCp . . . [77,82,84,88] which, surprisingly at first, was found not to be complementary to the 3' terminus of the genomic RNAs, . . . pypGpU$_{OH}$.[90-92] Explanation for this phenomenon came about through the work of Colonno and Banerjee,[92-94] who found that a small RNA molecule (approximately 70 nucleotides in length) is synthesized during in vitro transcription and contains the unblocked and unmethylated 5' terminal sequence ppApCpGp.... This sequence is complementary to the 3' terminal sequence of the VSV genome RNA and, since it appears to be involved in the initiation of RNA transcription in vitro, has been designated the "leader" RNA. Such a "leader" sequence has not yet been demonstrated in vivo.

2. Primary Transcription

Initial synthesis of intracellular rhabdovirus mRNA occurs by transcription from the parental genome utilizing the infecting virion polymerase (primary transcription). This synthesis occurs independently of protein synthesis and can be monitored using cycloheximide or puromycin to inhibit cellular and viral protein synthesis, thus preventing the synthesis and activity of new polymerase molecules.[7,44,95,96]

Primary transcription appears to be the main source of viral mRNA during the first 20 to 60 min of infection by VSV-Indiana. It has been found by Flamand and Bishop[7] that normally one out of five adsorbed VSV virions participate in primary transcription, and that the inactive particles remain inactive throughout the infection cycle.[45] For mammalian rhabdoviruses, the optimum temperature for primary in vivo transcription was found to be around 36 to 39°C,[7,44] which is much higher than the optimum (28 to 31°C) found for in vitro synthesis.[97-99] On the other hand, the temperature optimum for both in vivo and in vitro transcription of fish rhabdoviruses is strikingly low, 15 to 25°C,[8,62,100] most likely reflecting the temperature optimum for replication of these viruses in their respective hosts.

The rate, completeness, and extent of primary transcription can be measured by hybridization experiments. The in vivo transcription assay developed by Flamand and Bishop[7] enables one to detect very small amounts of viral complementary mRNA in infected cells. The procedure involves using [^{3}H] nucleoside-labeled virus preparations of high specific activity to infect cells, followed by incubation of the cells in the presence of inhibitors of protein synthesis (puromycin or cycloheximide). The infected-cell nucleic acids are then extracted at various times postinfection and the presence of viral mRNA contained therein is determined in a qualitative assay by self-annealing the extracted nucleic acids and calculating the increase in [^{3}H] RNase resistance. By annealing samples of the infected-cell nucleic acids to an excess of *added* [^{3}H] viral RNA and determining the maximum increase in [^{3}H] RNase resistance, the total quantity of intracellular viral mRNA can be determined in a quantitative assay.[45] Secondary transcription can be measured by applying the same procedures to cells infected with [^{3}H] virus in the absence of protein synthesis inhibitors. Using this technique, it was found that VSV-Indiana primary transcription occurs at a linear rate up to 6 hr postinfectin at 31 to 39.5°C, but is reduced with time at a temperature of 40°C. The transcription process is complete (i.e., the transcripts represent the entire viral genome) by 3 hr postinfection for all rhabdoviruses tested (Table 8). In addition, it was calculated that for VSV-Indiana, 1 genome mass equivalent of viral mRNA is synthesized every 90 sec.[7,45] This same reaction was estimated to be at least 5 times slower for rabies virus (A. Flamand, unpublished results).

Primary transcription has also been demonstrated in a variety of other rhabdoviruses such as two fish viruses, PFR[8] and SVCV,[100] two rabies-subgroup viruses, rabies HEP-Flury[42,101] and Mokola,[17] as well as Kern Canyon virus.[102] Comparable rates and amounts of viral mRNA were detected in cells infected with VSV-Indiana, PFR, or SVCV. A comparison of both the rate and overall accumulation of VSV-Indiana, KCV, and rabies virus mRNAs is shown in Figure 11. Although the overall accumulation of KCV mRNA synthesized during primary transcription is equal to that of VSV, its rate of synthesis is about 3-fold lower, most likely due to the lower specific activity of its polymerase as compared to that of VSV-Indiana.[97] Comparable rates and mRNA accumulation have been observed in cells infected with the two rabies-subgroup viruses (rabies-HEP and Mokola) but both are much lower by comparison to VSV-Indiana. This most likely reflects the low specific activity of the rabies virion transcriptase which has been estimated to be approximately 1/150 that of VSV-Indiana.[99]

The primary transcription process of VSV has been shown to be uninhibited by actinomycin D, cycloheximide, and puromycin,[7,44] while the effect of interferon [or of poly (rI):poly (rC)] upon primary transcription remains a controversial issue.[10,95,103-105] Data presented by Marcus et al.,[95] Marcus and Sekellick,[105] and Manders et al.,[103] indicate that interferon acts mainly by reducing viral primary transcription, presumably by effecting a block in the initiation of transcription.[105] Alternatively, data of Repik et al.,[10] and Baxt et al.,[104] lends itself to the interpretation that interferon does not anfect primary transcription, but instead most likely inhibits viral protein synthesis which is required for secondary transcription and RNA replication. It is true, however, that in certain experiments by the latter two investigators, approximately two fold fewer viral transcripts were recovered in cells treated with both interferon plus cycloheximide, as opposed to cells treated with cycloheximide alone. These results are in complete agreement with those of Marcus and Sekellick.[105] Nonetheless, one should bear in mind that 20 to 40-fold less RNA was transcribed in VSV-infected cells in the

TABLE 8

Completeness of Rhabdovirus Transcription in vivo in the Presence or Absence of Cycloheximide

	Percent ribonuclease resistance after self-annealing				
Time (hr)	VSV-Ind	VSV-NJ	Chandipura	Cocal	Piry
0	15	12	10	9	50
1	90	100	74	97	82
2	100	100	96	100	100
3	100	97	98	100	100
3 + cycloheximide	100	100	100	100	100

Note: Virus transcription in cells in the presence or absence of cycloheximide. BHK cells were incubated for 1 hr at 38°C, some in the presence of 100 μg cycloheximide/mℓ, and were then infected with highly [^{3}H] nucleoside-labeled rhabdoviruses in the presence of 10 μg DEAE-dextran/mℓ. Following adsorption at 4°C for 30 min, the inoculum was discarded and fresh media (± cycloheximide) added to the infected cell monolayers. The monolayers were incubated at 38°C for the specified amount of time, after which the media was discarded and the cellular nucleic acids were extracted as described by Flamand and Bishop.[7] The presence of viral mRNA was determined by self-annealing the extracted nucleic acids (1 to 2 × 10^3 ^{3}H-cpm per sample of 60 μℓ) in sealed capillary tubes for 48 hr at 60°C. After annealing, the samples were expelled into a high salt buffer (0.4 *M*NaCl, 0.01 *M* Tris-HCl, pH 7.4) and subsequently divided into two equal volumes. The samples were precipitated by the addition of cold 5% trichloroacetic acid before or after digestion with pancreatic RNase, and the percent RNase resistance determined.

presence of interferon or cycloheximide as compared to that in untreated control cells.[10] Taken in this light, the 2-fold difference in the amount of RNA synthesized in cycloheximide-treated cells with and without interferon is minimal and the RNA synthesized in the presence of interferon is comparable to that synthesized during primary transcription. Using a coupled transcription and translation system, Ball and White[106] found no difference in the rates of VSV mRNA synthesis or degradation, the size distribution, or in degrees of methylation or polyadenylation of the RNAs synthesized in extracts of control cells and interferon-treated cells. The rate of translation, however, was reduced up to 70% in the interferon-treated cells. From the above observations and from the results of Baxt et al.,[104] (who found that the synthesis of those RNA species [20 S and 42 S] which are completely dependent on viral protein synthesis were preferentially inhibited in interferon-treated cells) and others,[107-110] it appears that interferon has its major effect on viral RNA translation. However, since primary transcription of VSV is inhibited to a small extent,[10,104,105] "it seems prudent to recognize the possibility that the antiviral effects of interferon action, like those on the immune system, are multifaceted."[105]

3. Secondary Transcription

Excellent reviews of the VSV secondary transcription process have recently been published by Bishop and Smith[42] and Bishop.[111] Since little new data concerning secondary transcription has been published since the writing of the above reviews, this section will serve mainly to highlight certain aspects of secondary transcription contained in those reviews.

Secondary transcription of rhabdoviruses occurs in cells treated with actinomycin D, but is inhibited in cells treated with cycloheximide (see Figure 11), puromycin, in-

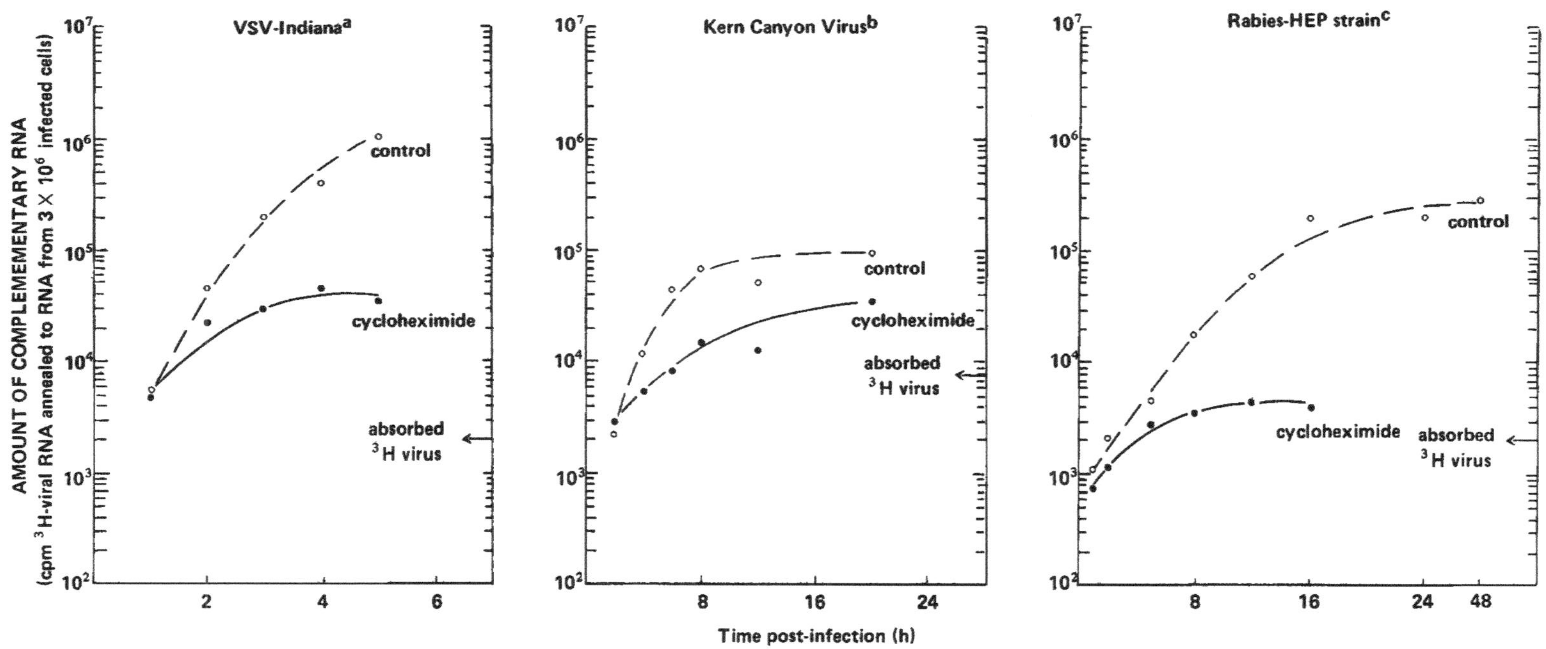

FIGURE 11. Viral mRNA synthesis in BHK cells by VSV-Indiana, KCV, and rabies viruses. The synthesis of mRNA by these viruses in BHK cells in the presence or absence of cycloheximide was determined as described by Flamand and Bishop. (Data modified and reprinted from (a) Bishop, D. H. L. and Flamand, A., *J. Mol. Biol.*, 87, 31, 1974; (b) unpublished data of Repik, P. and Reichmann, M. E., 1977; and (c) Bishop, D. H. L. and Flamand, A., *Control Processes in Virus Multiplication*, Burke, D. C. and Russell, W. C., Eds., Cambridge University Press, 1975, 95. With permission.)

terferon, or poly (rI) poly (rC).[7,10,44,45] These results, plus the fact that no secondary transcription or viral-like RNA is observed in cells infected with *ts* mutants having an RNA-negative phenotype (Table 9) indicate that secondary transcription and the synthesis of viral RNA is dependent on the synthesis of *de novo* viral proteins.

By equating the rates of RNA transcription in cycloheximide-treated cells at different times postinfection to the rate established by VSV-Indiana infecting active virions, Flamand and Bishop[45] have calculated that approximately 450 transcriptive intermediates are present in BHK infected cells by 5 hr postinfection. The greatest increase in transcriptive intermediates was found to occur between 1 and 3 hr postinfection. No increase was observed after 3 hr of infection even though progeny virus continued to be liberated. In amount, the total viral mRNA produced per infected BHK cell was calculated to be about 2×10^4 genome mass equivalent copies of viral mRNA per cell. In similar experiments using a higher MOI, it was found that the initial rate of mRNA synthesis was proportional to the MOI although the total amount of viral mRNA obtained was equivalent to that obtained at the lower MOI.[45] This suggests that there is a finite limit to the amount of mRNA which can be synthesized in vivo.

Experiments utilizing VSV-Indiana and different cell types (BHK, mouse L-cells, and primary or secondary chick embryo fibroblasts) have demonstrated that both the yield of infectious virus and the total amount of viral mRNA are dependent on the host cell type.[10,45,122] VSV appears to replicate most efficiently in BHK cells although the reason for this phenomenon is unknown. The low rate and overall accumulation of both KCV and rabies virus mRNAs (Figure 11) may therefore be a consequence of the host cell type (i.e., infections in BHK cells are only slightly cytopathic) and the fact that infections with these viruses do not result in a significant early inhibition of host cell macromolecular synthesis. The low specific activity of their respective virion polymerases most likely is one major cause of these phenomena.

Secondary transcription of VSV-Indiana in BHK cells starts at a time (1 to 2 hr postinfection) when the first progeny virus and newly synthesized viral-like RNA can be detected,[45] which indicates that the newly synthesized viral-like RNA species are the templates for secondary transcription. These observations also suggest that the viral-like RNA species have the potential of becoming progenitors of infectious virions or templates for further rounds of secondary transcription.

4. *Transcription of Temperature-Sensitive Mutants In Vivo at Nonpermissive Temperatures*

a. ***VSV-Indiana***

The more than 400 VSV-Indiana *ts* mutants have been categorized into five groups by complementation analyses (see review by Pringle[123]). Using a sensitive screening procedure, Rettenmier et al.,[127] has detected a sixth complementation group consisting of rare mutants, the natures of which have not yet been etsablished. Since synthesis of RNA at nonpermissive temperatures is inhibited for mutants of groups I and IV and for some mutants of group II,[114,124] these mutants are known as RNA-negative mutants. Mutants in groups III and V are deficient in assembly and maturation but do program the synthesis of intracellular viral mRNA[45,114,116] and are therefore RNA-positive.

The amount of viral mRNA-synthesized in BHK cells at both permissive and nonpermissive temperatures by certain VSV-Indiana mutants from all five complementation groups has been examined (Table 9).

The group I mutant (GI-114) assayed by Repik et al.,[112] was found to possess a thermolabile transcriptase when assayed in vivo for viral mRNA synthesis at 40°C by comparison to its activity at 31°C. These results agree with those of Szilágyi and Prin-

TABLE 9

Complementation Groups of VSV-Indiana and VSV-New Jersey Temperature-sensitive mutants

Virus	Complementation group	RNA phenotype	Protein affected	Transcription in vivo: Primary	Transcription in vivo: Secondary	Examples	Ref.
VSV-Indiana	I	−	L	−	−	GI—114	7,17,112,113
				+	−	OI—5, GI—16	
	II	−/+	N	+	−	GII—22	112,114,115
				+	+	OII—52	
	III	+	M	+	+	OIII—23, OIII—31	7,45,114,116
	IV	−	(N,NS)?	+	−	OIV—41, OIV—194	45,114,117,118
	V	+	G	+	+	OV—45, OV—110	112,116,119
VSV-New Jersey	A	−	—	+	−	A1, A6	61,120
	B	−	—	−	−	B1, B2	61,120
	C	+	—	+	+	C1, C2	61,120
	D	+	N	+	+	D1	61,120,121
	E	−/+	—	+	−	E1, E3	61,120
				+	+	E2	
	F	−	—	−	−	F1, F2	61,120

gle[125] who demonstrated that this mutant had little transcriptase activity when assayed in vitro at the nonpermissive temperature. Perlman and Huang[126] also found that this mutant was temperature-sensitive at 37°C in both primary and secondary transcription but not in the replication of virion RNA. With other group I mutants, a thermostable transcriptase has been demonstrated. For these mutants, intracellular primary transcription was observed at temperaures nonpermissive for viral production, in amounts essentially similar to those obtained at permissive temperatures.[7,42,45,112,114] However, no secondary transcription was detected in cells infected with these mutants at high temperature. Because some, but not all, *ts* mutants of group I so far investigated are able to perform primary transcription at nonpermissive temperatures, it seems that mutants in this group are pleiotropic in behavior, some having thermolabile polymerase activites and others possessing polymerase activities which, when synthesized at permissive temperatures, are functional (thermostable) at high termperatures.[42] Results obtained by many investigators and summarized in the reviews by Wagner[43] and Pringle[123] agree with this postulate. A thermolabile transcriptase component (e.g., for *ts* GI-114) can be best understood as a functional protein when formed at 31°C, but which is changed into a nonfunctional form when put at a high temperature (37 to 40°C). Translation at the nonpermissive temperature would result only in the nonfunctional form. A *thermostable* transcriptase component (e.g., *ts* OI-5) can best be understood as a functional form of the protein when synthesized at 31°C, which remains in that form when shifted to a nonpermissive temperature. For such mutants, translation at the nonpermissive would yield only a nonfunctional configuration of the protein.

The in vitro reconstitution experiments of Hunt et al.[113] have confirmed that thermosensitivity of the L protein accounts for the defectiveness of the transcriptase of the group I mutants. The in vitro transcriptase activity of three group I *ts* mutants (*ts* GI-13, *ts* GI-16, and *ts* GI-114) restricted at 39°C was restored only by the L-protein fractions derived from wild-type VSV nucleocapsids. Using a sensitive hybridization procedure, it was found by Repik et al.,[112] that the serologically related wild-type VSV-New Jersey virus is able to complement the transcription defect of VSV-Indiana *ts* GI-114 in vivo at 40°C. This complementation is inhibited by puromycin, indicating that newly synthesized proteins from the wild-type virus are needed by the mutant to correct its transcriptional defect. The serologically unrelated Mokola virus is unable to complement the *ts* GI-114 defect even though Mokola does syntheize viral mRNA in infected BHK cells[17] To determine the extent to which VSV-NJ proteins are able to complement the transcriptional defects of other VSV-Indiana mutants, similar assays were performed using wild-type VSV-NJ plus a different VSV-Indiana group I mutant (*ts* GI-16) or a group IV mutant (*ts* OIV-100).[17,112] No complementation of the transcriptional defects of these two mutants was detected at 39.5°C, so it appears that the transcription defect can be corrected only in mutants possessing a thermolabile enzyme.

As seen in Table 9, primary transcription has been demonstrated for mutants of complementation groups II, III, IV, and V at the nonpermissive temperature. Secondary transcription was observed for mutants of groups II, III, and V at the nonpermissive temperature in amounts equal to or slightly less than that at the permissive temperature.

It should be mentioned that one group II mutant (*ts* GII-22) has been classifed as RNA-negative.[125] However, Unger and Reichmann[114] did detect RNA synthesis at the nonpermissive temperature in cells infected with this mutant. Only primary transcription was observed, so that presumably RNA replication, and therefore secondary transcription, does not occur. In vitro transcription assays performed on *ts* GII-22 also demonstrated that it synthesizes RNA at 39°C.[125] Mutants of complementation group II appear to have a lesion in their N protein.[115]

No secondary transcription is detected at the nonpermissive temperature for group IV mutants, and although they have been classified as RNA-negative, they are not strictly so since they are capable of primary transcription.[7,45,112,114] It therefore seems that mutants of this group are defective in a protein needed for a replicative and/or transcriptive function, although no group IV thermolabile transcriptase mutants have yet been identified. It is not yet known which protein is defective in these mutants, although the N or NS are likely candidates.[117,118]

b. VSV-New Jersey

Approximately 50 chemically induced *ts* mutants of the VSV-New Jersey M strain have been classified into six complementation groups designated A, B, C, D, E, and F.[120] Since mutants in groups A, B, and F are unable to induce significant amounts of RNA synthesis, they are classified as RNA-negative. Mutants in groups C and D were determined to be RNA-positive, while two out of three mutants in group E are RNA-positive and the other is RNA-negative (Table 9).

The data of Lesnaw and Reichmann[61] indicate that the *ts* lesions in Groups A(A1) and E(E1) are not in the virion transcriptase, since these mutants were able to carry out primary transcription at the nonpermissive temperature (39.5°C). Temperature-shift experiments suggest that both mutants are defective in replication. Mutants B1 and F1 were inhibited in both primary and secondary transcription at the high temperature and so failed to synthesize any RNA. In temperature-shift experiments, mutant B1 continued to transcribe and replicate following a shift up to the nonpermissive temperature, while mutant F1 remained inhibited from synthesizing any RNA. This data indicates that there is a functional link between transcription and replication. Mutants C1 and D1 were able to carry out both primary and secondary transcription normally at the nonpermissive temperature and so must contain a defect in a structural protein. Wunner and Pringle[121] have shown that mutant D1 contains a lesion in its N protein.

Lesnaw and Reichmann[61] suggested that the complementation groups of VSV-New Jersey can be equated with those of VSV-Indiana, such that groups A, B, and E of VSV-New Jersey correspond to groups IV, I, and II of VSV-Indiana, while groups C and D parallel those of groups III and V. Complementation group F of VSV-NJ appears to be unique and as yet does not have an analogue in the VSV-Indiana groups.

D. Proposed Mechanism of VSV-Indiana Transcription

Prior to 1976, it was not known whether transcription of the individual mRNAs resulted from independent initiations at five different sites on the genome or from a single initiation followed by sequential read-through of the genes. Low doses of ultraviolet light has a major effect on RNA by producing covalent uridine dimers,[128,129] with the effect that transcription (and subsequent gene expression) beyond such dimers is blocked.[130] It stands to reason, therefore, that if the mRNAs are synthesized independently from five different sites, then the sensitivity of each mRNA to UV irradiation should be directly proportional to its size (i.e., the target size for each mRNA species will be directly related to its physical size). However, if the mRNAs are synthesized from a single initiation point, the transcription of a particular mRNA species will depend upon its position in the genome and on the prior transcription of all preceding genes. This approach was originally developed by Hackett and Sauerbier[131,132] to map ribosomal RNA genes and was the approach used by Ball and White[133] and Ball et al.[134] to resolve the initiation site(s) question of VSV transcription. By using a coupled transcription and translation reaction in a cell-free system, these investigators determined that VSV-Indiana transcription initiates at a single point on the genome RNA and proceeds sequentially in the order 3′-N-NS-M-G-L-5′. The same conclusions were reached independently by Abraham and Banerjee[135] using similar techniques, and by Reddy et al.,[136] using various VSV-Indiana defective interfering particles as probes to determine the order of in vitro transcription. Ball[137] subsequently demonstrated that the identical situation occurs in vivo — during primary transcription in VSV-infected mouse L cells, RNA synthesis initiates at a single site on the genome and proceeds in an obligatory order through the genes.

What is the mechanism of VSV transcription? Although the precise transcription process has not yet been established, models of transcription have been proposed by Bishop,[42,111] Banerjee and associates,[138,139] and Ball and associates,[134] involving either independent mRNA initiation[140] or processing of plus stranded RNA to form mRNA species. Such models are discussed in Chapters 12 and 13.

An intact 42 S plus strand copy of the VSV genome is a necessary intermediate in replication and has been demonstrated in vivo.[49,141,142] With the proposed transcription cleavage models,[138,139] such a product would be generated in the absence of cleavage, and would represent a relatively easy switch from transcription to replication. With a model involving a "stop-start" mechanism,[42,111] generation of the 42 S plus strand would be brought about by the inhibition of all initiation events except the onset initiation of the "leader RNA".

ACKNOWLEDGMENTS

I wish to thank Robert Futrelle, Ellen Kordik, and Paula Bay for reading portions of the manuscript and for their helpful suggestions. I also gratefully acknowledge the assistance of Marj Gillett and Roxanne Flowers in the preparation of this manuscript.

The personal research reported in this chapter was supported by PHS grants CA 05234 and CA 09069 from the National Cancer Institute and PHS grant AI 05492 from the National Institute of Allergy and Infectious Diseases.

REFERENCES

1. **Walter, G. and Mudd, J. A.,** Iodination of vesicular stomatitis virus with lactoperoxidase, *Virology,* 52, 574, 1973.
2. **Moore, N. F., Kelley, J. M., and Wagner, R. R.,** Envelope proteins of vesicular stomatitis virions: accessibility to iodination, *Virology,* 61, 292, 1974.
3. **Simpson, R. W., Hauser, R. E., and Dales, S.,** Viropexis of vesicular stomatitis virus by L cells, *Virology,* 37, 285, 1969.
4. **Dahlberg, J. E.,** Quantitative electron microscopic analysis of the penetration of VSV into L cells, *Virology,* 58, 250, 1974.
5. **Hummeler, K., Koprowski, H., and Wiktor, T. J.,** Structure and development of rabies virus in tissue culture, *J. Virol.,* 1, 152, 1967.
6. **Iwasaki, Y., Wiktor, T. J., and Koprowski, H.,** Early events of rabies virus replication in tissue cultures: an electron microscopic study, *Lab. Invest.,* 28, 142, 1973.
7. **Flamand, A. and Bishop, D. H. L.,** Primary *in vivo* transcription of vesicular stomatitis virus and temperature-sensitive mutants of five vesicular stomatitis virus complementation groups, *J. Virol.,* 12, 1238, 1973.
8. **Roy, P., Clark, H F., Madore, H. P., and Bishop, D. H. L.,** RNA polymerase associated with virions of pike fry rhabdovirus, *J. Virol.,* 15, 338, 1975.

9. **Wagner, R. R., Levy, A. H., Snyder, R. M., Ratcliff, G. A., Jr., and Hyatt, D. E.,** Biologic properties of two plaque variants of vesicular stomatitis virus (Indiana serotype), *J. Immunol.*, 91, 112, 1963.
10. **Repik, P., Flamand, A., and Bishop, D. H. L.,** Effect of interferon upon the primary and secondary transcription of vesicular stomatitis and influenza viruses, *J. Virol.*, 14, 1169, 1974.
11. **Mudd, J. A.,** Glycoprotein fragment associated with vesicular stomatitis virus after proteolytic digestion, *Virology*, 62, 573, 1974.
12. **Schloemer, R. H. and Wagner, R. R.,** Association of vesicular stomatitis virus glycoprotein with virion membrane: characterization of the lipophilic tail fragment, *J. Virol.*, 16, 237, 1975.
13. **Cartwright, B., Smale, C. J., and Brown, F.,** Surface structure of vesicular stomatitis virus, *J. Gen. Virol.*, 5, 1, 1969.
14. **Bishop, D. H. L., Repik, P., Obijeski, J. F., Moore, N. F., and Wagner, R. R.,** Restitution of infectivity to spikeless vesicular stomatitis virus by solubilized viral components, *J. Virol.*, 16, 75, 1975.
15. **Bussereau, F., Cartwright, B., Doel, T. R., and Brown, F.,** A comparative study of the surface of different strains of vesicular stomatitis virus, *J. Gen. Virol.*, 29, 189, 1975.
16. **Helenius, A. and Simons, K.,** Solubilization of membranes by detergents, *Biochim. Biophys. Acta*, 415, 29, 1975.
17. **Repik, P.,** Restoration of Glycoproteins and Infectivity to Spikeless Virus Particles, and the *in vivo* Transcription Processes of Animal RNA Viruses, Ph. D. thesis, Rutgers University, 1977.
18. **Kelley, M. J., Emerson, S. U., and Wagner, R. R.,** The glycoprotein of vesicular stomatitis virus is the antigen that gives rise to and reacts with neutralizing antibody, *J. Virol.*, 10, 1231, 1972.
19. **Wiktor, T. J., György, E., Schlumberger, H. D., Sokol, F., and Koprowski, H.,** Antigenic properties of rabies virus components, *J. Immunol.*, 110, 269, 1973.
20. **Dales, S.,** Early events in cell-animal virus interactions, *Bacteriol. Rev.*, 37, 103, 1973.
21. **Lonberg-Holm, K. and Philipson, L.,** Early interactions between animal viruses and cells, *Monogr. Virol.*, 9, 1, 1974.
22. **Dietzschold, B., Schneider, L. G., and Cox, J. H.,** Serological characterization of the three major proteins of vesicular stomatitis virus, *J. Virol.*, 14, 1, 1974.
23. **Tignor, G. H., Murphy, F. A., Clark, H F., Shope, R. E., Madore, P., Bauer, S. P., Buckley, S. M., and Meredith, C. D.,** Duvenhage virus: morphological, biochemical, histopathological and antigenic relationships to the rabies serogroup, *J. Gen. Virol.*, 37, 595, 1977.
24. **Toyoshima, K. and Vogt, P. K.,** Enhancement and inhibition of avian sarcoma viruses by polycations and polyanions, *Virology*, 38, 414, 1969.
25. **Booth, J. C.,** Enhancement by diethylaminoethyl-dextran of the plaque-forming activity of foot-and-mouth disease virus-antibody complexes in pig kidney IB-RS-2 cells, *Arch. Virol.*, 55, 251, 1977.
26. **Hesse, J., Ebbesen, P., and Kristensen, G.,** Correlation between polyion effect on cell susceptibility to *in vitro* infection with murine C-type viruses and polyion effect on some membrane-related functions, *Intervirology*, 9, 1973, 1978.
27. **Kaplan, M. M., Wiktor, T. J., Maes, R. F., Campbell, J. B., and Koprowski, H.,** Effect of polyions on the infectivity of rabies virus in tissue culture: construction of a single-cycle growth curve, *J. Virol.*, 1, 145, 1967.
28. **Schloemer, R. H. and Wagner, R. R.,** Cellular adsorption function of the sialoglycoprotein of vesicular stomatitis virus and its neuraminic acid, *J. Virol.*, 15, 882, 1975.
29. **Cartwright, B., Smale, C. J., and Brown, F.,** Dissection of vesicular stomatitis virus into infective ribonucleoprotein and immunizing components, *J. Gen. Virol.*, 7, 19, 1970.
30. **Collins, M.,** Electrokinetic properties of dissociated chick embryo cells. I. pH-surface charge relationships and the effect of calcium ions, *J. Exp. Zool.*, 163, 23, 1966.
31. **Breindl, M. and Koch, G.,** Competence of suspended Hela cells for infection by inactivated poliovirus particles and by isolated viral RNA, *Virology*, 48, 136, 1972.
32. **Cohn, Z. A. and Parks, E., 1967,** The regulation of pinocytosis in mouse macrophages. II. Factors inducing vesicle formation, *J. Exp. Med.*, 125, 213, 1967.
33. **Schloemer, R. H. and Wagner, R. R.,** Sialoglycoprotein of vesicular stomatitis virus: role of the neuraminic acid in infection, *J. Virol.*, 14, 270, 1974.
34. **Atanasiu, P., Tsiang, H., Perrin, P., and Favre, S.,** Mise en évidence de l'acide sialique sur le virus rabique. Conséquences de sa suppression sur les pouvoirs infectieux et hémagglutinant, *C. R. Acad. Sci. Ser. D.*, 283, 111, 1976.
35. **Cartwright, B. and Brown, F.,** Role of sialic acid in infection with vesicular stomatitis virus, *J. Gen. Virol.*, 35, 197, 1977.
36. **Heine, J. W. and Schnaitman, C. A.,** Fusion of vesicular stomatitis virus with the cytoplasmic membrane of L cells, *J. Virol.*, 3, 619, 1969.

37. **Heine, J. W. and Schnaitman, C. A.**, Entry of vesicular stomatitis virus into L cells, *J. Virol.*, 8, 786, 1971.
38. **Dal Canto, M. C., Rabinowitz, S. G., and Johnson, T. C.**, *In vivo* assembly and maturation of vesicular stomatitis virus, *Lab. Invest.*, 35, 515, 1976.
39. **Bishop, D. H. L., Emerson, S. U., and Flamand, A.**, Reconstitution of infectivity and transcriptase activity of homologous and heterologous viruses: vesicular stomatitis (Indiana serotype), Chandipura, vesicular stomatitis (New Jersey serotype), and Cocal viruses, *J. Virol.*, 14, 139, 1974.
40. **Szilágyi, J. F. and Uryvayev, L.**, Isolation of an infectious ribonucleoprotein from vesicular stomatitis virus containing an active RNA transcriptase, *J. Virol.*, 11, 279, 1973.
41. **Fazekas de St. Groth, S.**, Viropexis: the mechanism of influenza virus infection, *Nature (London)*, 162, 294, 1948.
42. **Bishop, D. H. L. and Smith, M. S.**, Rhabdoviruses, in *The Molecular Biology of Animal Viruses*, Vol. 1, Nayak, D. P., Ed., Marcel Dekker, New York, 1977, 176.
43. **Wagner, R. R.**, Reproduction of rhabdoviruses, in *Comprehensive Virology*, Vol. 4, Fraenkel-Conrat, H. and Wagner, R. R., Eds., Plenum Press, New York, 1975, 1.
44. **Huang, A. S. and Manders, E. K.**, Ribonucleic acid synthesis of vesicular stomatitis virus. IV. Transcription by standard virus in the presence of defective intefering particles, *J. Virol.*, 9, 909, 1972.
45. **Flamand, A. and Bishop, D. H. L.**, *In vivo* synthesis of RNA by vesicular stomatitis virus and its mutants, *J. Mol. Biol.*, 87, 31, 1974.
46. **Huang, A. S., Baltimore, D., and Stampfer, M.**, Ribonucleic acid synthesis of vesicular stomatitis virus. III. Multiple complementary messenger RNA molecules, *Virology*, 42, 946, 1970.
47. **Mudd, J. A. and Summers, D. F.**, Polysomal ribonucleic acid of vesicular stomatitis virus-infected HeLa cells, *Virology*, 42, 958, 1970.
48. **Schincariol, A. L. and Howatson, A. F.**, Replication of vesicular stomatitis virus II. Separation and characterization of virus-specific RNA species, *Virology*, 49, 766, 1972.
49. **Morrison, T. G., Stampfer, M., Lodish, H. F., and Baltimore, D.**, *In vitro* translation of vesicular stomatitis virus messenger RNAs and the existence of a 40S "plus" strand, in *Negative Strand Viruses*, Vol. 1, Mahy, B. W. J. and Barry, R. D., Eds., Academic Press, New York, 1975, chap. 18.
50. **Moyer, S. A., Grubman, M. J., Ehrenfeld, E., and Banerjee, A. K.**, Studies on the *in vivo* and *in vitro* messenger RNA species of vesicular stomatitis virus, *Virology*, 67, 463, 1975.
51. **Rose, J. K. and Knipe, D.**, Nucleotide sequence complexities, molecular weights, and polyadenylic-acid content of the vesicular stomatitis virus messenger RNA species, *J. Virol.*, 15, 994, 1975.
52. **Villarreal, L. P., Breindl, M., and Holland, J. J.**, Determination of molar ratios of vesicular stomatitis virus induced RNA species in baby hamster kidney BHK-21 cells, *Biochemistry*, 15, 1663, 1976.
53. **Rhodes, D. P., Moyer, S. A., and Banerjee, A. K.**, *In vitro* synthesis of methylated messenger RNA by the virion-associated RNA polymerase of vesicular stomatitis virus, *Cell*, 3, 327, 1974.
54. **Schaffer, F. L., Hackett, A. J., and Soergel, M. E.**, Vesicular stomatitis virus autointerference, *Fed. Proc.*, 28, 1867, 1969.
55. **Schaffer, F. L. and Soergel, M. E.**, Molecular weight estimates of vesicular stomatitis virus ribonucleic acids from virions, defective particles, and infected cells, *Arch. Gesamte Virusforsch.*, 39, 203, 1972.
56. **Morrison, T., Stampfer, M., Baltimore, D., and Lodish, H. F.**, Translation of vesicular stomatitis messenger RNA by extracts from mammalian and plant cells, *J. Virol.*, 13, 62, 1974.
57. **Knipe, D., Rose, J. K., and Lodish, H. F.**, Translation of individual species of vesicular stomatitis viral mRNA, *J. Virol.*, 15, 1004, 1975.
58. **Both, G. W., Moyer, S. A., and Banerjee, A. K.**, Translation and identification of the viral mRNA species isolated from subcellular fractions of vesicular stomatitis virus-infected cells, *J. Virol.*, 15, 1012, 1975.
59. **Freeman, G. J., Rose, J. K., Clinton, G. M., and Huang, A. S.**, RNA synthesis of vesicular stomatitis virus. VII. Complete separation of the mRNA's of vesicular stomatitis virus by duplex formation, *J. Virol.*, 21, 1094, 1977.
60. **Perrault, J. and Holland, J. J.**, Absence of transcriptase activity or transcription-inhibiting ability in defective interfering particles of vesicular stomatitis virus, *Virology*, 50, 159, 1972.
61. **Lesnaw, J. A. and Reichmann, M. E.**, RNA synthesis by the temperature-sensitive mutants of vesicular stomatitis virus, New Jersey serotype, *Virology*, 63, 492, 1975.
62. **McAllister, P. E. and Wagner, R. R.**, Virion RNA polymerases of two salmonid rhabdoviruses, *J. Virol.*, 22, 839, 1977.
63. **Ermine, A. and Flamand, A.**, RNA synthesis in BHK-21 cells infected by rabies virus, *Ann. Microbiol. Paris*, 128A, 477, 1977.
64. **Grubman, M. J., Moyer, S. A., Banerjee, A. K., and Ehrenfeld, E.**, Sub-cellular localization of vesicular stomatitis virus messenger RNAs, *Biochem. Biophys. Res. Commun.*, 62, 531, 1975.

65. **Stamminger, G. and Lazzarini, R. A.**, Analysis of the RNA of defective VSV particles, *Cell*, 3, 85, 1974.
66. **Ehrenfeld, E. and Summers, D. F**, Adenylate-rich sequences in vesicular stomatitis virus messenger ribonucleic acid, *J. Virol.*, 10, 683, 1972.
67. **Soria, M. and Huang, A. S.**, Association of polyadenylic acid with messenger ribonucleic acid of vesicular stomatitis virus, *J. Mol. Biol.*, 77, 449, 1973.
68. **Ehrenfeld, E.**, Polyadenylation of vesicular stomatitis virus messenger RNA, *J. Virol.*, 13, 1055, 1974.
69. **Galet, H. and Prevec, L.**, Polyadenylate synthesis by extracts from L cells infected with vesicular stomatitis virus, *Nature London New Biol.*, 243, 200, 1973.
70. **Villarreal, L. P., and Holland, J. J.**, Synthesis of poly(A) *in vitro* by purified virions of vesicular stomatitis virus, *Nature London New Biol.*, 246, 17, 1973.
71. **Banerjee, A. K. and Rhodes, D. P.**, *In vitro* synthesis of RNA that contains polyadenylate by virion-associated RNA polymerase of vesicular stomatitis virus, *Proc. Natl. Acad. Sci. U.S.A.*, 70, 3566, 1973.
72. **Banerjee, A. K., Moyer, S. A., and Rhodes, D. P.**, Studies on the *in vitro* adenylation of RNA by vesicular stomatitis virus, *Virology*, 61, 547, 1974.
73. **Murphy, M. F. and Lazzarini, R. A.**, Synthesis of viral mRNA and polyadenylate by a ribonucleoprotein complex from extracts of VSV-infected cells, *Cell*, 3, 77, 1974.
74. **Reichmann, M. E., Villarreal, L. P., Kohne, D., Lesnaw, J., and Holland, J. J.**, RNA polymerase activity and poly(A) synthesizing activity in defective T particles of vesicular stomatitis virus, *Virology*, 58, 240, 1974.
75. **Naito, S. and Ishihama, A.**, Function and structure of RNA polymerase from vesicular stomatitis virus, *J. Biol. Chem.*, 251, 4307, 1976.
76. **Preston, C. M. and Szilágyi, J. F.**, Cell-free translation of RNA synthesized *in vitro* by a transcribing nucleoprotein complex prepared from purified vesicular stomatitis virus, *J. Virol.*, 21, 1002, 1977.
77. **Franze-Fernandez, M. T. and Banerjee, A. K.**, *In vitro* RNA transcription by the New Jersey serotype of vesicular stomatitis virus. I. Characterization of the mRNA species, *J. Virol.*, 26, 179, 1978.
78. **Darnell, J. E., Philipson, L., Wall, R., and Adesnik, M.**, Polyadenylic acid sequences: role in conversion of nuclear RNA into messenger RNA, *Science*, 174, 507, 1971.
79. **Rottman, F., Shatkin, A. J., and Perry, R. P.**, Sequences containing methylated nucleotides at the 5′ termini of messenger RNAs: possible implications for processing, *Cell*, 3, 197, 1974.
80. **Shatkin, A. J.**, Capping of eucaryotic mRNAs, *Cell*, 9, 645, 1976.
81. **Rose, J. K.**, Nucleotide sequences of ribosome recognition sites in mRNAs of vesicular stomatitis virus, *Proc. Natl. Acad. Sci. U.S.A.*, 74, 3672, 1977.
82. **Rose, J. K.**, Heterogeneous 5′-terminal structures occur on vesicular stomatitis virus mRNAs, *J. Biol. Chem.*, 250, 8098, 1975.
83. **Shatkin, A. J., Banerjee, A. K., Both, G. W., Furuichi, Y., and Muthukrishnan, S.**, Dependence of translation on 5′-terminal methylation of mRNA, *Fed. Proc.*, 35, 2214, 1976.
84. **Moyer, S. A. and Banerjee, A., K.**, *In vivo* methylation of vesicular stomatitis virus and its host-cell messenger RNA species, *Virology*, 70, 339, 1976.
85. **Moyer, S. A., Abraham, G., Adler, R., and Banerjee, A. K.**, Methylated and blocked 5′ termini in vesicular stomatitis virus *in vivo* mRNAs, *Cell*, 5, 59, 1975.
86. **Testa, D. and Banerjee, A. K.**, Two methyltransferase activities in the purified virions of vesicular stomatitis virus, *J. Virol.*, 24, 786, 1977.
87. **Abraham, G., Rhodes, D. P., and Banerjee, A. K.**, The 5′-terminal structure of the methylated mRNA synthesized *in vitro* by vesicular stomatitis virus, *Cell*, 5, 51, 1975.
88. **Rhodes, D. P. and Banerjee, A. K.**, 5′ terminal sequence of vesicular stomatitis virus mRNAs synthesized *in vitro*, *J. Virol.*, 17, 33, 1976.
89. **Blobel, G. and Dobberstein, B.**, Transfer of proteins across membranes. I. Presence of proteolytically processed and unprocessed nascent immunoglobulin light chains on membrane-bound ribosomes of murine myeloma, *J. Cell Biol.*, 67, 835, 1975.
90. **Banerjee, A. K. and Rhodes, D. P.**, 3′-terminal sequence of vesicular stomatitis virus genome RNA, *Biochem. Biophys. Res. Commun.*, 68, 1387, 1976.
91. **Keene, J. D., Rosenberg, M., and Lazzarini, R. A.**, Characterization of the 3′ terminus of RNA isolated from vesicular stomatitis virus and from its defective interfering particles, *Proc. Natl. Acad. Sci. U.S.A.*, 74, 1353, 1977.
92. **Colonno, R. J. and Banerjee, A. K.**, *In vitro* RNA transcription by the New Jersey serotype of vesicular stomatitis virus. II. Characterization of the leader RNA, *J. Virol.*, 26, 188, 1978.
93. **Colonno, R. J. and Banerjee, A. K.**, A unique RNA species involved in initiation of vesicular stomatitis virus RNA transcription *in vitro*, *Cell*, 8, 197, 1976.

94. **Colonno, R. J. and Banerjee, A. K.**, Mapping and initiation studies on the leader RNA of vesicular stomatitis virus, *Virology*, 77, 260, 1977.
95. **Marcus, P. I., Engelhardt, D. L., Hunt, J. M., and Sekellick, M. J.**, Interferon action: inhibition of vesicular stomatitis virus RNA synthesis induced by virion-bound polymerases, *Science*, 174, 593, 1971.
96. **Wertz, G. W. and Levine, M.**, RNA synthesis by vesicular stomatitis virus and a small plaque mutant: effects of cycloheximide, *J. Virol.*, 12, 253, 1973.
97. **Aaslestad, H., G., Clark, H F., Bishop, D. H. L., and Koprowski, H.**, Comparison of the ribonucleic acid polymerases of two rhabdoviruses, Kern Canyon virus and vesicular stomatitis virus, *J. Virol.*, 7, 726, 1971.
98. **Huang, A. S., Baltimore, D., and Bratt, M. A.**, Ribonucleic acid polymerase in virions of Newcastle disease virus: comparison with the vesicular stomatitis virus polymerase, *J. Virol.*, 7, 389, 1971.
99. **Kawai, A.**, Transcriptase activity associated with rabies virion, *J. Virol.*, 24, 826, 1977.
100. **Roy, P. and Clewley, J. P.**, Spring viremia of carp virus RNA and virion associated transcriptase activity, *J. Virol.*, 25, 912, 1978.
101. **Villarreal, L. P. and Holland, J. J.**, Transcribing complexes in cells infected by vesicular stomatitis virus and rabies virus, *J. Virol.*, 14, 441, 1974.
102. **Repik, P.**, unpublished data, 1977.
103. **Manders, E. K., Tilles, J. G., and Huang, A. S.**, Interferon-mediated inhibition of virion-directed transcription, *Virology*, 49, 573, 1972.
104. **Baxt, B., Sonnabend, J. A., and Bablanian, B.**, Effects of interferon on vesicular stomatitis virus transcription and translat ion, *J. Gen. Virol.*, 35, 325, 1977.
105. **Marcus, P. I. and Sekellick, M. J.**, Interferon action. III. The rate of primary transcription of vesicular stomatitis virus is inhibited by interferon action, *J. Gen. Virol.*, 38, 391, 1978.
106. **Ball, L. A. and White, C. N.**, Effect of interferon pretreatment on coupled transcription and translation in cell-free extracts of primary chick embryo cells, *Virology*, 84, 496, 1978.
107. **Falcoff, E., Falcoff, R., Lebleu, B., and Revel, M.**, Correlation between the antiviral effect of interferon treatment and the inhibition of *in vitro* mRNA translation in noninfected L cells, *J. Virol.*, 12, 421, 1973.
108. **Gupta, S. L., Sopori, M. L., and Lengyel, P.**, Inhibition of protein synthesis directed by added viral and cellular messenger RNAs in extracts of interferon-treated Ehrlich ascites tumor cells. Location and dominance of the inhibitor(s), *Biochem. Biophys. Res. Commun.*, 54, 777, 1973.
109. **Samuel, C. E. and Joklik, W. K.**, A protein synthesizing system from interferon-treated cells that discriminates between cellular and viral messenger RNAs, *Virology*, 58, 476, 1974.
110. **Content, J., Lebleu, B., Nudel, U., Zilberstein, A., Berissi, H., and Revel, M.**, Blocks in elongation andnitiation of protein synthesis induced by interferon treatment of mouse L cells, *Eur. J. Biochem.*, 54, 1, 1975.
111. **Bishop, D. H. L.**, Virion polymerases, in *Comprehensive Virology*, Vol. 10, Fraenkel-Conrat, H. and Wagner, R. R., Eds., Plenum Press, New York, 1977, Chap. 3.
112. **Repik, P., Flamand, A., and Bishop, D. H. L.**, Synthesis of RNA by mutants of vesicular stomatitis virus (Indiana serotype) and the ability of wild-type VSV New Jersey to complement the VSV-Indiana *ts*G I-114 transcription defect, *J. Virol.*, 20, 157, 1976.
113. **Hunt, D. M., Emerson, S. U., and Wagner, R. R.**, RNA^- temperature-sensitive mutants of vesicular stomatitis virus: L-protein thermosensitivity accounts for transcriptase restriction of group I mutants, *J. Virol.*, 18, 596, 1976.
114. **Unger, J. T. and Reichmann, M. E.**, RNA synthesis in temperature-sensitive mutants of vesicular stomatitis virus, *J. Virol.*, 12, 570, 1973.
115. **Combard, A., Printz-Ané, C., Martinet, C., and Printz, P.**, Temperature-sensitive defect of vesicular stomatitis virus in complementation group II, *J. Virol.*, 21, 913, 1977.
116. **Lafay, F.**, Envelope proteins of vesicular stomatitis virus: effect of temperature-sensitive mutations in complementation groups III and V, *J. Virol.*, 14, 1220, 1974.
117. **Combard, A., Martinet, C., Printz-Ane, C., Friedmann, A., and Printz, P.**, Transcription and replication of vesicular stomatitis virus: effects of temperature-sensitive mutations in complementation group IV, *J. Virol.*, 13, 922, 1974.
118. **Ngan, J. S. C., Holloway, A. F., and Cormack, D. V.**, Temperature sensitive mutants of vesicular stomatitis virus: comparison of the *in vitro* RNA polymerase defects of group I and group IV mutants, *J. Virol.*, 14, 765, 1974.
119. **Deutsch, V.**, Parental G protein reincorporation by a vesicular stomatitis virus temperature-sensitive mutant of complementation group V at nonpermissive temperature, *Virology*, 69, 607, 1976.
120. **Pringle, C. R., Duncan, I. B., and Stevenson, M.**, Isolation and characterization of temperature-sensitive mutants of vesicular stomatitis virus, New Jersey serotype, *J. Virol.*, 8, 836, 1971.

121. **Wunner, W. H. and Pringle, C. R.,** A temperature-sensitive mutant of vesicular stomatitis virus with two abnormal virion proteins, *J. Gen. Virol.*, 23, 97, 1974.
122. **Wertz, G. W. and Youngner, J. S.,** Interferon production and inhibition of host synthesis in cells infected with vesicular stomatitis virus, *J. Virol.*, 6, 476, 1970.
123. **Pringle, C. R.,** Genetics of rhabdoviruses, in *Comprehensive Virology,* Vol. 9, Fraenkel-Conrat, H. and Wagner, R. R., Eds., Plenum Press, New York, 1977, chap. 6.
124. **Printz-Ané, C., Combard, A., and Martinet, C.,** Study of the transcription and replication of vesicular stomatitis virus by using temperature-sensitive mutants, *J. Virol.*, 10, 889, 1972.
125. **Szilágyi, J. F. and Pringle, C. R.,** Effect of temperature-sensitive mutations on the virion-associated RNA transcriptase of vesicular stomatitis virus, *J. Mol. Biol.*, 71, 281, 1972.
126. **Perlman, S. M. and Huang, A. S.,** Ribonucleic acid synthesis of vesicular stomatitis virus. V. Interactions between transcription and replication, *J. Virol.*, 12, 1395, 1973.
127. **Rettenmier, C. W., Dumont, R., and Baltimore, D.,** Screening procedure for the complementation-dependent mutants of vesicular stomatitis virus, *J. Virol.*, 15, 41, 1975.
128. **Wacker, A.,** Molecular mechanisms of radiation effects, in *Prog. Nucleic Acid Res. Mol. Biol.*, Vol. 1, Davidson, J. N. and Cohn, W. E., Eds., Academic Press, New York, 1963, 369.
129. **Miller, R. L. and Plagemann, P. G. W.,** Effect of ultraviolet light on mengovirus: formation of uracil dimers, instability and degradation of capsid, and covalent linkage of protein to viral RNA, *J. Virol.*, 13, 729, 1974.
130. **Michalke, H. and Bremer, H.,** RNA synthesis in *Escherichia coli* after irradiation with ultraviolet light, *J. Mol. Biol.*, 41, 1, 1969.
131. **Hackett, P. B. and Sauerbier, W.,** Radiological mapping of the ribosomal RNA transcription unit in *E. coli, Nature,* 251, 639, 1974.
132. **Hackett, P. B. and Sauerbier, W.,** The transcriptional organization of the ribosomal RNA genes in mouse L-cells, *J. Mol. Biol.*, 91, 235, 1975.
133. **Ball, L. A. and White, C.N.,** Order of transcription of genes of vesicular stomatitis virus, *Proc. Natl. Acad. Sci. U.S.A.*, 73, 442, 1976.
134. **Ball, L. A., White, C. N., and Collins, P. L.,** A transcriptional map of vesicular stomatitis virus, in *Animal Virology,* Vol. 4, ICN-UCLA Symp. Molecular Cellular Biol., Baltimore, D., Huang, A. S. and Fox, C. F., Eds., Academic Press, New York, 1976, chap. 27.
135. **Abraham, G. and Banerjee, A. K.,** Sequential transcription of the genes of vesicular stomatitis virus, *Proc. Natl. Acad. Sci. U.S.A.*, 73, 1504, 1976.
136. **Reddy, D. V. R., Schnitzlein, W.M., and Reichmann, M. E.,** Use of defective interfering particle RNA probes in the determination of the order of *in vitro* transcription of vesicular stomatitis virus genes, *J. Virol.*, 21, 432, 1977.
137. **Ball, L. A.,** Transcriptional mapping of vesicular stomatitis virus *in vivo, J. Virol.*, 21, 411, 1977.
138. **Colonno, R. J., Abraham, G., and Banerjee, A. K.,** Blocked and unblocked 5′-termini in vesicular stomatitis virus product RNA *in vitro:* their possible role in mRNA biosynthesis, in *Prog. Nucleic Acid Res. Mol. Biol.*, 19, 83, 1976.
139. **Banerjee, A. K., Abraham, G., and Colonno, R. J.,** Vesicular stomatitis virus: mode of transcription, *J. Gen. Virol.*, 34, 1, 1977.
140. **Hefti, E. and Bishop, D. H. L.,** The 5′ sequence of VSV *in vitro* transcription product RNA (± SAM), *Biochem. Biophys. Res. Commun.*, 68, 393, 1976.
141. **Chang, S. H., Hefti, E., Obijeski, J. F., and Bishop, D. H. L.,** RNA transcription by the virion polymerases of five rhabdoviruses, *J. Virol.*, 13, 652, 1974.
142. **Soria, M., Little, S. P., and Huang, A. S.,** Characterization of vesicular stomatitis virus nucleocapsids. I. Complementary 40 S RNA molecules in nucleocapsids, *Virology,* 61, 270, 1974.
143. **Bishop, D. H. L. and Flamand, A.,** Transcription processes of animal RNA viruses, in *Control Processes in Virus Multiplication,* Burke, D. C. and Russell, W. C., Eds., Cambridge University Press, 1975, 95.
144. **Repik, P. and Bishop, D. H. L .,** unpublished observations, 1975.
145. **Repik, P. and Reichmann, M. E.,** unpublished data, 1978.

Chapter 2

THE IN VITRO mRNA TRANSCRIPTION PROCESS

Amiya K. Banerjee

TABLE OF CONTENTS

I. INTRODUCTION

Consistent with the observation that the genome RNA of purified vesicular stomatitis virus (VSV) is noninfectious,[1] the presence of an RNA-dependent RNA polymerase associated with the virions was discovered.[2] This activity is elicited when the virions are disrupted with a nonionic detergent and following incubation with ribonucleoside triphosphates, RNA is synthesized in vitro that is complementary to the viral genome.[2,3] Thus VSV, like other viruses that contain virion RNA polymerases, initiate infection in the host cells by first transcribing its genome into mRNA using its resident RNA polymerase. Subsequent events involve translation of these mRNAs for the eventual replication of the virus. As discussed in Chapter 10, the RNA synthesizing activity in the purified virions of VSV is carried out by the L and NS proteins using RNA-N protein as the template-bound complex.[4] The precise functions of each protein in the transcriptive processes are still unclear. Since the genome of VSV is a linear single-stranded RNA of negative polarity and the in vitro mRNAs consist of several species of different sizes, the VSV system is a model for studying the processes involved in RNA transcription and replication in vitro. Moreover, it is an excellent system to explore the intricacies of RNA-protein interactions.

In the present chapter, mRNA synthesis by the Indiana Serotype of VSV will be discussed since this serotype has been extensively used for transcription studies in vitro. References to other serotypes will be made whenever necessary.

II. IN VITRO mRNA SYNTHESIS

A. Reaction Conditions

The in vitro RNA synthesis by VSV has been studied using the following transcribing complexes: (1) triton-disrupted purified VSV,[3,5] (2) ribonucleoprotein particle purified from virus,[6] (3) ribonucleoprotein particle isolated from infected cells,[7] and (4) purified L and NS proteins, using the N protein-RNA complex as the template.[8,9]

A typical reaction condition for mRNA synthesis by triton-disrupted purified VSV in vitro is as follows:[5] A 0.2 m*ℓ* incubation mixture contains 0.1 *M* NaCl, 0.05 *M* Tris-HCl(pH 8.0), 5 m*M* $MgCl_2$, 4 m*M* dithiothreitol, 1 m*M* ATP, GTP, and CTP, 0.1 m*M* UTP, labeled UTP, 0.05% Triton N101®, and 20 to 50 μg of purified VSV. The temperature optimum for the reaction is 30°C[10] and RNA synthesis continues linearly for hours. RNA synthesis decreases with increase in temperature but all size classes of mRNA species are synthesized at 37°C and 40°C.[11] A specific activity as high as 30 nmol of GMP incorporation/hr/mg of protein has been reported.[12] The presence of salt is essential although Na^+ can be replaced by other monovalent cations. The optimum salt concentrations appear to be similar for different rhabdoviruses.[13] The reaction is also dependent on the presence of reducing agents and a nonionic detergent is required if the intact virus is used for RNA synthesis. Although the requirement for UTP is one tenth of the other three ribonucleoside triphosphates, recent experiments show that the concentrations of GTP, CTP, and UTP can be reduced to 0.02 m*M* without any significant loss in the rate of the RNA synthesis.[14] However, for maximal RNA synthesis, the concentration of ATP cannot be decreased below 0.5 m*M*.[14] These results indicate that the polymerase has a Km for ATP significantly higher than for the other triphosphates apparently due to a requirement of ATP for initiation of RNA synthesis. (See conclusion).

Inhibition of in vitro RNA synthesis was demonstrated by some nonspecific reagents such as aurintricaboxylic acid and polyethelene sulfonate.[15] These reagents appear to act at or near the transcription initiation site. Recently, antibodies prepared against

NS protein and N protein-RNA complex were shown to inhibit VSV transcription in vitro.[16] In vitro RNA trascription is inhibited at concentrations of virions higher than 200 μg of virus protein/mℓ. The putative inhibitor that is present within the virions seems to be an internal viral protein which shows some degree of serotype specificity.[17] Stimulation of VSV RNA synthesis in vitro by a variety of mammalian and plant cell-free extracts has been shown recently.[18—20] The precise component(s) responsible for this effect on transcriptase and its mode of action are still unclear.

B. Nature of the RNA Products

The mRNA products synthesized in vitro by purified VSV fall into four distinct size classes when analyzed by velocity sedimentation:[5] 31S, 17S, 14.5S, and 12S (Figure 1). By competition annealing experiments, these mRNA species were shown to be identical to the corresponding mRNA species isolated from infected cells.[5,21] Each of the mRNA species has been identified by translation into virus-specific proteins in in vitro protein synthesizing systems.[22] Such analyses indicate that the 17S mRNA codes for the G protein, the 14.5S mRNA codes for N protein, and the 12S mRNA, which contains two mRNA species of very similar molecular weight[22,23] codes for both the M and NS proteins. To date, the 31S mRNA synthesized in vitro has not been translated in vitro. However, the molecular weight of this mRNA is compatible with its serving as the template for the L protein and the corresponding mRNA species isolated from infected cells code for the L protein in vitro.[24] Recently the five distinct VSV mRNA species were resolved by gel electrophoresis after hybridization to the viral genome RNA followed by RNase T_2 digestion[25,26] (Figure 2). The molecular weights of the individual mRNA species are shown in Table 1. On the basis of the total molecular weights of the several species of VSV mRNA, it seems that almost the entire viral genome is transcribed in vitro. In fact, VSV genome RNA hybridized with 12 to 18S mRNA or with 31S RNA is protected against RNase digestion to the extent of approximately 45%.[5] Thus almost 90% of the genome RNA can be accounted for by the mRNA species synthesized in vitro. Whether the remaining portion of the genome RNA is untranscribed or is a template for transcription of a hitherto undiscoverd mRNA is still unclear.

One striking feature of the in vitro RNA synthesis is that the individual mRNA species are consistently synthesized with unequal frequencies in the order N > M + NS > G > L (see Figure 1). These results indicate that there may be a transcriptional control operative under these in vitro conditions. Interestingly, the synthesis of in vivo mRNA species also falls in the same order.[27] In fact, it was later shown that the initiation of RNA synthesis occurs at a single site at the 3′-terminus of the genome RNA and the individual mRNAs are formed sequentially. (See below).

III. MODIFICATION OF mRNAs IN VITRO

A. Polyadenylation

mRNAs synthesized in vitro by purified VSV undergo several modifications, one of which is polyadenylation. Polyadenylic acid, or Poly(A), consisting of approximately 200 bases is added to the 3′-termini of the RNA chains during transcription, as also found in the viral mRNAs isolated from infected cells.[28-32] The polyadenylation process is strictly transcription-dependent since no Poly(A)-polymerase activity could be detected in the absence of transcription.[33] Cordycepin triphosphate (2′-deoxy adenosine 5′-triphosphate) did not inhibit polyadenylation or RNA synthesis or replace ATP in the RNA synthesis reaction.[33] Pulse-chase experiments indicated that Poly(A) sequences are added to the mRNAs during and after their release from the genome RNA

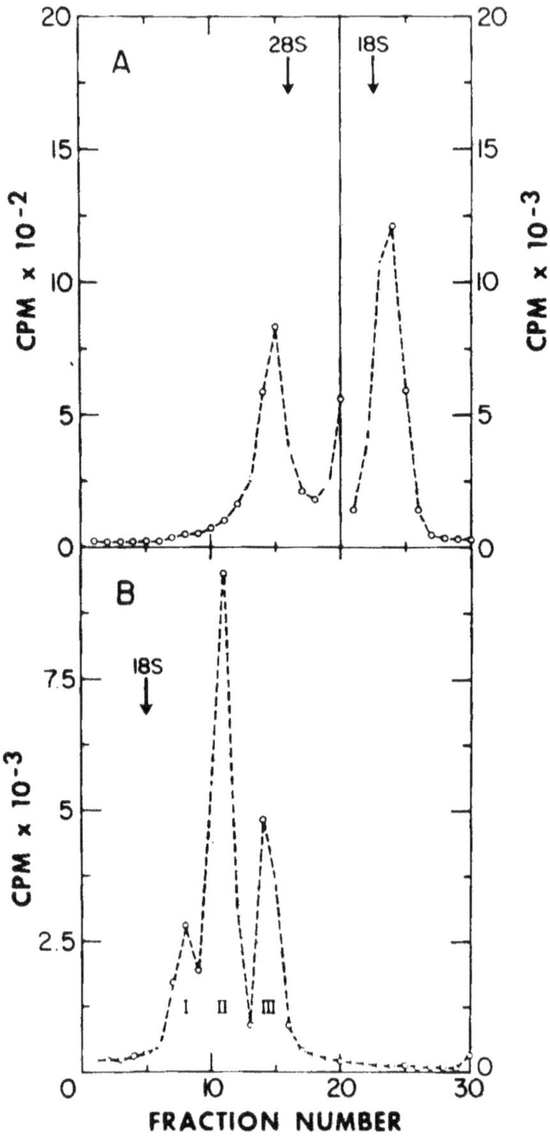

FIGURE 1. Velocity sedimentation analysis of the in vitro VSV product RNAs. RNA was synthesized in vitro with α-^{32}P UTP.[63] At the end of 2 hr synthesis at 30°C, the reaction mixture was adjusted to 0.5% SDS and the RNA purified by extraction with phenol, denatured with $(CH_3)_2SO$, followed by binding to and elution from oligo(dT)-cellulose column. Equal aliquotes of purified RNA were sedimented through 12 ml linear sucrose gradients. The gradients were centrifuged in a spinco SW41 rotor at 23°C for 17 hr at 20,500 rpm(A) or at 33,000 rpm(B). Fractions were collected from a hole pierced at the bottom of the tube, and acid-precipitable radioactivity in each fraction was determined. The $S_{20}w$ for peaks I, II, III, are 17S, 14.5S, and 12S, respectively.

template. In the presence of S-adenosylhomocysteine — an inhibitor of methyltransferase — the mRNA species synthesized in vitro contain giant heterogeneous polyaden-

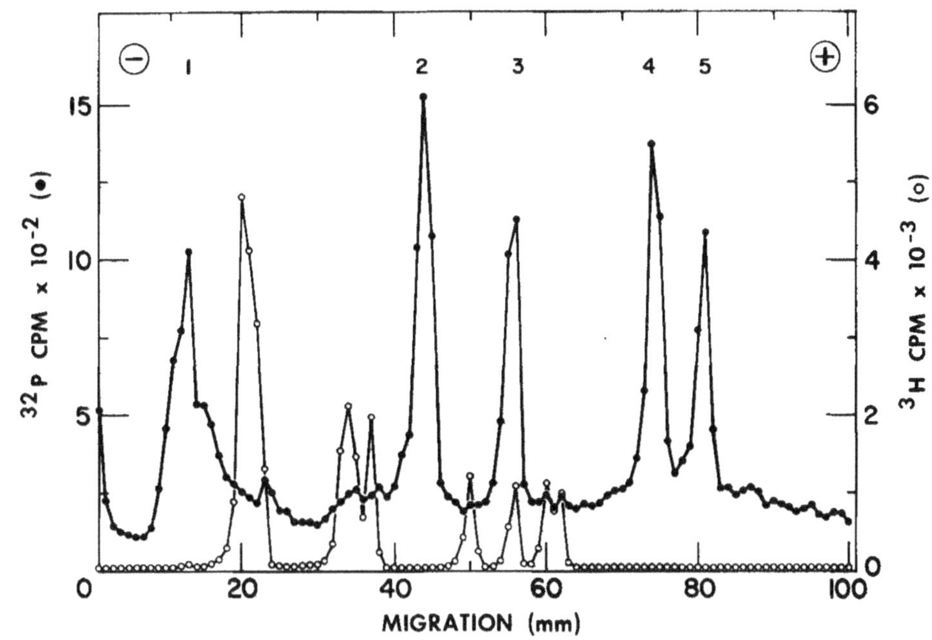

FIGURE 2. Separation of RNA duplexes derived from five VSV mRNA species. VSV mRNAs were synthesized in vitro using α-^{32}P CTP as the labeled precursor, and the product was purified by oligo(dT)-cellulose column, chromatography and SDS-sucrose gradient sedimentation to isolate the four size classes of Poly(A) containing mRNA.[25] Approximately 10^4 cpm each of the 31S, 17S, 14.5S, and 2×10^4 cpm of ^{3}H-uridine-labeled reovirus double-stranded RNA, the mixture was digested with RNAase T_1 and analyzed by gel electrophoresis. •——•, P^{32}-labeled VSV RNA duplexes; O——O, ^{3}H labeled reovirus double-stranded RNA.[25] (From Rhodes, D. P., Abraham, G., Colonno, R. J., Jelinek, W., and Banerjee, A. K., *J. Virol.*, 21, 1105, 1977. With permission.)

TABLE 1

Molecular Weights of VSV mRNA Species

Species[a]	$S_{20,w}$[b]	Gene[c] assignment	Molecular weight ($\times 10^{-6}$)[d]
1	31	L	1.7
2	17	G	0.7
3	14.5	N	0.5
4	12	NS	0.28
5	12	M	0.26

[a] For assignment of number, see Figure 2.
[b] See Moyer, S. A. and Banerjee, A. K., *Cell*, 4, 37, 1975.
[c] See Both, G. W., Moyer, S. A., and Banerjee, A. K., *Proc. Nat. Acad. Sci. U.S.A.*, 72, 274, 1975.
[d] Averages of the data from References 22, 24—26.

ylic acid ranging from 200 to 2000 nucleotides in length. Although the reason for this phenomenon is not clear, it appears that S-adenosylhomocysteine may have a direct effect on the polyadenylation reaction in vitro.[19]

The precise mechanism(s) of polyadenylation of VSV mRNAs is not known. That the Poly(A) stretches are not coded by the genome RNA comes from the observation that the genome RNA contains no corresponding stretch of Poly(U), although the genome RNA does contain an unusually high content of uridine (31 mol%).[28,31,34] An attractive hypothesis is that short oligo (U) sequences located at the end of each gene serve as mRNA termination signals, with the transcriptase using these sequences for synthesis of Poly(A) possibly by a slippage mechanism.[33] A similar mechanism has been shown in the procaryotic RNA polymerase system.[35] Determination of nucleotide sequences of the genome at the specific termination sites of the transcriptase may shed some light on the polyadenylation process.

B. 5′-Terminal Capping and Methylation

Unlike the usual triphosphates that are present at the 5′-termini of procaryotic mRNAs the VSV mRNAs synthesized in vitro contain the capped structure as G(5′)ppp(5′)Ap . . . .[36] A guanosine residue blocks the 5′-terminal adenosine residue of the RNA by a 5′-5′ linkage containing three phosphate groups. The phosphate bonds are resistant to the action of alkaline phosphatase but are sensitive to the action of nucleotide pyrophosphatase. This kind of novel structure has been found to be present in most eucaryotic mRNAs.[37] In the VSV system, the α phosphate of ATP and α and β phosphates of GTP are incorporated into the blocked structure.[33] This is in contrast to the corresponding structures in reo, vaccinia, and cytoplasmic polyhedrosis virus mRNAs where the α phosphate of the blocking nucleotide and the α and β phosphates of the 5′-terminal nucleotide are present in the blocked structures.[37] Since the blocking guanosine base contains free 2′ and 3′ hydroxyl groups (similar to the 3′-terminal residue), the 5′-terminal portion of the RNA can be selectively bound to dihydroxyboryl cellulose.[38]

In addition the capping reaction, purified VSV contains methyltransferase activities which transfer methyl groups from S-adenosylmethionine to the 7 position of the blocking G and to the 2′-hydroxyl position of the penultimate nucleotide yielding ^{7}mG(5′)ppp(5′)Ap^{m} . . . .[39] Unlike many eucaryotic mRNAs,[37] VSV mRNAs do not contain N^{6}-methyl adenosine residues (N^{6}A^{m}) located internally in the RNA chain. Moreover, the virion-associated methylase does not additionally methylate the 5′-penultimate base (N^{6}A^{2m}) as found in the VSV mRNAs isolated from the infected cells.[40,41] The methylation reaction is totally inhibited by the inhibitor, S-adenosylhomocysteine. Recently it has been possible to demonstrate two distinct methyltransferase activities in purified VSV.[42] Limited concentrations of the methyl donor ($<$ 0.1 m*M*) result in a single methylation of the penultimate base in the 2′-hydroxyl position, i.e., G(5′)ppp(5′)Ap^{m} . . . , whereas saturating concentrations of the methyl donor ($>$ 7 m*M*) also result in methylation of the blocking guanosine residue giving rise to the dimethylated cap structure, ^{7}mG(5′)ppp(5′)A^{m}. Pulse-chase experiments demonstrate that the monomethylated cap structure is the precursor of the dimethylated cap. In this respect, VSV is quite different from vaccinia or reovirus where 7-methylation of guanosine in mRNA is followed by 2′-0 methylation of the 5′-penultimate base.[37]

It is still unclear which of the viral proteins are involved in the capping and methylation reactions. These activities are discernible in highly purified virus cores free from the G and the M proteins. Moreover, by reconstitution experiments using separated L and NS proteins and N protein RNA complex, it was shown that the capping, methylating, and polyadenylating activities are transcription dependent.[43] Thus all these activities seem to be mediated by a maximum of three viral proteins, i.e., N, NS, and L.

IV. NUCLEOTIDE SEQUENCES AT THE 5′-TERMINI OF mRNAs

As described earlier, all the VSV mRNA species synthesized in vitro contain the novel structure G(5′)ppp(5′)Ap . . . .[36] The 5′-terminal analysis was then extended and all the mRNA species were found to contain a common hexanucleotide sequence, G(5′)ppp(5′)A–A–C–A–G.[35] In the presence of the methyl donor, S-adenosylmethionine, the RNA transcripts contained the identical hexanucleotide sequence except that methylation occurred on the guanosine and the penultimate adenosine bases. Thus it appears that the virion polymerase recognizes, for biosynthesis of the mRNA species, a common sequence 3′U–U–G–U–C 5′ located at the beginning of the five genes on the genome RNA.

Bishop and his colleagues have reported that in vitro RNA products contain as many as six different initiation sequences. These include: pppA–C–G, pppA–A–py, GpppA–A, pppG–C, pppG–A, pppG–G. The predominant sequences are pppA–C–G (25 to 35%) and GpppA–A (30 to 40%).[13,44,45] As will be shown later, the sequence (p)ppA–C–G is present in the VSV product RNA but not in any of the mRNA species. The RNA containing this sequence is a small leader RNA which is complementary to the 3′-terminal region of the genome RNA. The reasons for the discrepancies in results in the two laboratories regarding the other 5′-terminal sequences are unclear. Since the VSV mRNAs isolated from the infected cells also contain the same 5′-terminal hexanucleotide sequence,[46] it can be generally accepted that the predominant, if not the only, 5′-terminal sequence of VSV mRNAs is G(5′)ppp(5′)A–A–C–A–G.

It is interesting to note that four out of the five mRNA species synthesized in vitro by the New Jersey Serotype of VSV contain a common 5′-terminal sequence G(5′)ppp(5′)A–A–C–A–G, the same as found in the mRNAs of Indiana serotype.[47] The results suggest that these sequences have been conserved during evolution of the serotypes from a common ancestor, probably because they are important for biosynthesis of virus-specific mRNAs.

V. MECHANISM OF mRNA SYNTHESIS IN VITRO

A. Multiple Initiation vs. Single Initiation

As mentioned earlier, γ-^{32}PATP is not incorporated into VSV mRNA and only the α phosphate of ATP is retained in the 5′-blocked triphosphate structure as G(5′)$p^{\alpha}p^{\beta}p^{\alpha}$(5′)Ap . . . .[36] Thus it was difficult to ascertain whether the mRNAs are initiated molecules or are generated by a cleavage mechanism. Moreover, it was found that the 5′-terminal sequence of the mRNAs, G(5′)ppp(5′)A–A–C–A–G, does not base pair with the 3′-terminal sequence . . . PypGpU of the genome RNA.[48] Consequently, it seems that transcription of VSV genome RNA may occur by either of the two processes shown in Figure 3. By the multiple initiation model, the virion-associated RNA polymerase initiates transcription at five distinct promoter sites distal to the 3′-terminus of the template where the complementary sequence 3′UUCUG5′ on the genome RNA may serve as the recognition site for the polymerase. The mRNAs are presumably initiated with pppAp . . . which, in the presence of GTP, is converted into a blocked structure G(5′)ppp(5′)Ap . . . by a "capping" enzyme present in the virion. The β and γ phosphates of ATP and γ phosphate of GTP are removed during this process. On the other hand, in the single initiation model, the polymerase initiates transcription at the 3′-terminus of the genome RNA with a 5′-terminal pppA–C–Pu . . . sequence in the newly synthesized RNA. As the polymerase transcribes the genome RNA, an endonuclease in the virion cleaves the nascent RNA into cistron-sized mRNAs containing 5′-monophosphorylated sequence pA–A–C–A–G. Subsequent "capping" with GTP occurs at these 5′-termini, and the 3′-termini of the released mRNAs are polyadenylated. This model predicts that a "leader" RNA sequence complementary to the

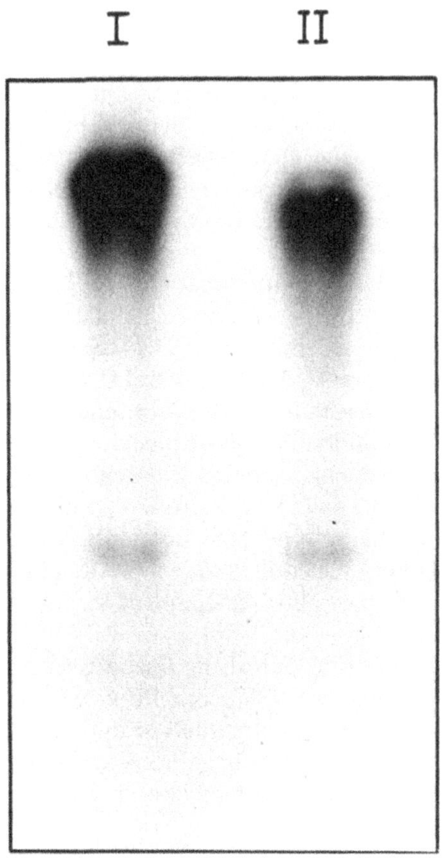

FIGURE 3. Models for VSV mRNA synthesis. (From Banerjee, A. K. and Rhodes, D. P., *Biochem. Biophys. Res. Commun.*, 68, 1387, 1976. With permission.)

3′-terminal portion of the genome RNA should be present in the RNA product. In addition, since the synthesis of the mRNAs is sequential, a genetic map of the genome RNA can be constructed.

B. Ultraviolet Effect on Transcription

In order to distinguish between the two models, the so-called "ultraviolet transcription mapping" originally developed by Sauerbier and his colleagues,[49] was applied to the VSV system. This method has successfully identified the single promotor and the direction of the transcription for "early" T7 bacteriophage mRNA, as well as the single precursor of 23S and 16S rRNA in *E. coli.*[49] The technique relies on ultraviolet light (UV) induced lesions in DNA or RNA, which cause premature chain termination without reinitiation during RNA synthesis. If ultraviolet lesions occur essentially randomly in long stretches of DNA or RNA, the irradiation should cause an exponential decrease in transcription with increasing distance from the promotor. Thus, in the VSV system, if the mRNAs are synthesized by independent initiation events the sensitivity of each gene to UV irradiation should be proportional to the size of the gene, i.e., the target-sizes of the genes should be proportional to the molecular weight of the corresponding mRNA species. On the other hand, if the mRNAs are synthesized sequentially from a single promotor at the 3′-terminus of the genome, the transcription of a particular gene will depend on the prior transcription of its 3′-proximal genes. Moreover, the target size of a gene will be the sum of molecular weights of its 3′-proximal genes or the corresponding mRNAs.

Using this technique, Ball and White[18] and Abraham and Banerjee[50] came to the conclusion that in the VSV system, UV irradiation caused a polar effect on the synthesis of mRNA *in vitro*. Consequently a single promotor is present at or near the 3′-terminus of the VSV genome and the mRNAs are synthesized sequentially. Ball and White[18] used UV irradiated purified VSV in a coupled transcription-translation system and measured the rate of decrease of synthesis of individual VSV proteins with increasing doses of irradiation. Abraham and Banerjee,[50] on the other hand, measured directly the synthesis of individual VSV mRNA species following increasing amounts of UV irradiation of purified virus. Target-size calculations in either approach showed that the target size of only the N protein gene corresponded to the molecular weight of its mRNA. The target-size of the other genes appeared to be cumulative, indicating a single start for the virion polymerase at the 3′-terminus of the genome RNA. In the coupled system,[18] L protein is not detected. However, from a consideration of its size and from genetic data, the gene coding for L protein can be placed at the 5′ half of the genome RNA. In the in vitro RNA synthesis system[50] the target size of the L protein mRNA was measured directly, but M and NS protein mRNAs are not easily separated. Consequently, genes coding for M and NS polypeptides were placed contiguously towards the 3′ end of the genome flanked by the N and G genes. Thus by combining the two approaches the complete gene order of VSV genome is established. as 3′−N−NS−M−G−L−5′.

C. The Leader RNA

The strategy of the virion polymerase is consistent with transcription of the genome RNA occurring at a single initiation site as outlined in Figure 3.

Although the cleavage of precursor mRNAs has not been demonstrated, the sequential synthesis of mRNAs is clearly operative in vitro. As predicated by the processing model, a leader RNA complementary to the 3′-terminal portion of the genome RNA was discovered among the RNA products synthesized in vitro.[51] (Figure 4). The characteristic properties of the leader RNA[52] include:

1. It is an initiated RNA molecule since it has a polyphosphorylated 5′-terminus (p)ppA−C−G which appears to be complementary to the 3′-terminal sequence of the genome RNA, . . . Py—G—C_{OH}. Consistent with this notion, purified leader RNA anneals to the 3′-terminal region of the genome RNA and protects it from RNase digestion.
2. The leader RNA has an estimated length of 48 nucleotides and is rich in adenosine (48 mol %) although it does not bind to oligo (dT)-cellulose.
3. (α-^{32}P)GMP labeled leader RNA after complete T_1 digestion yields five unique spots by two-dimensional electrophoresis and homochromatography. The tentative sequences of these spots are: ppACG, AG, AAG, CUCAG, $AC(A_3C)_2C(AUU)_2AUCAUUA_4G$, with some of the smaller sequences occurring in multiple molar yields.[53]
4. Kinetic studies show that the leader RNA and the mRNA coding for N protein anneal to 42S genome RNA at the same rate, indicating that the leader RNA is coded for only once on the genome RNA.
5. In the absence of UTP, an oligonucleotide is synthesized in vitro which contains the same 5′-terminal sequence as the leader RNA indicating that the synthesis of

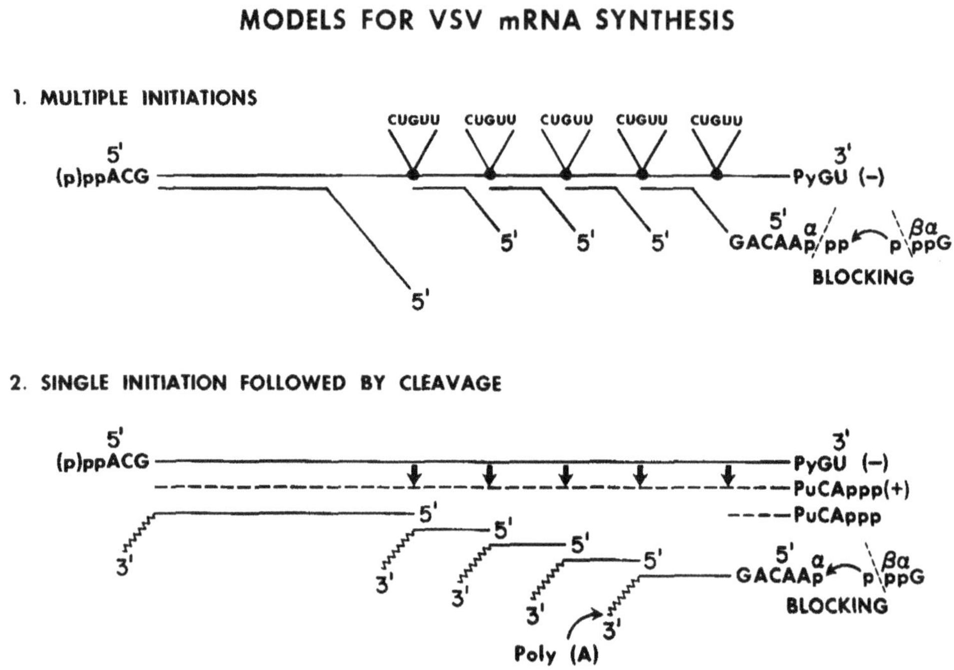

FIGURE 4. Polyacrylamide gel electrophoresis of leader RNA. Product RNA synthesized in vitro in the presence of α-^{32}P CTP by both the Indiana (I) and New Jersey (II) serotypes of VSV were purified by phenol-SDS extraction, sephadex G-50 chromatography. After fractionation by oligo(dT)-cellulose, the nonpoly(A)-containing RNA was electrophoresed on a 20% polyacrylamide slab gel and autoradiography was carried out after electrophoresis for 16 hr at 140 V.[54] (From Colonno, R. J. and Banerjee, A. K., *J. Virol.*, 26, 188, 1978. With permission.)

the leader RNA is the first event during in vitro RNA transcription, presumably followed by the synthesis of mRNA species.

6. Leader RNA synthesized in vitro by the New Jersey serotype of VSV is identical in size as the corresponding leader RNA from VSV Indiana.[54] (See Figure 3). The leader RNAs of Indiana and New Jersey serotypes have extensive sequence homology including the 5′-terminal ppA−C−G sequence yet the leader RNAs will not cross-hybridize to the heterologons genome RNA.[53,54]
7. It was shown that a defective interfering particle of VSV isolated from a heat-resistant strain of VSV can transcribe the genome RNA in vitro to synthesize the leader RNA and the mRNAs for N, NS, M,and G, but not the 31S mRNA species.[20] This is the only defective particle which has retained the 3′-terminal half of the wild type genome RNA[55] and transcribes in vitro.

The above results thus indicate that the 3′-terminal region of VSV genome RNA is rich in uridine and is used as template for initiation of viral mRNA synthesis by the virion-associated transcriptase.

D. A Model for the Biosynthesis of VSV mRNAs

A model for the mechanism of biosynthesis of VSV mRNAs in vitro has been recently proposed[56] (Figure 4). The virion RNA of negative polarity contains the 5′ terminal sequence (p)ppA−C−G and the 3′-terminal . . . PyG−U_{OH} sequence, and is enclosed within the nucleocapsid protein N. The initiation of RNA synthesis occurs at

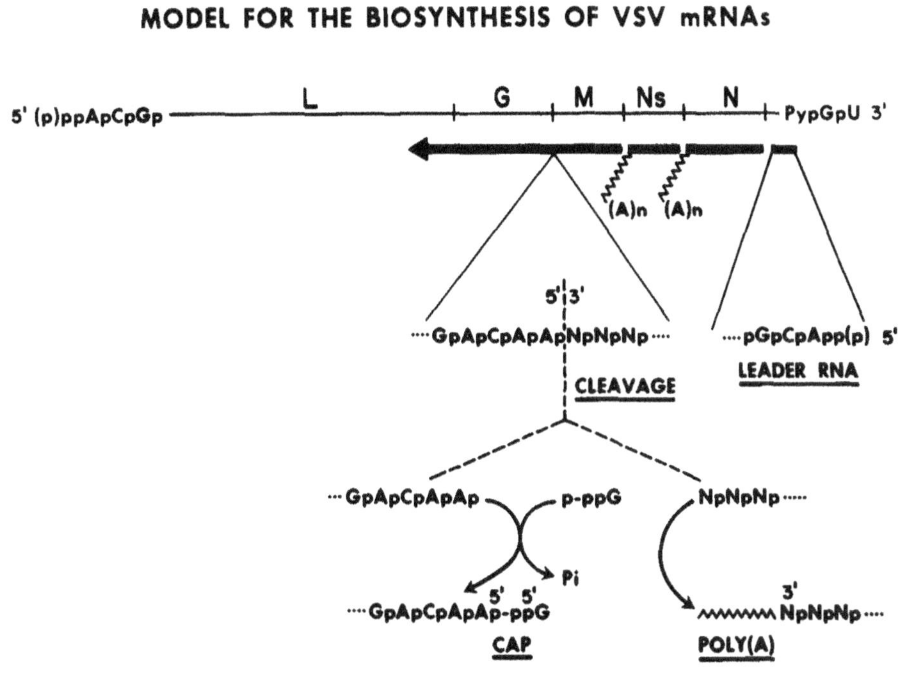

FIGURE 5. A model for the biosynthesis of VSV mRNAs.[56] (From Banerjee, A. K., Abraham, G., and Colonno, R. J., et al., *J. Gen. Virol.*, 34, 1,1977. With permission.)

the 3′-terminus of the genome RNA presumably mediated by the structural proteins L and NS. The first RNA product synthesized in vitro is the leader RNA which is 48 nucleotides long and complementary to the corresponding 3′-terminal protion of the VSV genome RNA. As the transcribing complex completes the synthesis of the leader RNA and enters into the adjacent N protein gene, a putative nuclease present in the virion cleaves the newly synthesized RNA at a specific site and releases the leader RNA. The 5′-terminal pA–A–C–A–G seqence of the nascent N protein mRNA is then capped by GTP to form the sequence G(5′)ppp(5′)AACAG. The transcribing complex continues to move along the genome RNA and the rest of the mRNAs are synthesized sequentially by similar cleavages of the growing RNA molecule followed by capping of the 5′-termini and polyadenylation of the 3′-termini of the released mRNAs. Thus, during complete transcription of the genome RNA, the RNA species are synthesized sequentially in the order 3′ leader—N—NS—M—G—L 5′.

VI. ADDENDUM

The complete nucleotide sequences of the leader RNA synthesized in in vitro by the Indiana and New Jersey serotype of VSV have been determined[64,65] by the technique described by Donis-Keller, Maxam, and Gilbert[66] in combination with the standard two dimensional fingerprint techniques described by Barrell.[67] The complementary sequences of the leader RNA representing the 3′-terminal sequence of the genome RNAs of both the serotypes are also shown in Table 2.

There are some striking similarities in the sequences between the leader RNAs synthesized by two serologically distinct VSV; (a) both contain 48 nucleotides which are adenine-rich (50%) and terminate variably at the 3′-terminus with cytosine (68%) and adenosine (32%) in Indiana and cytosine (40%) and adenosine (60%) in New Jersey leader RNA; (b) the first 24 nucleotides from the 5′-termini of the two RNAs are identical except for one base change at the 12th nucleotide with A for Indiana and C for New Jersey leader RNA. Beyond the 24th nucleotide, there are also extensive homologies except for the bases 25 and 26, 32 and 33, 37 and 38, 41 and 42, 44 and 45 (Table 2); (c) the locations of the guanosine residues are such that both leader RNAs generate identical sized T_1 oligonucleotides including the large RNase T_1 oligonucleotide containing 28 bases. Moreover, the T_1 oligonucleotides occur at identical locations in the nucleotide sequence through the first 40 nucleotides. (d) Both the leader RNAs terminate with a polypurine sequence $AGAGGGAAAC_{OH}$ for New Jersey and $AGGAGAAAAC_{OH}$ for Indiana leader RNA.

The above results show that the leader RNAs synthesized by two serotypes of VSV share extensive sequence homology, and correspondingly similar homologies exist between the 3′-terminal sequence of their genome RNAs. It is important to mention that previous hybridization studies with the leader RNA of New Jersey with Indiana genome RNA showed virtually no sequence homology.[54] It is now apparent from the sequences of two leader RNAs that few base differences between these RNAs are spaced such that RNase digestion at the mismatched positions after hybridization under the conditions employed, would result in melting of the hybrids into smaller lengths with complete digestion of the resulting single-stranded RNA.

The sequence homology at the 3′-terminal region of the genomes of both viruses suggests that they possibly share common binding and initiation sequence at their 3′-terminal region for the virion-associated RNA polymerase. In fact, the homology of the polymerase-binding site may account for the fact that the New Jersey serotype of VSV interfere with the replication of VSV Indiana.[68] In addition, by in vivo complementation studies, it has been shown that the L protein of VSV New Jersey can complement the transcription defect of a VSV Indiana mutant that is temperature sensitive in the L protein gene.[69]

It is interesting to note that the leader RNAs of both serotypes of VSV contain a polypurine sequence at their 3′-termini. These sequences have been implicated in the recognition site for the processing enzyme present in the virion.[64] Thus, it appears that both viruses may share not only a common strategy for biosynthesis of leader RNA and mRNAs, but also they may indeed share a common processing mechanism.

Finally, it appears from the extensive sequence homologies between the 3′-terminal portion of the genome RNA that these viruses may have evolved from a common ancestor. Since this portion of the genome RNA is involved in the binding and initiation by the RNA polymerase, the conservation of this sequence during evolution may be essential.

VII. CONCLUSIONS

The available data does not preclude a stop-start model where the virion polymerase intiates transcription from a promotor site at the 3′-terminus and after the synthesis and release of the leader RNA it initiates at a second promotor site to synthesize N-mRNA.[56] Following termination of the N-mRNA, the polymerase initiates at a third promotor site and so on. This model would suggest that the mRNA species are initiated with pppA ends and during the capping reaction with GTP and α and β phosphates of ATP are removed to give 5′ ends: $G(5')p^{\alpha}p^{\beta}\text{-}p^{\alpha}(5')A$ . . . . Hefti and Bishop[45,46] have reported that a small portion (5 to 10%) of the RNA species synthesized in vitro contain sequences like pppA–A–Py. This result, if confirmed, would be indicative of a stop-start model operating in vitro.

The capping reaction in VSV, however, is quite distinct from that observed in other viral systems (e.g., reo, vaccinia, and CPV) where the nascent RNA molecules are capped with GTP but retain the α and β phosphates of the initiating nucleotide as $G(5')p^{\alpha}\text{-}p^{\beta}p^{\alpha}(5')^{A}G$.[36] Moreover, the capping reaction in the latter viruses can be prevented by PP_i, resulting in the formation of the 5' polyphosphorylated RNA molecules. In the VSV system, P_i or PP_i does not inhibit the capping reaction.[57] Using ribonucleoside triphosphate analogues, e.g., GMPPNHP, it has been shown in the reo and vaccinia virus systems that this analogue of GTP failed to cap the 5'-termini of the mRNAs indicating that the cleavage between the β and γ phosphates of GTP is essential for capping.[58,59] n the VSV system, the same analogue is incorporated into the RNA chains and the 5'-terminal structure appears to be identical with the normal cap.[60] From the above observations, it appears that the processes leading to the formation of cap in the VSV system are unique. This may reflect a novel mode of mRNA synthesis by VSV, e.g., processing.

More recent experiments have suggested that mRNAs may indeed be formed by a processing mechanism. Using inosine-5' triphosphate (ITP) in place of GTP, it has been shown that RNA products as large as 14.5S contain a 5'-terminal sequence pp-A–C–(I) . . . . Since normally 5'-ppA–C–G sequence is only observed in the leader RNA, the results indicate that the GTP analogue permits synthesis in vitro of a larger RNA that may be an mRNA precursor.[61] Also, it has recently been shown the AMPPNHP will not support initiation of VSV RNA synthesis in vitro, but the analogue can be used for elongation of RNA that was previously initiated with ATP.[62] During this elongation process, mRNAs are synthesized with the normal 5'-terminal cap structure.[63] Since AMPPNHP fails to initiate RNA synthesis, the above results indicate that during elongation in the presence of AMPPNHP, no further initiation has occurred. This would indicate that the formation of mRNAs may have been mediated by a cleavage process. Further studies along this line should throw some light on the precise mechanism of mRNA synthesis.

Finally, the proposed cleavage model, also relates to the replication aspect of the VSV genome. If the putative processing event is blocked, the polymerase will read-through and sythesize a 42S plus strand — a required intermediate of replication. So far, the complement of the genome RNA has not been shown to be synthesized in vitro. Thus it remains to be seen whether cellular components or modfication of L, NS, or N proteins are needed for switching transcription to replication.

REFERENCES

1. **Huang, A. S. and Wagner, R. R.**, Comparative sedimentation co-efficients of RNA extracted from plaque-forming and defective particles of vesicular stomatitis virus, *J. Mol. Biol.*, 22, 381, 1966.
2. **Baltimore, D., Huang, A. S., and Stampfer, M.**, Ribonucleic acid synthesis of vesicular stomatitis virus. II. An RNA polymerase in the virion, *Proc. Nat. Acad. Sci. U.S.A.*, 66, 572, 1970.
3. **Bishop, D. H. L. and Roy, P.**, Properties of the product synthesized by vesicular stomatitis virus particles, *J. Mol. Biol.*, 58, 799, 1971.
4. **Emerson, S. U., Hunt, M., and Mellon, M.**, Viral transcriptase, Volume I.
5. **Moyer, S. A. and Banerjee, A. K.**, Messenger RNA species synthesized *in vitro* by the virion-associated RNA polymerase of vesicular stomatitis virus, *Cell*, 4, 37, 1975.
6. **Szilágyi, J. F. and Uryvayev, L.**, Isolation of an infectious ribonucleoprotein from vesicular stomatitis virus containing an active transcriptase, *J. Virol.*, 11, 279, 1973.
7. **Toneguzzo, F. and Ghosh, H. P.**, Characterization and translation of methylated and unmethylated vesicular stomatitis virus mRNA synthesized *in vitro* by ribonucleoprotein particles from vesicular stomatitis virus-infected L cells, *J. Virol.*, 17, 477, 1976.

8. **Emerson, S. U. and Wagner, R. R.**, L protein requirement for *in vitro* RNA synthesis by vesicular stomatitis virus, *J. Virol.*, 12, 1325, 1973.
9. **Naito, S. and Ishihama, A.**, Function and structure of RNA polymerase from vesicular stomatitis virus, *J. Biol. Chem.*, 251, 4037, 1976.
10. **Aaslestad, H. G., Clark, H F., Bishop, D. H. L., and Koprowski, H.**, Comparison of the ribonucleic acid polymerase of two rhabdoviruses, Kern Canyon virus and vesicular stomatitis virus, *J. Virol.*, 7, 726, 1971.
11. **Deutsch, V. and Banerjee, A. K.**, Effect of temperature on the enzymatic activities present in purified virions of vesicular stomatitis virus, *Biochem. Biophys. Res. Commun.*, 88, 1360, 1979.
12. **Bishop, D. H. L., Obijeski, J. F., and Simpson, R. W.**, Transcription of the influenza ribonucleic acid genome by a virion polymerase. I. Optimal conditions for *in vitro* activity of the ribonucleic acid-dependent ribonucleic acid polymerase, *J. Virol.*, 8, 66, 1971.
13. **Chang, S. H., Hefti, E., Obijeski, J. F., and Bishop, D. H. L.**, RNA transcription by the virion polymerases of five rhabdoviruses, *J. Virol.*, 13, 652, 1974.
14. **Testa, D. and Banerjee, A. K.**, *Biochim. Biophys. Res. Commun.*, 88, 1360, 1979.
15. **Hunt, D. M. and Wagner, R. R.**, Inhibition by aurintricarboxylic acid and polyethylene sulfonate of RNA transcription of vesicular stomatitis virus, *J. Virol.*, 16, 1146, 1975.
16. **Imblum, R. L. and Wagner, R. R.**, Inhibition of viral transcriptase by immunoglobulin directed against the nucleocapsid NS protein of vesicular stomatitis virus, *J. Virol.*, 15, 1357, 1975.
17. **Perrault, J. and Kingsbury, D. T.**, Inhibitor of vesicular stomatitis virus transcriptase in purified virions, *Nature (London)*, 248, 45, 1974.
18. **Ball, L. A. and White, C. N.**, Order of transcription of genes of vesicular stomatitis virus, *Proc. Nat. Acad. Sci., U.S.A.*, 73, 442, 1976.
19. **Rose, J. K., Lodish, H. F., and Brock, M. L.**, Giant heterogenous polyadenylic acid on vesicular stomatitis virus mRNA synthesized *in vitro* in the presence of S-adenosylhomocysteine, *J. Virol.*, 21, 683, 1977.
20. **Colonno, R. J., Lazzarrini, R., Keene, J. D., and Banerjee, A. K.**, *In vitro* synthesis of messenger RNA by a defective interfering particle of vesicular stomatitis virus, *Proc. Nat. Acad. Sci. U.S.A.*, 74, 1884, 1977.
21. **Moyer, S. A., Grubman, M. J., Ehrenfeld, E., and Banerjee, A. K.**, Studies on the *in vivo* and *in vitro* messenger RNA species of vesicular stomatitis virus, *Virology*, 67, 463, 1975.
22. **Both, G. W., Moyer, S. A., and Banerjee, A. K.**, Translation and identification of the mRNA species synthesized *in vitro* by the virion-associated RNA polymerase of vesicular stomatitis virus, *Proc. Nat. Acad. Sci. U.S.A.*, 72, 274, 1975.
23. **Rose, J. K. and Knipe, D.**, Nucleotide sequence complexities, molecular weights, and poly(A) content of the vesicular stomatitis virus mRNA species, *J. Virol.*, 15, 994, 1975.
24. **Morrison, T., Stampfer, M., Baltimore, D., and Lodish, H. F.**, Translation of vesicular stomatitis messenger RNA by extracts from mammalian and plant cells, *J. Virol.*, 13, 72, 1974.
25. **Rhodes, D. P., Abraham, G., Colonno, R. J., Jelinek, W., and Banerjee, A. K.**, Characterization of vesicular stomatitis virus mRNA species synthesized *in vitro*, *J. Virol.*, 21, 1105, 1977.
26. **Freeman, G. J., Rose, J. K., Clinton, G. M., and Huang, A. S.**, RNA synthesis of vesicular stomatitis virus by duplex formation, *J. Virol.*, 21, 1094, 1977.
27. **Villarreal, L. P., Breindl, M., and Holland, J. J.**, Determination of molar ratios of vesicular stomatitis virus induced RNA species in BHK_{21} cells, *Biochemistry*, 15, 1663, 1976.
28. **Banerjee, A. K. and Rhodes, D. P.**, *In vitro* synthesis of RNA that contains polyadenylate by virion-associated RNA polymerase of vesicular stomatitis virus, *Proc. Nat. Acad. Sci. U.S.A.*, 70, 3566, 1973.
29. **Villarreal, L. P. and Holland, J. J.**, Synthesis of Poly(A) *in vitro* by purified virions of vesicular stomatitis virus, *Nature (London) New Biol.*, 245, 17, 1973.
30. **Galet, H. and Prevec, L.**, Polyadenylate synthesis by extracts from L cells infected with vesicular stomatitis virus, *Nature (London) New Biol.*, 243, 200, 1973.
31. **Ehrenfeld, E.**, Polyadenylation of vesicular stomatitis virus messenger RNA, *J. Virol.*, 13, 1055, 1974.
32. **Ehrenfeld, E. and Summers, D. F.**, Adenylate-rich sequences in vesicular stomatitis virus messenger ribonucleic acid, *J. Virol.*, 12, 683, 1972.
33. **Banerjee, A. K., Moyer, S. A., and Rhodes, D. P.**, Studies on the *in vitro* adenylation of RNA by vesicular stomatitis virus, *Virology*, 61, 547, 1974.
34. **Marshall, S. and Gillespie, D.**, Poly(U) tracts absent from viral RNA, *Nature (London) New Biol.*, 240, 43, 1972.
35. **Chamberlain, M. and Berg, P.**, Mechanism of RNA polymerase action: characterization of the DNA-dependent synthesis of polyadenylic acid, *J. Mol. Biol.*, 8, 708, 1964.

36. **Abraham, G., Rhodes, D. P., and Banerjee, A. K.**, Novel initiation of RNA synthesis *in vitro* by vesicular stomatitis virus, *Nature (London)*, 255, 37, 1975.
37. **Shatkin, A. J.**, Capping of eucaryotic mRNAs, *Cell*, 9, 645, 1976.
38. **Rhodes, D. P. and Banerjee, A. K.**, 5'-Terminal sequence of vesicular stomatitis virus mRNAs synthesized *in vitro*, *J. Virol.*, 17, 33, 1976.
39. **Abraham, G., Rhodes, D. P., and Banerjee, A. K.**, 5'-Terminal structure of methylated mRNA synthesized *in vitro* by vesicular stomatitis virus, *Cell*, 5, 51, 1975.
40. **Moyer, S. A. and Banerjee, A. K.**, *In vivo* methylation of vesicular stomatitis virus and its host-cell messenger RNA species, *Virology*, 70, 339, 1976.
41. **Rose, J. K.**, Heterogenous 5'-terminal structures occur on vesicular stomatitis virus mRNAs, *J. Biol. Chem.*, 250, 8098, 1975.
42. **Testa, D. and Banerjee, A. K.**, Two methyl transferase activities in the purified virions of vesicular stomatitis virus, *J. Virol.*, 24, 786, 1977.
43. **Abraham, G. and Banerjee, A. K.**, The nature of the RNA products synthesized *in vitro* by subviral components of vesicular stomatitis virus, *Virology*, 71, 230, 1976.
44. **Hefti, E. and Bishop, D. H. L.**, The 5' nucleotide sequence of vesicular stomatitis viral RNA, *J. Virol.*, 15, 90, 1975.
45. **Hefti, E. and Bishop, D. H. L.**, The 5' sequence of VSV viral RNA and its *in vitro* transcription product RNA, *Biochem. Biophys. Res. Commun.*, 66, 785, 1975.
46. **Rose, J. K.**, Nucleotide sequences of ribosome recognition sites in messenger RNAs of vesicular stomatitis virus, *Proc. Nat. Sci. U.S.A.*, 74, 3672, 1977.
47. **Franze-Fernandez, M. T. and Banerjee, A.K.**, *In vitro* transcription by the New Jersey Serotype of Vesicular stomatitis virus. I. Characterization of the mRNA species, *J. Virol.*, 26, 179, 1978.
48. **Banerjee, A. K. and Rhodes, D. P.**, 3'-Terminal sequence of vesicular stomatitis virus genome RNA, *Biochem. Biophys. Res. Commun.*, 68, 1387, 1976.
49. **Sauerbier, W.** UV damage at the transcriptional level, *Adv. Radiat. Biol.*, 6, 49, 1976.
50. **Abraham, G. and Banerjee, A. K.**, Sequential transcription of the genes of vesicular stomatitis virus, *Proc. Nat. Acad. Sci. U.S.A.*, 73, 1504, 1976.
51. **Colonno, R. J. and Banerjee, A. K.**, A unique RNA species involved in initiation of vesicular stomatitis virus RNA transcription *in vitro*, *Cell*, 8, 197, 1976.
52. **Colonno, R. J. and Banerjee, A. K.**, Mapping and initiation studies on the leader RNA of vesicular stomatitis virus, *Virology*, 77, 260, 1977.
53. **Colonno, R. J. and Banerjee, A. K.**, unpublished observations, 1978.
54. **Colonno, R. J. and Banerjee, A. K.**, *In vitro* transcription of mRNA by the New Jersey Serotype of vesicular stomatitis Virus II. Characterization of the leader RNA, *J. Virol.*, 26, 188, 1978.
55. **Schnitzlein, W. M. and Reichmann, E.**, The size and the cistronic origin of defective vesicular stomatitis virus particle RNAs in relation to homotypic and heterotypic interference, *J. Mol. Biol.*, 101, 307, 1976.
56. **Banerjee, A. K., Abraham, G., and Colonno, R. J.**, Vesicular stomatitis virus: mode of transcription, *J. Gen. Virol.*, 34, 1, 1977.
57. **Abraham, G. and Banerjee, A. K.**, unpublished observations, 1976.
58. **Furuichi, Y. and Shatkin, A. J.**, 5'-Termini of reovirus mRNA: ability of viral cores to form caps post-transcriptionally, *Virology*, 77, 566, 1977.
59. **Moss, B., Gershowitz, A., Wei, C. M., and Boone, R.**, Formation of the guanylyated and methylated 5'-terminus of vaccinia virus mRNA, *Virology*, 72, 341, 1976.
60. **Banerjee, A. K., Colonno, R. J., Testa, D., and Franze-Fernandez, M. T.**, Mechanism of RNA synthesis *in vitro* by vesicular stomatitis virus, in *Negative Strand Viruses and the Host Cells*, Barry, R. D. and Mahy, B. W. J., Eds. Academic Press, New York, 1978, 249.
61. **Testa, D. and Banerjee, A. K.**, *In vitro* synthesis of a possible precursor mRNA by purified vesicular stomatitis virus, *Biochem. Biophys. Res. Commun.*, 82, 655, 1978.
62. **Testa, D. and Banerjee, A. K.**, Initiation of RNA synthesis in vitro by vesicular stomatitis virus; role of ATP, *J. Biol. Chem.*, 254, 2053, 1979.
63. **Rhodes, D. P., Moyer, S. A., and Banerjee, A. K.**, *In vitro* synthesis of methylated messenger RNA by the virion-associated RNA polymerase of vesicular stomatitis virus, *Cell*, 3, 327, 1974.
64. **Colonno, R. J. and Banerjee, A. K.**, Complete nucleotide sequence of the leader RNA synthesized *in vitro* by vesicular stomatitis virus, *Cell*, 15, 93, 1978.
65. **Colonno, R. J. and Banerjee, A. K.**, Nucleotide sequence of the leader RNA of the New Jersey serotype of vesicular stomatitis virus, *Nucleic Acids Res.*, 1978, in press.
66. **Donis-Keller, H., Maxam, A. M., and Gilbert, W.**, Mapping adenines, guanines and pyrimidines in RNA, *Nucleic Acids Res.*, 4, 2527, 1977.

67. **Barrell, G. G.,** Fractionation and sequence analysis of radio-active nucleotides, in *Procedures in Nucleic Acids Research,* Vol.2, Cantoni, G. L. and Davis, D. R., Eds., Harper & Row, New York, 1971, 751.
68. **Legault, D., Takayesu, D., and Prevec, L.,** Heterotypic exclusion between vesicular stomatitis viruses of the New Jersey and Indiana serotypes, *J. Gen. Virol.,* 35, 53, 1977.
69. **Repik, P., Flamand, A., and Bishop, D. H. L.,** Synthesis of RNA by mutants of vesicular stomatitis virus (Indiana serotype) and the ability of wild-type VSV New Jersey to complement the VSV Indiana ts GI-114 transcription defect, *J. Vitrol.,* 20, 157, 1976.

Chapter 3

RIBOSOME RECOGNITION AND TRANSLATION OF VESICULAR STOMATITIS VIRUS MESSENGER RNA

John K. Rose

TABLE OF CONTENTS

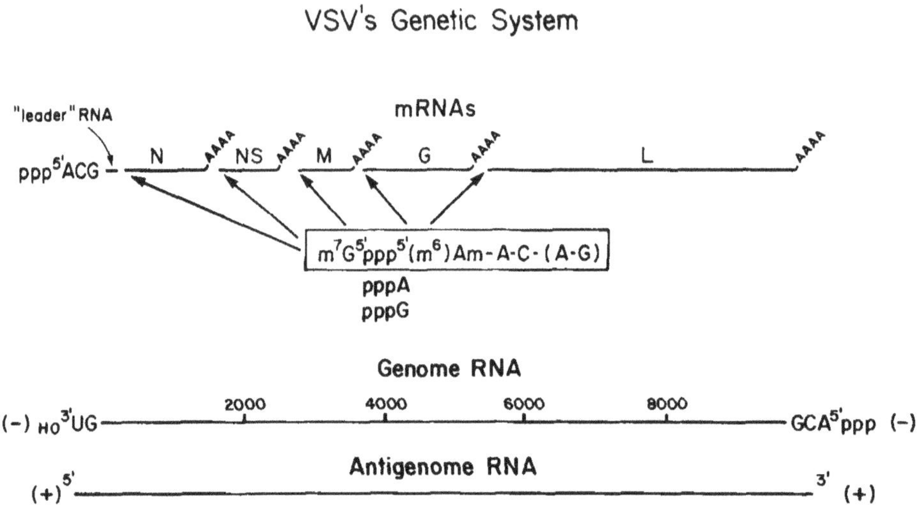

FIGURE 1. Diagram of VSVs genetic system. The diagram shows the positive stranded mRNAs and leader RNA that are transcribed from the negative strand genome RNA. Approximate lengths in nucleotides are indicated on the genome RNA.

I. INTRODUCTION

During the past few years, work on vesicular stomatitis virus (VSV) has led to an extensive characterization of the viral genetic system. Much of this work has focused on the five viral mRNAs that encode the five viral proteins. The purpose of this chapter is first to review the basic work on purification of the VSV mRNAs as well as the assignment of these mRNAs to the viral proteins that they encode by in vitro translation. In the second and third sections, some of the recent work on in vivo and in vitro translation of VSV mRNAs is reviewed. The final section is devoted exclusively to a review of work on the isolation and nucleotide sequencing of the ribosome binding sites from all of the VSV mRNAs.

Figure 1 is a diagram showing the VSV mRNAs, the negative strand genome RNA from which they are transcribed, and the genome-length positive strand RNA that serves as a replication intermediate. The diagram is included for completeness, and other chapters have details of the system that are not reviewed here.

II. VSV mRNAS AND PROTEINS

Initial work showed that the VSV mRNA species could be fractionated into two size classes (termed 28S and 13 to 15S) by sucrose gradient sedimentation.[1] Both size classes contained mRNA complementary to the single stranded RNA genome of VSV.[1] Subsequent in vitro translation studies[2] using both wheat germ and reticulocyte systems showed that the small mRNA size class could be translated to form the N, NS, M, and probably G proteins, while the large-size class could be translated to form the L protein.

Further work on fractionation of the VSV mRNA species employed formamide-polyacrylamide gel electrophoresis[3,4] or high-resolution gradient sedimentation.[5,6] Both of these procedures resolved the smaller mRNA size class into three bands or peaks (12S, 15S, and 18S). Fingerprint analysis[3] indicated that the 12S mRNA band was a mixture of mRNAs, while the 15S and 18S bands were pure mRNA species. In vitro

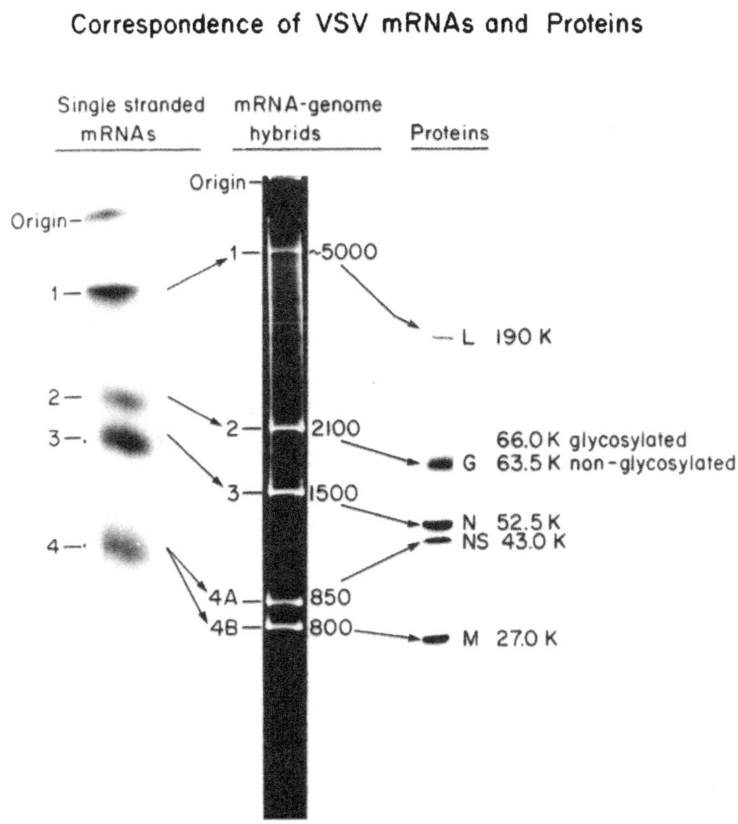

FIGURE 2. Correspondence of VSV mRNAs and proteins. Left lane: autoradiogram of ^{32}P-labeled VSV mRNAs separated by formamide-polyacrylamide gel electrophoresis. Middle lane: ethidium bromide stained mRNA-genome RNA hybrids obtained after ribonuclease treatment of VSV mRNAs that had been hybridized to the VSV genome RNA. Separation was by polyacrylamide gel electrophoresis; approximate lengths are given in nucleotides. The assignment of the smaller mRNA-genome bands to the NS and M proteins was by the relative sensitivities of their transcription to UV inactivation. Right lane: autoradiogram of VSV proteins that were labeled in infected cells with ^{35}S-methionine and separated by SDS-polyacrylamide gel electrophoresis. Molecular weights are from Knipe et al.[4]

translation of the mRNA bands eluted from gels,[4] or pur fied on velocity gradients,[5,6] showed that the 12S mRNA band contained a mixture of two mRNAs encoding the NS and M proteins, while the 15S and 18S mRNAs encoded the N and G proteins respectively.

The correspondence between the mRNAs and the proteins encoded is indicated in Figure 2. The left lane shows an autoradiogram of the VSV mRNAs separated by formamide-polyacrylamide gel electrophoresis. Complete separation of the two smaller mRNAs (these normally run together on gels or gradients) as mRNA-genome RNA hybrids has been achieved recently[7,8] and is shown in the central lane. The assignment of the mRNAs to the specific proteins encoded is indicated by the arrows between the central lane and the right lane, which shows an autoradiogram of the VSV proteins separated by gel electrophoresis.

One anomalous feature of the mRNA-protein assignments should be noted. The NS mRNA is about 850 nucleotides long (minus poly[A]) by chemical length determination as well as gel mobilty[3,7] and should only be capable of encoding a protein of about

31,000 daltons. However, the NS protein appears to have a molecular weight of 43,000 or greater.[4] This discrepancy is presumably due to anomalous migration of the NS protein.

III. IN VIVO TRANSLATION STUDIES

A. Relative Efficiencies of Translation of VSV mRNAs

Villarreal et al.[9] found that the molar ratios of the VSV mRNAs extracted from infected cells are very similar to the molar ratios of the VSV proteins being synthesized in those cells. From these results, it was concluded that VSV mRNAs were all translated with equal efficiencies in vivo. It was also found that the molar amounts of the mRNA species were not equal, and decreased in the order of transcription of the genome, N–NS–M–G–L.[10] Lodish and Froshauer,[11] in a more detailed study which included analysis of polysome size in vivo, found that the G and N mRNAs were translated with equal efficiency in vivo.

It is clear from the experiments of Nuss and Koch[12] that VSV mRNAs are not translated with equal efficiency under conditions of hypertonic shock in vivo. These experiments showed that translation of the G and M mRNAs was inhibited considerably more than translation of N, NS, and L mRNAs when in fected cells were subjected to increasing hypertonic shock. A possible explanation for these differential effects of hypertonic shock is suggested by the structures of the ribosome binding sites from VSV mRNAs (see below).

There is no information available on the efficiency of translation of VSV mRNA compared to host mRNA, although it is well known that VSV mRNA translation occurs in the absence of host mRNA translation late during infection.[13] This host shut-off could be a passive process involving synthesis of large amounts of very efficient viral mRNAs that compete efficiently against host mRNAs for the translation apparatus. Alternatively, it could be a more active process involving alteration of the host translational apparatus. There is one report of a temperature-sensitive mutation in VSV that reduces the extent of host shut-off,[14] but the molecular nature of the effect is not yet known.

B. Posttranslational Processing In Vivo

Several groups have found that the VSV N, NS, M, and L proteins synthesized in vitro in several systems are the same size as those found in virions,[2,4-6,15-17] suggesting no major processing of these proteins. The G protein is an exception because the in vitro product appears to be about 3000 daltons smaller than the G protein found in virions.[2,4-6,15-17] As discussed below under in vitro processing, this difference in gel mobility results from the absence of glycosylation in vitro. Experiments in vivo[18] indicated that G protein with mobility as fast as the in vitro product could not be detected even by very short pulse labeling, suggesting that the G protein is glycosylated concomitantly with or shortly after synthesis. These pulse-labeling experiments were able to detect two intracellular forms of the glycoprotein, G_1 and G_2. The G_2 form apparently differs from the G_1 form by the presence of N-acetylneuraminic acid, which is added 10 to 20 min before the G_2 form appears on the cell surface.

IV. IN VITRO TRANSLATION STUDIES

A. Role of Terminal Modifications in VSV mRNA Translation

VSV has proved to be an especially useful system for studying the role of 5′-terminal modifications in mRNA translation. The mRNAs can be synthesized in vitro by the virion-bound polymerase[19] and have the capped 5′-termini $m^7G^{5'}rppp^{5'}rAmpAp$ or $G^{5'}rppp^{5'}rApAp$ depending on the presence or absence of the methyl donor S-adenosylmethionine[20] during synthesis. In vivo, the functional mRNAs have the 5′-terminus $m^7G^{5'}rppp^{5'}r(m^6)Amp(m^6)$ Amp with parentheses indicating partial methylations.[21-23]

It was reported initially that in vitro translation of VSV mRNAs in the wheat germ system required the presence of a 5′-terminal m^7G on the mRNAs.[24] These experiments were performed with mRNA synthesized in vitro. Somewhat different results were obtained in the reticulocyte translation system using VSV mRNA (synthesized in vivo) from which the m^7G was removed by a chemical procedure.[25] In this system, the uncapped mRNA was translated to form the correct VSV proteins, but with 25 to 30% efficiency, compared to control RNA. Even totally unmethylated VSV mRNA synthesized in vitro is translated with 12 to 13% efficiency in the reticulocyte system.[26]

It is now clear that the effect of the m^7G on translocation depends on the translation system employed and that the wheat germ system shows a very strong dependence on the m^7G.[26] A range of effects of methylation on VSV mRNA translation have been found in other systems.[16,17] These differences may, at least in part, be due to differences in salt concentrations employed in the various systems.[27]

There are no data available on the effect of 5′-modifications on VSV mRNA translation in vivo, where ionic conditions and competition for the translation apparatus may be very different than in in vitro systems. A small fraction of RNA having the same size as the VSV mRNA species is synthesized in vivo with 5′-triphosphate termini ($ppp^{5'}A$ and $ppp^{5'}G$), and these RNAs are not found on polyribosomes.[22] Until the nucleotide sequences at these triphosphate ends can be compared to the mRNA sequences, the significance of this finding will not be known.

In addition to effects of 5′-terminal modifications on translation, there is one report that the VSV mRNA made in vitro without poly(A) is translated poorly in a fractionated reticulocyte system.[17] This could possibly be due to degradation of the poly(A) minus RNA in the system.

B. Coupled Transcription and Translation

Ball and White[10] have reported the development of an in vitro system in which VSV mRNAs and proteins are synthesized concomitantly. Using this system, they were able to analyze the effect of preirradiation of virus with UV-light on the rate of transcription of each gene, because the proteins encoded by each gene are readily separated, while the mRNAs are not. From this study, it was determined that the VSV mRNAs are transcribed in the order N–NS–M–G–L. In this study, it was also noted that the presence of the cell extract greatly enhanced the synthesis of proper mRNA transcripts, a result obtained by others also.[28,29]

C. Posttranslational Processing In Vitro

Toneguzzo and Ghosh[30] reported that glycosylation of the G protein will occur in vitro if a membrane fraction from HeLa cells is present during protein synthesis. Similar results were obtained by Katz et al.[31] using membranes from dog pancreas. Rothman and Lodish[32] have reported a more detailed analysis of this process, showing that the glycosylation occurs in two stages on the nascent chain of the G protein.

V. ISOLATION AND SEQUENCE ANALYSIS OF RIBOSOME RECOGNITION SITES IN VSV mRNAS

VSV mRNAs are ideal for examination of mRNA structure because they are readily

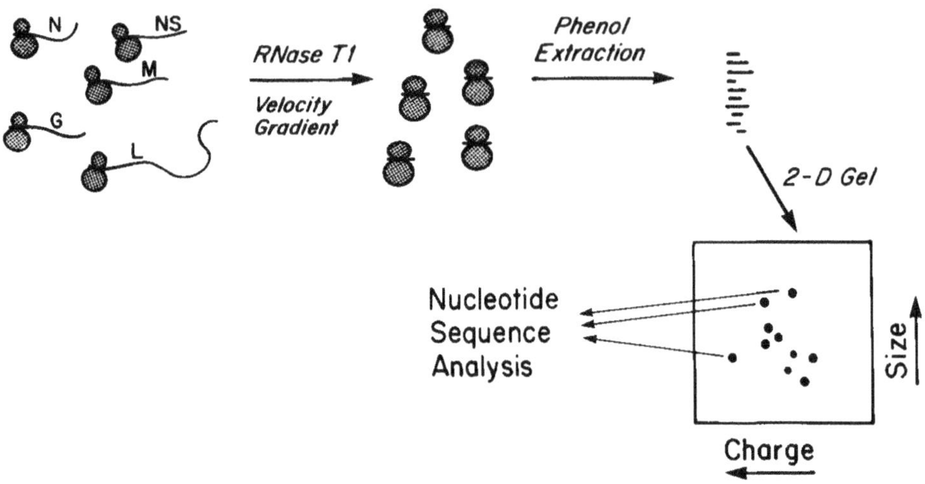

FIGURE 3. Isolation of ribosome protected translation initiation sites. The diagram shows the procedure used for isolation of ribosome protected translation initiation sites from VSV mRNAs. Details are given in the text and in References 34 and 35.

labeled with ^{32}P in vivo and in vitro, and can be isolated in pure form. The procedure that was used to isolate the translation initiation sites from VSV mRNAs is diagrammed in Figure 3. Total ^{32}P-labeled mRNA (synthesized in vivo or in vitro) was incubated in a reticulocyte cell-free translation system in the presence of anisomycin, an inhibitor that blocks translation elongation. The initiation complex was then treated with RNase to digest all sequences except those protected by the ribosome. The ribosome-bound fragments were then purified by sucrose gradient sedimentaton, followed by separation of the fragments on the two-dimensional gel system of de Wachter and Fiers.[33] Purified mRNAs were used to determine which mRNA gave rise to which spot on the two-dimensional gel. The details of the isolation procedure and the sequence analysis of the sites have been reported.[34,35]

The sequences determined for the protected sites from each of the five VSV mRNAs are shown in Figure 4. Each site contains a single AUG initiation codon, and the amino acid sequences shown are predicted from the genetic code. The sequences do not all share common features, indicating that a variety of structures can result in equal translation efficiencies in vivo.[9] The capped 5′-end of the mRNA only appears in the site when it is close to the initiation codon. Homologies among some of the sites can be seen (dashed boxes), but their significance is not known.

The sequences shown (Figure 4) are protected in the 80S initiation complex. Protection at the earlier stage of the 40S initiation complex has been examined also.[35] It was found in some of the VSV mRNAs that the 40S complex gave partial protection of larger sequences than those protected in the 80S complex. These larger sites always contained the initiation codon, but not always the 5′-end of the mRNA. Thus, if there is a recognition of the 5′ end of the mRNA by the 40S complex,[36] presumably some rearrangement of the mRNA structure or ribosome movement occurs so that the 40S initiation complex protects only the initiation site and not the 5′-end of the mRNA in some cases.

Consideration of the structures of the protected sites (Figure 4) suggests that the differential effects of hypertonic shock on VSV mRNA translation in vivo[12] may be related to the distance between the 5′-end of the mRNA and the initiation codon. The 5′-end is close to the initiation site of those mRNAs (N, NS, and L) with translation

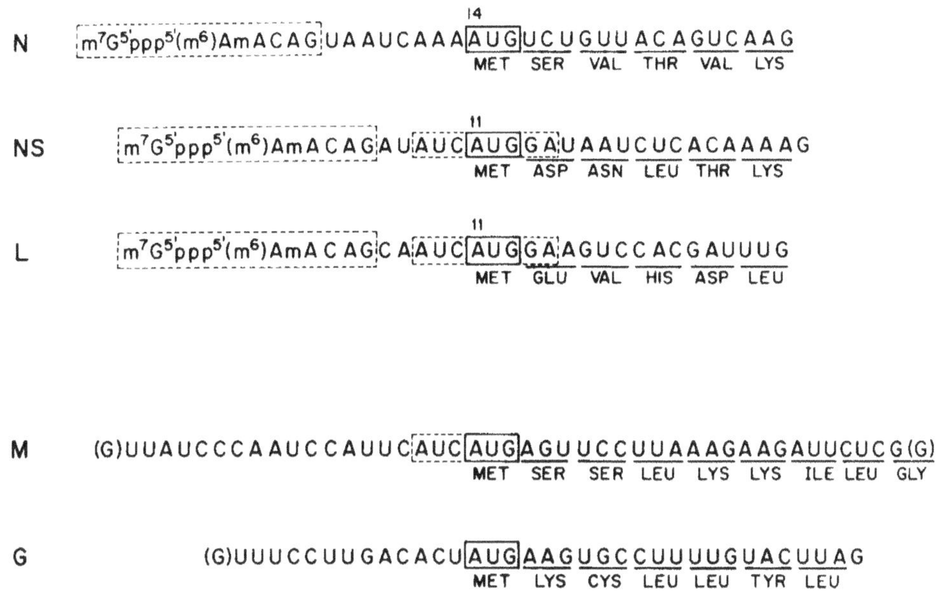

FIGURE 4. Complete sequences of the ribosome recognition sites from the VSV mRNAs. The sequences corresponding to the major ribosome protected sites from each of the indicated VSV mRNAs are shown. Regions of homology are indicated by dashed boxes. The sequences of the N, NS, and G sites are from Reference 34 and the sequences from the L and M sites are from Reference 35. In the previously published N sequence,[34] the sequence AAAAU which occurs in the noncoding region was reported incorrectly as AAAU, and the sequence AAAAG which occurs at the 3′-end of the NS sequence was reported incorrectly as AAAG.

resistant to hypertonic shock, and distant from the initiation site in those mRNAs (M and G) with translation sensitive to hypertonic shock. Thus, hypertonic shock might inhibit a transition of the 40S complex from the 5′-end of the mRNA to the initiation site.

Although there is a notable lack of homology among the VSV initiation sites, a strong homology between the VSV M mRNA binding site and the alfalfa mosaic virus (AMV) coat mRNA binding site[37] has been found.[35] The two sites show a 12-nucleotide perfect homology, C–A–U–C–A–U–G–A–G–U–U–C, as well as correspondence of one half of the remaining positions (17 out of 34). In addition to these homologies, these mRNAs are about the same length, have long leader sequences preceding the initiation codons, and encode proteins of about the same size. Thus, it seems likely that these two genes from apparently totally unrelated viruses have been derived from a common ancestral gene. Further sequence analysis of RNAs encoding proteins of similar size and function may reveal relatedness of the genes from other apparently unrelated viruses.

REFERENCES

1. **Huang, A. S., Baltimore, D., and Stampfer, M.,** Ribonucleic acid synthesis of vesicular stomatitis virus. III. Multiple complementary messenger RNA molecules, *Virology*, 42, 946, 1970.
2. **Morrison, T., Stampfer, M., Baltimore, D., and Lodish, H. F.,** Translation of vesicular stomatitis messenger RNA by extracts from mammalian and plant cells, *J. Virol.*, 13, 62, 1974.

3. **Rose, J. K. and Knipe, D.**, Nucleotide sequence complexities, molecular weights, and poly(A) content of the vesicular stomatitis virus mRNA species, *J. Virol.*, 15, 994, 1975.
4. **Knipe, D., Rose, J. K., and Lodish, H. F.**, Translation of individual species of vesicular stomatitis viral mRNA, *J. Virol.*, 15, 1004, 1975.
5. **Both, G. W., Moyer, S. A., and Banerjee, A. K.**, Translation and identification of the mRNA species synthesized *in vitro* by the virion-associated RNA polymerase of vesicular stomatitis virus, *Proc. Natl. Acad. Sci. U.S.A.*, 72, 274, 1975.
6. **Both, G. W., Moyer, S. A., and Banerjee, A. K.**, Translation and identification of the viral mRNA species isolated from subcellular fractions of vesicular stomatitis virus-infected cells, *J. Virol.*, 15, 1012, 1975.
7. **Freeman, G. J., Rose, J. K., Clinton, G. M., and Huang, A. S.**, RNA synthesis of vesicular stomatitis virus. VII. Complete separation of the mRNAs of VSV by duplex formation, *J. Virol.*, 21, 1094, 1977.
8. **Rhodes, D. P., Abraham, G., Colonno, R. J., Jelinek, W., and Banerjee, A. K.**, Characterization of vesicular stomatitis virus mRNA species synthesized *in vitro*, *J. Virol.*, 21, 1105, 1977.
9. **Villarreal, L. P., Breindl, M., and Holland, J. J.**, Determination of molar ratios of VSV induced RNA species in BHK_{21} cells, *Biochemistry*, 15, 1663, 1976.
10. **Ball, L. A. and White, C. N.**, Order of transcription of genes of vesicular stomatitis virus, *Proc. Natl. Acad. Sci. U.S.A.*, 73, 442, 1976.
11. **Lodish, H. F. and Froshauer, S.**, Rates of initiation of protein synthesis by two purified species of VSV mRNA, *J. Biol. Chem.*, 252, 8804, 1977.
12. **Nuss, D. and Koch, G.**, Differential inhibition of vesicular stomatitis virus polypeptide synthesis by hypertonic initiation block, *J. Virol.*, 17, 283, 1976.
13. **Yamazaki, S. and Wagner, R. R.**, Action of interferon: kinetics and differential effects on viral functions, *J. Virol.*, 6, 421, 1970.
14. **Stanners, C. P., Francoeur, A. M., and Lam, T.**, Analysis of a VSV mutant with attenuated cytopathogenicity: mutation in viral function, P, for inhibition of protein synthesis, *Cell*, 11, 273, 1977.
15. **Morrison, T. G. and Lodish, H. F.**, Site of synthesis of membrane and non-membrane proteins of vesicular stomatitis virus, *J. Biol. Chem.*, 250, 6955, 1975.
16. **Toneguzzo, F. and Ghosh, H. P.**, Characterization and translation of methylated and unmethylated VSV mRNA synthesis *in vitro* by ribonucleoprotein particles from VSV-infected L-cells, *J. Virol.*, 17, 477, 1976.
17. **Preston, C. M. and Szilágyi, J. F.**, Cell-free translation *in vitro* by a transcribing nucleoprotein complex prepared from purified vesicular stomatitis virus, *J. Virol.*, 21, 1002, 1977.
18. **Knipe, D., Lodish, H. F., and Baltimore, D.**, Localization of two cellular forms of the VSV glycoprotein, *J. Virol.*, 21, 1121, 1977.
19. **Baltimore, D., Huang, A. S., and Stampfer, M.**, Ribonucleic acid synthesis of vesicular stomatitis virus, *Proc. Natl. Acad. Sci. U.S.A.*, 66, 572, 1970.
20. **Abraham, G., Rhodes, D. P., and Banerjee, A. K.**, The 5'-terminal structure of the unmethylated mRNA synthesized *in vitro* by vesicular stomatitis virus, *Nature (London)*, 255, 37, 1975.
21. **Moyer, S. A., Abraham, G., Adler, R., and Banerjee, A. K.**, *In vivo* methylation of VSV mRNA, *Cell*, 5, 59, 1975.
22. **Rose, J. K.**, Heterogeneous 5'-terminal structures occur on vesicular stomatitis virus mRNAs, *J. Biol. Chem.*, 250, 8098, 1975.
23. **Moyer, S. A. and Banerjee, A. K.**, *In vivo* methylation of VSV and its host cell mRNA species, *Virology*, 70, 339, 1976.
24. **Both, G. W., Banerjee, A. K., and Shatkin, A. J.**, Methylation-dependent translation of viral messenger RNAs *in vitro*, *Proc. Natl. Acad. Sci. U.S.A.*, 72, 1189, 1975.
25. **Rose, J. K. and Lodish, H. F.**, Translation *in vitro* of vesicular stomatitis virus mRNA lacking 5'-terminal 7-methylguanosine, *Nature (London)*, 262, 32, 1976.
26. **Lodish, H. F. and Rose, J. K.**, Relative importance of 7-methylguanosine in translation and ribosome binding of VSV mRNA in wheat germ and reticulocyte cell-free systems, *J. Biol. Chem.*, 252, 1181, 1977.
27. **Weber, L. A., Hickey, E. D., Nuss, D. L., and Baglioni, C.**, 5'-terminal 7-methylguanosine and mRNA function: influence of potassium concentration on translation *in vitro*, *Proc. Natl. Acad. Sci. U.S.A.*, 74, 3254, 1977.
28. **Breindl, M. and Holland, J. J.**, Studies on the *in vitro* transcription and translation of vesicular stomatitis virus mRNA, *Virology*, 73, 106, 1976.
29. **Rose, J. K., Lodish, H. F., and Brock, M. L.**, Giant heterogeneous poly(A) on VSV mRNA synthesized *in vitro* in the presence of S-adenosyl homocysteine, *J. Virol.*, 21, 683, 1977.

30. **Toneguzzo, F. and Ghosh, H. P.**, Synthesis and glycosylation *in vitro* of glycoprotein of vesicular stomatitis virus, *Proc. Natl. Acad. Sci. U.S.A.*, 74, 1516, 1977.
31. **Katz, F. N., Rothman, J. E., Lingappa, V. R., Blobel, G., and Lodish, H. F.**, Membrane assembly *in vitro:* synthesis, glycosylation and assymetric insertion of a transmembrane protein, *Proc. Natl. Acad. Sci. U.S.A.*, 73, 3278, 1977.
32. **Rothman, J. E. and Lodish, H. F.**, Synchronized transmembrane insertion and glycosylation of a nascent membrane protein, *Nature (London)*, 269, 775, 1977.
33. **de Wachter, R. and Fiers, W.**, Preparative two-dimensional gel electrophoresis of ^{32}P-labeled RNA on polyacrylamide slab gels, *Anal. Biochem.*, 49, 184, 1972.
34. **Rose, J. K.**, Nucleotide sequences of ribosome recognition sites in messenger RNAs of vesicular stomatitis virus, *Proc. Natl. Acad. Sci. U.S.A.*, 74, 3672, 1977.
35. **Rose, J. K.**, Complete sequences of the ribosome recognition sites from VSV mRNAs: recognition by the 40S and 80S complexes, *Cell*, 14, 345, 1978.
36. **Shatkin, A. J.**, Capping of eucaryotic mRNAs, *Cell*, 9, 645, 1976.
37. **Koper-Zwarthoff, E. C., Lockard, R. E., de Weerd, B., RajBhandary, U. L., and Bol, J. F.**, Nucleotide sequence of the 5′-terminus of alfalfa mosaic virus RNA 4 leading into the coat protein cistron, *Proc. Natl. Acad. Sci. U.S.A.*, 74, 5504, 1977.

Chapter 4

GENE ORDER

L. Andrew Ball

TABLE OF CONTENTS

TABLE 1

Determinations of the Gene Order of VSV[a]

Gene Order	Method	Ref.
	Transcriptional mapping by:	
3′... –N — NS — M — G — ...5′		17
–N — (NS , M) — G — L		18
–N — NS — M — G — L	UV irradiation,	26
Leader –N — (NS , M) — G — L	in vitro	13,14
— NS — M —		46
–N — M — G — L	UV irradiation, in vivo	45
–N — (NS , M) —		30
Leader –N — (NS , M , G) —	Kinetics of RNA synthesis,	31
–(N , NS , M , G) — L	in vitro	32
	Physical mapping by:	
–(N , NS , M , G) — L	Hybridization to defective	4,5,38,39
Leader –	particle and genome RNAs	14
–N — (NS , M) — G — L	Electron microscopy of mRNA-genome RNA duplexes	40,41

[a] Throughout this chapter, the structural genes are referred to by the letters that designate the viral proteins for which they code, i.e., N, NS, M, G and L.[53] Genes shown in brackets were not mapped relative to one another. The genome RNA, being of negative polarity, is represented in a 3′ to 5′ orientation.

I. INTRODUCTION

The five known structural genes of vesicular stomatitis virus (VSV), Indiana serotype, are arranged within the viral genome in the order 3′–N–NS–M–G–L–5′. The evidence will be reviewed in this chapter. In addition, a nucleotide sequence has been located at the extreme 3′ terminus of the genome which codes for a small transcript, the "leader" RNA. Strictly speaking, this nucleotide sequence does not constitute a gene, but its position in the viral RNA will also be discussed. The evidence for the locations of the structural genes and the leader RNA is based on two separate approaches, transcriptional and physical mapping. The first method measures the distances which the viral RNA polymerase must traverse in order to transcribe each gene, i.e., the distances between the genes and their promotor site(s). The second method measures the gene locations directly, in relation to the termini of the viral RNA. Since there appears to be only one promotor site, which is situated at the 3′ end of the genome, the two methods yield similar results. Data from both approaches and from several laboratories support the proposed gene order and are summarized in Table 1.

The genome of VSV (Indiana serotype) is a single-stranded RNA molecule of molecular weight 3.6 to 4.0 × 10^6 which is complementary to the viral messenger RNAs.[1-3] Five mRNA species have been identified in the cytoplasm of infected cells and each corresponds to one of the five known structural genes of the virus (Table 2). The sum of their molecular weights approximates that of the genome RNA, and they contain nucleotide sequences which, together, are complementary to more than 90% of the genome.[4-6] Low concentrations of nonionic detergents can be used to activate the RNA-dependent RNA polymerase of purified virus particles in vitro.[7] Under optimum

TABLE 2

Messenger RNAs of VSV

S Value	Molecular weight (× 10^{-6})									Average	Encoded Protein
28—31S	1.77	2.07	1.82	—	≥ 1.62	1.2	1.76	1.89	2.12	1.78	L
17S	0.72	0.67	0.92	0.71	0.67	0.63	0.58	0.57	0.71	0.69	G
14.5S	0.56	0.49	0.61	0.48	0.52	0.50	0.41	0.37	0.52	0.50	N
12S	0.32	0.34	0.34	0.19	0.25	0.29	0.23	0.19	0.27	0.27	NS
12S	0.32	0.34	0.34	0.19	0.25	0.29	0.20	0.15	0.27	0.26	M
Sum	3.59	3.81	3.93	—	≥ 3.31	2.91	3.18	3.17	3.89	3.47	
Method	a	b	c	d	c	e	f	f	e		
Reference	47	48	6	6	49	34	35	36	8		9,26,50—52

Note: Where appropriate, the molecular weights have been corrected for a poly(A) content of 100 nucleotides.[49]

[a] Polyacrylamide gel electrophoresis in the presence of sodium dodecyl sulfate.
[b] Sucrose density gradient centrifugation.
[c] Polyacrylamide gel electrophoresis in the presence of 98% formamide.
[d] Polyacrylamide gel electrophoresis in the presence of 8 *M* urea.
[e] Polyacrylamide gel electrophoresis in the presence of 98% formamide and 6 *M* urea.
[f] Polyacrylamide gel electrophoresis of mRNA-genome RNA duplexes.

conditions, the products of transcription closely resemble the authentic viral mRNAs; all five species are synthesized as intact, capped, methylated, polyadenylated, and translatable molecules.[6,8-12] In addition, transcription in vitro results in the synthesis of a small RNA molecule about 48 nucleotides long, which is complementary to a part of the genome RNA, but is not capped, methylated, or polyadenylated.[13-16] The nucleotide sequences which code for these six distinct products of transcription have been located on the viral RNA by the methods described below.

II. TRANSCRIPTIONAL MAPPING

A. UV Target-size Analysis

Transcriptional mapping means determining the spatial relationship between genes and their promotor sites. In situations where each gene is transcribed from a separate promotor, such information is worthless for mapping the genes relative to one another. However, where several genes are transcribed by read-through from a single promotor site, transcriptional mapping can reveal the order of genes within the transcriptional unit. In the case of VSV, transcriptional mapping data were used to determine both the number of promotor sites and the gene order.[17,18]

The technique was developed by Hackett and Sauerbier to map the ribosomal RNA genes of *E. coli*[19] and mouse L-cells.[20] It depends on the fact that irradiation of DNA with UV light results in the formation of covalent dimers between adjacent thymine residues.[21] Such dimers block the passage of DNA-dependent RNA polymerase and thus result in premature termination of transcription.[22] Assuming that the distribution of thymine residues is random, and that every TT sequence has an equal chance of dimerization, then the probability of UV-induced termination is directly proportional to the distance transcribed. Under these circumstances, the UV-sensitivity of transcription of a gene provides a measure of the proximity of its promotor site.

Fewer studies have been made of the effects of UV light on RNA templates. Thus, while it is known that irradiation of RNA results in the formation of uracil dimers,[23] the ability of such dimers to block transcription is not so firmly established. In fact, the success of the transcriptional mapping studies of VSV, Sendai virus,[24] and Newcastle disease virus[25] provides the strongest evidence that the technique is applicable to RNA as well as to DNA genomes.

Measurements of the UV-sensitivities of VSV gene expression are summarized in Table 3. Ball and White[17] used a cell-free system in which detergent-activated VSV-directed coupled transcription and translation of four of the five viral genes. Purified virus preparations were first exposed to doses of radiation sufficient to introduce up to 10 UV hits per genome, and then used to program the coupled system. Synthesis of individual viral proteins was assayed by polyacrylamide gel electrophoresis of the cell-free system products, and their sensitivites (or target-sizes) were calculated relative to that of the entire viral genome. Its target-size was measured from the UV-sensitivity of infectivity. This approach yielded target-size estimates for the N, NS, M and G protein genes (Table 3). Rather than approximating the individual physical sizes of the genes, the measured target-sizes were closely similar to the cumulative physical sizes, summed in the order of increasing UV-sensitivity, i.e., N < NS < M < G (Table 3). The simplest interpretation of this result was that the order of UV-sensitivity reflected the physical order of genes within a single transcriptional unit.

Although no L protein synthesis was detected in the coupled system, its putative mRNA was present among the products of transcription in vitro. Analysis of the response of the individual mRNAs to UV irradiation of the template therefore enabled Abraham and Banerjee[18] and Ball et al.[26] to measure the target-size of the L protein

TABLE 3

Gene	Relative target-size (daltons of RNA × 10^{-6})				RNA molecular weight (× 10^{-6})	
					Individual[a]	Cumulative[b]
Leader	—	—	0.02	—	0.02	0.02
N	0.55	0.6	0.55	0.48	0.50	0.52
NS	0.83	—	0.85	0.96	0.27	0.79
M	1.12	1.2			0.26	1.05
G	1.76	2.1	2.06	1.72	0.69	1.74
L	—	3.9	4.06	3.90	1.78	3.52
Method	a	b	c	d		
Reference	17	45	13,14,18	26		

Note: The letters below code the "Method" column:

a. From measurements of protein synthesis in vitro, target-sizes were calculated relative to that of viral infectivity, which was taken to correspond to an RNA of 3.82 × 10^6 molecular weight.[54]
b. From measurements of protein synthesis in vivo, target-sizes were calculated as in (a).
c. From measurements of RNA synthesis in vitro, target-sizes were calculated relative to that of 14.5s (N protein) mRNA, which was taken to correspond to an RNA of 0.55 × 10^6 molecular weight.
d. From measurements of RNA synthesis in vitro, target-sizes were calculated as in (a).

[a] Calculated as an average of the molecular weight measurements shown in Table 2.

[b] Summed in the order of increasing UV-sensitivity, i.e., leader, N, NS, M, G, L.

gene, and to confirm the previous measurements for the N and G protein genes. By this method of analysis, however, the NS and M protein genes could not be studied separately because their mRNAs were not resolved by the methods used.

The target-sizes derived by Abraham and Banerjee[18] were calculated relative to the UV-sensitivity of the N protein mRNA rather than that of the entire RNA genome. This approach depends on the assumption that the N protein gene, being the most resistant to irradiation, is close to the promotor site, and thus has a target-size equivalent to its physical size. While this assumption was substantiated by the other transcriptional mapping studies, and is the only simple way to rationalize the target-size data, it nevertheless involves a circular argument which can be avoided by using instead the UV-sensitivity of infectivity as the standard.

The genes of VSV displayed the same cumulative target-sizes when their UV-sensitivity was assayed during primary transcription in infected cells (Table 3). This eliminated the possibility that the sequential mechanism was peculiar to detergent-activated virus, or limited to transcription in vitro. Moreover, these results provided confirmation of the target-size of the L protein gene, which otherwise depended only on measurements of the 31S putative mRNA.

The discovery that the five genes were transcribed sequentially suggested that the initiation site for RNA synthesis might be at the 3′ terminus of the viral RNA. However, it was clear from the nucleotide sequence at the 5′ ends of the mRNAs, all of which were $^{m7}G^{5'pr}ppp^{5'pr}A_mpA_{(m)}pCpApGp$ . . . ,[27,28] that none of them were transcribed from the extreme 3' end of the genome, which had the sequence . . . Pyp GpU_{OH}.[29] Abraham et al.[13] therefore sought and found a small RNA with the properties expected for the product of initial transcription. Its 5' terminal sequence, ppApCpGp . . . , was not capped or methylated, but was complementary to the 3' end of the genome. Moreover, it could be isotopically labeled by incorporation of (β,γ-^{32}P) ATP, suggesting that the diphosphorylated terminus was derived from a triphosphate-mediated initiation event. This "leader" RNA was about 48 nucleotides long and was not polyadenylated. Transcriptional mapping studies[13,14] showed that its UV target was about equivalent to its physical size (0.02×10^6 daltons) which suggested that it lay very close to a promotor site and was transcribed before the five structural genes (Table 3). However, its target-size was so small that it made no detectable contribution to those of the distal genes, and could not, therefore, be definitely assigned to the same transcriptional unit. Nevertheless, the simplest interpretation of all the UV mapping studies is that transcription initiates at the 3' terminus of the genome, and proceeds in an order which reflects the physical arrangement of the viral genes: leader, N, NS, M, G, L. An alternative possibility which cannot be eliminated by transcriptional mapping data alone is that there exists a hierarchy of six interdependent promotor sites, which may be scattered throughout the genome,and that transcription of each gene activates the next promotor in the hierarchy. However, the physical mapping studies which are discussed below (section III) support the more straightforward interpretation.

B. Kinetics of RNA Synthesis and Abundance of Products

As well as being reflected in the UV target-sizes of the genes, sequential transcription from a single promotor site has two other consequences for RNA synthesis. If transcription is synchronized, the products of proximal genes will appear sooner and perhaps in greater molar abundance than those of more distal genes. Using a coupled transcription-translation system in which they halted the synthesis of translatable mRNA by adding S-adenosylhomocysteine, Breindl and Holland showed that the N protein mRNA was transcribed before those of the NS or M proteins.[30] Rose et al. were able to detect T1 oligonucleotides derived from the N protein mRNA before those from the NS, M, or G protein messages.[31] At even earlier times during transcription they detected three characteristic T1 oligonucleotides which they tentatively assigned to the leader RNA. Correspondingly, sequences which, like the L protein mRNA, were complementary to the RNA of defective VSV particles that mapped at the 5' end of the genome, were synthesized only at late times during transcription in vitro.[32,33]

The relative abundances of the RNA products also reflect the sequence of transcription. Villarreal et al.[34] reported that the molar ratios of the mRNA species made in infected cells were N:(NS + M):G:L:: 100:107:36:5·6. The five viral proteins were synthesized in almost identical ratios: N:NS:M:G:L:: 100:(48):53:34:6·9. (The figure for the NS protein is approximate because of uncertainty about its true molecular weight.) The gradient of molar abundance of the products of transcription in vitro followed the same gene order,[35,36] although measurements of the relative molar yield of the leader RNA have not been reported.

As is discussed in detail elsewhere in this volume, it is not clear how the six discrete RNA products arise during sequential transcription of the genome, although precursor processing by nucleolytic cleavage is an attractive possibility. However, the gradient of molar abundances presumably reflects a progressive attenuation of transcription by dissociation of the polymerase from its template as the reaction proceeds.

III. PHYSICAL MAPPING

A. Hybridization to Defective Particle and Genome RNAs

Repeated passage of VSV at high multiplicity gives rise to defective particles which lack regions of the viral genome.[37] Analysis of the RNA-RNA hybridization between the 12 to 18S and 31S mRNAs on the one hand, and the subgenomic RNAs from a number of such defective particles on the other, made possible the derivation of a low-resolution physical map of the genome. Thus Leamnson and Reichmann[4] located the four smaller genes at one (unidentified) end of the viral RNA and the L protein gene at the other. Stamminger and Lazzarini[5] originally located the L protein gene as being one gene removed from the 5′ end of the genome, but these authors have recently modified their map in light of subsequent hybridization data.[38] They now locate the L protein gene at the 5′ end and the other four genes at the 3′ end of the genome, in agreement with more recent results from Reichmann's laboratory.[39]

The three-base complementarity which has been shown between the 5′ end of the leader RNA and the 3′ end of the genome is insufficient evidence from which to conclude that this is, indeed, its map location. However, Colonno and Banerjee[14] showed that annealing of leader RNA to the genome protected the 3′ end of VSV RNA from nucleolytic degradation. Conversely, annealing of the N protein mRNA failed to protect the 3′ end. Moreover, the leader RNA appeared to hybridize only once per genome, indicating that it has a unique location at the extreme 3′ terminus.

B. Electron Microscopic Analysis of mRNA-genome RNA Duplexes

Single- and double-stranded RNA can be distinguished by electron microscopy, so annealing of individual mRNA molecules to VSV genome RNA results in observable duplex regions at the positions of the corresponding genes. Contour length measurements of the single- and double-stranded regions permit accurate physical mapping of the genes. Moreover, since the poly (A) tracts on the mRNA molecules are not encoded in the genome, they form single-stranded tails at one end of the annealed regions and thus reveal the orientation of the duplexes.

Herman et al.[40] used the products of in vitro transcription as a source of viral mRNA, which was then selected by oligo-dT cellulose chromatography and resolved into 12S, 14.5S, and 17S fractions by sucrose gradient centrifugation. Although far from pure, these fractions were enriched for NS + M, N and G protein mRNAs, respectively.[9] Each fraction was annealed separately to VSV genome RNA and the resulting partial duplexes were examined by electron microscopy. Two examples of such partial duplexes, formed with 14.5S mRNA (upper panel) and 17S mRNA (lower panel) are shown in Figure 1. The N protein mRNA (14.5 S) clearly anneals to the 3′ end of the genome; the region at the extreme 3′ terminus which is expected to anneal only to the leader RNA is probably too small to be readily visualized as a single-stranded tail by this technique. The G protein mRNA (17S), on the other hand, anneals to a region near the middle of the genome.

Using the same approach, Herman and Ball[41] examined partial duplexes formed between VSV genome RNA and the 31S putative L protein mRNA, synthesized in vitro[8] (Figure 2). The L protein message spans about 59% of the genome, corresponding to a molecular weight of about 2.2×10^6. The duplex region usually extends to one end of the genome, although a short single-stranded tail is occasionally visible (see Figure 2). The presence of a poly(A) tract at this position has been demonstrated by hybridization with poly(dT) tails covalently linked to SV40 DNA (data not shown). This establishes the orientation of the duplexes, and locates the L protein gene at the 5′ terminus of the genome.

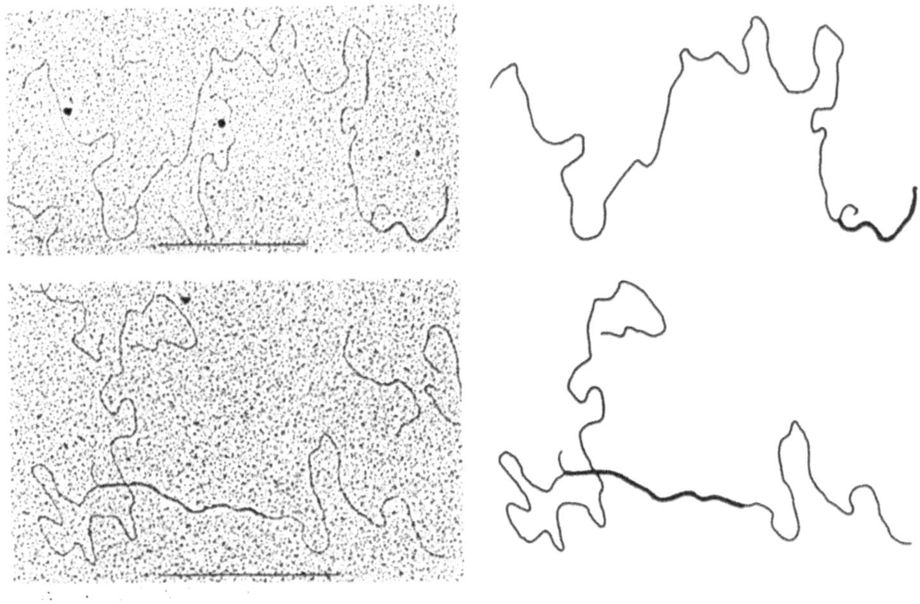

FIGURE 1. Electron micrographs (left) and interpretative line drawings (right) of partial duplexes formed between VSV genome RNA and individual species of viral messenger RNA. Top: 14.5S (N protein) mRNA; bottom: 17S (G protein) mRNA. The single-stranded poly(A) tails are clearly visible and indicate that the 3′ end of the genome RNA is to the right in both cases. The bars represent 1 μm. (From Herman, R. C., Adler, S., Lazzarini, R. A., Colonno, R. J., Banerjee, A. K., and Westphal, H., *Cell*, 15, 587, 1978. With permission.)

Contour length measurements of the single- to double-strand transition points of a number of partially duplexed molecules are summarized in Figure 3. They show that the physical order of the genes is 3′–N–(NS, M)–G–L–5′ and that the approximate positions of the intercistronic boundaries agree well with the results of transcriptional mapping (Table 4).

IV. DISCUSSION

The correspondence between the transcriptional and physical maps of VSV firmly establishes that transcription proceeds from one end of the genome to the other. The map location of the leader RNA further indicates that the initiation site is located precisely at the 3′ terminus. However, the termination site has not been located. Whether the 3′ end of the L protein mRNA (within the poly[A]) is complementary to the extreme 5′ terminus of the genome, or whether there is a short region beyond the L protein gene that is analogous to the region coding for the leader RNA remains to be established. Examination of partial duplexes formed with L protein mRNA indicates that if such a region exists, it is probably less than about 100 nucleotides long, based on the estimated limits of electron microscopic resolution.

The sequential nature of transcription suggests that the mechanism may share some of the features of positive strand synthesis during replication. However, the relationships between the two processes, and, in particular, the mechanism of transition from one to the other, are unclear. In one case, the information in the genome is copied into six discrete RNAs, five of which undergo extensive posttranscriptional processing to become functional messengers. In the other case, the genome information is copied into an intact complementary RNA which forms a nucleoprotein complex with the viral N protein, and functions as a template for virion RNA synthesis. It has been

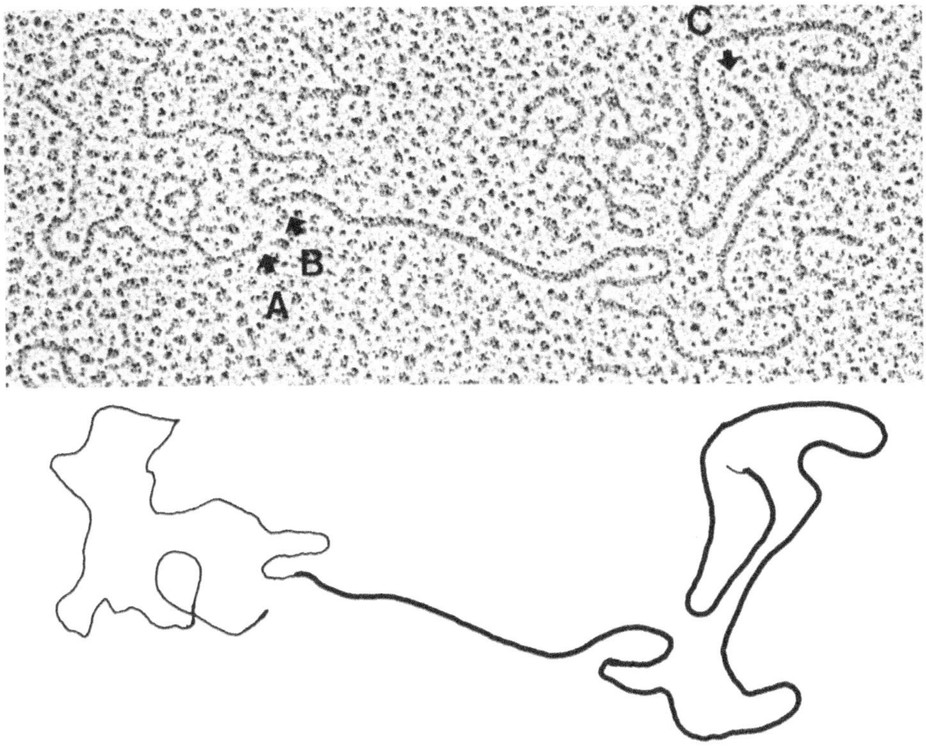

FIGURE 2. Electron micrograph (top) and interpretative line drawing (bottom) of a partial duplex formed between VSV genome RNA and 31S (putative L protein) mRNA.[41] A indicates the 3′ end of the genome; A → B is a region of single-stranded genome RNA; B indicates the position of the single- to double-strand transition and the 5′ end of the 31S mRNA; B → C is the mRNA-genome RNA duplex; C indicates the 5′ end of the genome and the 3′ end of the 31S mRNA. The poly(A) tail can perhaps be discerned.

suggested[13,16,26] that the mRNAs may be generated by nucleolytic cleavage of the nascent positive strand, with subsequent capping, methylation, and polyadenylation of the cleavage products. Such a scheme would account for the unusual capping mechanism of VSV, whereby two of the three phosphate groups in the cap are contributed by the capping GTP residue.[42] In the absence of cleavage, the product of synthesis would be an intact positive copy of the genome RNA, which is a necessary intermediate in replication. Thus, according to this model, the difference between trancription and replication would be the presence or absence of cleavage of the nascent positive strand at the intercistronic junctions.

The N protein can protect the viral RNA completely against nucleolytic action,[43] so it is tempting to speculate that it may be responsible for inhibiting cleavage of the positive RNA strand also. If its primary binding site was the leader RNA, the N protein might form there a nucleation site similar to the initial complex between tobacco mosaic virus RNA and the viral coat protein disc.[44] In the case of this helical ribonucleoprotein, the addition of further protein molecules (elongation) is a much faster reaction resembling a linear crystallization process. Without the correct nucleation site, the kinetics of RNA-protein association are drastically inhibited.

According to this model, therefore, VSV RNA synthesis in the absence of an excess of N protein would result in naked nascent RNA chains that were susceptible to nucleolytic processing to form leader RNA and the five mRNAs. However, when the

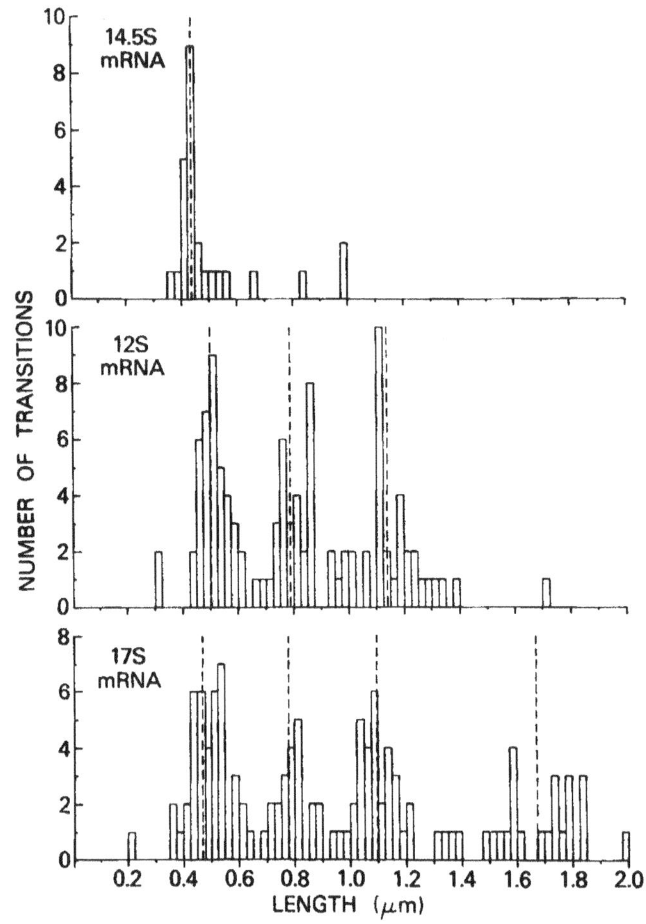

FIGURE 3. Histograms of single- to double-strand transitions observed in mRNA-genomic RNA partial duplexes. Duplexes were prepared by annealing genomic RNA to the 14.5S (N protein) mRNA (top panel), 12S (NS + M proteins) mRNA (middle panel) and 17S (G protein) mRNA (bottom panel). The broken vertical lines indicate the mean transition points, expressed in terms of the contour length from the 3′ terminus of the genome. The length of the viral RNA was 4.1 μm so 1 μm corresponds to about 0.95×10^6 daltons of RNA. (From Herman, R. C., Adler, S., Lazzarini, R. A., Colonno, R. J., Banerjee, A. K., and Westphal, H., *Cell*, 15, 587, 1978. With permission.)

viral N protein had accumulated sufficiently it would bind to a primary recognition site within the leader RNA to form a nucleus for the rapid binding of more N protein molecules. These would be added in a 5′ to 3′ direction as the nascent chain grew, and would protect the RNA against cleavage. The result would be an intact positive copy of the genome, complexed with its full complement of N protein molecules and thus in the right form to act as a template for genome RNA synthesis.

Although it is totally speculative, this model accounts for many of the observed facts concerning VSV transcription and replication. It is already providing a useful stimulus for the design of new experiments.

TABLE 4

Correlation of the Physical and Transcriptional Maps of VSV

Intercistronic boundary	Distance of the intercistronic boundary from the 3′ terminus of the genome (daltons of RNA × 10^{-6}) from EM analysis[a]	from UV target-size[b]
Leader/N	—	0.02
N/NS	0.44	0.55
NS/M	0.73	0.83
M/G	1.04	1.16
G/L	1.55	1.91
L terminus	—	3.95

[a] Calculated as averages of the data shown in Figure 3. The mean contour length of VSV genome RNA was 4.1 μm, and was taken to correspond to a molecular weight of 3.82×10^6.[54]

[b] Calculated as averages of the data shown in Table 3.

ACKNOWLEDGMENTS

I am grateful to Drs. R. Herman, S. Adler, R. A. Lazzarini, R. Colonno, A. K. Banerjee, and H. Westphal for their generosity in letting me present and discuss some of their unpublished data, shown in Figures 1 and 3. My own work[8,17,25,26,45] was supported by grant number CA-14733 from the National Cancer Institute.

REFERENCES

1. **Wagner, R. R.,** Reproduction of rhabdoviruses, in *Comprehensive Virology,* Vol. 4, Fraenkel-Conrat, H. and Wagner, R. R., Eds., Plenum Press, New York, 1976, 1.
2. **Huang, A.S., Baltimore, D., and Stampfer, M.,** Ribonucleic acid synthesis of vesicular stomatitis virus. III. Multiple complementary messenger RNA molecules, *Virology,* 42, 946, 1970.
3. **Mudd, J. A. and Summers, D. F.,** Polysomal ribonucleic acid of vesicular stomatitis virus-infected HeLa cells, *Virology,* 42, 958, 1970.
4. **Leamnson, R. N. and Reichmann, M. E.,** The RNA of defective vesicular stomatitis virus particles in relation to viral cistrons, *J. Mol. Biol.,* 85, 551, 1974.
5. **Stamminger, G. and Lazzarini, R. A.,** Analysis of the RNA of defective VSV particles, *Cell,* 3, 85, 1974.
6. **Moyer, S. A., Grubman, M. J., Ehrenfeld, E., and Banerjee, A. K.,** Studies on the *in vivo* and *in vitro* messenger RNA species of vesicular stomatitis virus, *Virology,* 67, 463, 1975.
7. **Baltimore, D., Huang, A. S., and Stampfer, M.,** Ribonucleic acid synthesis of vesicular stomatitis virus. II. An RNA polymerase in the virion, *Proc. Natl. Acad. Sci. U.S.A.,* 66, 572, 1970.
8. **Ball, L. A. and White, C. N.,** Coupled transcription and translation in mammalian and avian cell-free systems, *Virology,* 84, 479, 1978.
9. **Both, G. W., Moyer, S. A., and Banerjee, A. K.,** Translation and identification of the mRNA species synthesized *in vitro* by the virion-associated RNA polymerase of vesicular stomatitis virus, *Proc. Natl. Acad. Sci. U.S.A.,* 72, 274, 1975.
10. **Moyer, S. A. and Banerjee, A. K.,** Messenger RNA species synthesized *in vitro* by the virion-associated RNA polymerase of vesicular stomatitis virus, *Cell,* 4, 37, 1975
11. **Banerjee, A. K., Moyer, S. A., and Rhodes, D. P.,** Studies on the *in vitro* adenylation of RNA by vesicular stomatitis virus, *Virology,* 61, 547, 1974.
12. **Rhodes, D. P. and Banerjee, A. K.,** 5′-terminal sequence of vesicular stomatitis virus mRNAs synthesized *in vitro,* *J. Virol.,* 17, 33, 1976.

13. **Abraham, G., Colonno, R. J., and Banerjee, A. K.**, Evidence for the synthesis of a "leader" RNA segment followed sequential transcription of the genes of vesicular stomatitis virus, in *Animal Virology*, Vol. 4, ICN-UCLA Symposia on Molecular and Cellular Biology, Baltimore, D., Huang, A. S., and Fox, C. F., Eds., Academic Press, New York, 1976, 439.
14. **Colonno, R. J. and Banerjee, A. K.**, Mapping and initiation studies on the leader RNA of vesicular stomatitis virus, *Virology*, 77, 260, 1977.
15. **Colonno, R. J. and Banerjee, A. K.**, A unique RNA species involved in initiation of vesicular stomatitis virus RNA transcription *in vitro, Cell*, 8, 197, 1976.
16. **Colonno, R. J., Abraham, G., and Banerjee, A. K.**, Blocked and unblocked 5′-termini in vesicular stomatitis virus product RNA *in vitro*: their role in mRNA biosynthesis, in *Progress in Nucleic Acid Research and Molecular Biology*, Vol. 19, Volkin, E. and Cohn, W. E., Eds., Academic Press, New York, 1976.
17. **Ball, L. A. and White, C. N.**, Order of transcription of genes of vesicular stomatitis virus, *Proc. Natl. Acad. Sci. U.S.A.*, 73, 442, 1976.
18. **Abraham, G. and Banerjee, A. K.**, Sequential transcription of the genes of vesicular stomatitis virus, *Proc. Natl. Acad. Sci. U.S.A.*, 73, 1504, 1976.
19. **Hackett, P. B. and Sauerbier, W.**, Radiological mapping of the ribosomal RNA transcription unit in *E. coli, Nature (London)*, 251, 637, 1974.
20. **Hackett, P. B. and Sauerbier, W.**, The transcriptional organization of the ribosomal RNA genes in mouse L-cells, *J. Mol. Biol.*, 91, 235, 1975.
21. **Wacker, A.**, Molecular mechanisms of radiation effects, *Prog. Nuclec Acid Res. Mol. Biol.*, 1, 369, 1963.
22. **Michalke, H. and Bremer, H.**, RNA synthesis in *Escherichia coli* after irradiation with ultraviolet light, *J. Mol. Biol.*, 41, 1, 1969.
23. **Miller, R. L. and Plagemann, P. G. W.**, Effects of ultraviolet light on mengovirus: formation of uracil dimers, instability and degradation of capsid, and covalent linkage of protein to viral RNA, *J. Virol.*, 13, 729, 1974.
24. **Glazier, K., Raghow, R., and Kingsbury, D. W.**, Regulation of Sendai virus transcription: evidence for a single promotor *in vivo, J. Virol.*, 21, 863, 1977.
25. **Ball, L. A., Collins, P. L., and Hightower, L. E.**, Transcription, translation and mapping of the genes of Newcastle disease virus, in *Negative Strand Viruses and the Host Cell*, Mahy, B. W. J. and Barry, R. D., Eds., Academic Press, London, 1978, 367.
26. **Ball, L. A., White, C. N., and Collins, P. L.**, A transcriptional map of vesicular stomatitis virus, in *Animal Virology*, Vol. 4, ICN-UCLA Symposia on Molecular and Cellular Biology, Baltimore, D., Huang, A. S., and Fox, C. F., Eds., Academic Press, New York, 1976, 419.
27. **Rhodes, D. P. and Banerjee, A. K.**, 5′-terminal sequence of vesicular stomatitis virus mRNAs synthesized *in vitro, J. Virol.*, 17, 33, 1976.
28. **Rose, J. K.**, Heterogeneous 5′-terminal structures occur on vesicular stomatitis virus mRNAs, *J. Biol. Chem.*, 250, 8048, 1975.
29. **Banerjee, A. K. and Rhodes, D. P.**, 3′ -terminal sequence of vesicular stomatitis virus genome RNA, *Biochem. Biophys. Res. Commun.*, 68, 1387, 1976.
30. **Breindl, M. and Holland, J. J.**, Studies on the *in vitro* transcription and translation of vesicular stomatitis virus mRNA, *Virology*, 73, 106, 1976.
31. **Rose, J. K., Lodish, H. F., and Brock, M. L.**, Giant heterogeneous polyadenylic acid on vesicular stomatitis virus mRNA synthesized *in vitro* in the presence of S-adenosylhomocysteine, *J. Virol.*, 21, 683, 1977.
32. **Reddy, D. V. R., Schnitzlein, W. M., and Reichmann, M. E.**, Use of defective interfering particle RNA probes in the determination of the order of *in vitro* transcription of vesicular stomatitis virus genes, *J. Virol.*, 21, 432, 1977.
33. **Roy, P. and Bishop, D. H. L.**, Genome homology of vesicular stomatitis virus and defective T particles and evidence for the sequential transcription of the virion ribonucleic acid, *J. Virol.*, 9, 946, 1972.
34. **Villarreal, L. P., Breindl, M., and Holland, J.J.**, Determination of molar ratios of vesicular stomatitis virus induced RNA species in BHK-21 cells, *Biochemistry*, 15, 1663, 1976.
35. **Freeman, G. J., Rose, J. K., Clinton, G. M., and Huang, A. S.**, RNA synthesis of vesicular stomatitis virus. VII. Complete separation of the mRNAs of vesicular stomatitis virus by duplex formation, *J. Virol.* 21, 1094, 1977.
36. **Rhodes, D. P., Abraham, G., Colonno, R. J., Jelinek, W., and Banerjee, A. K.**, Characterization of vesicular stomatitis virus mRNA species synthesized *in vitro, J. Virol.*, 21, 1105, 1977.
37. **Huang, A. S.**, Defective interfering viruses, *Ann. Rev. Microbiol.*, 27, 101, 1973.
38. **Stamminger, G. and Lazzarini, R. A.**, Unpublished data, 1977.

39. **Schnitzlein, W. M. and Reichmann, M. E.**, The size and cistronic origin of defective vesicular stomatitis virus particle RNAs in relation to homotypic and heterotypic interference, *J. Mol. Biol.*, 101, 307, 1976.
40. **Herman, R. C., Adler, S., Lazzarini, R. A., Colonno, R. J., Banerjee, A. K., and Westphal, H.**, Intervening polyadenylate sequences in RNA transcripts of vesicular stomatitis virus, *Cell*, 15, 587, 1978.
41. **Herman, R. C. and Ball, L. A.**, Unpublished results, 1978.
42. **Abraham, G., Rhodes, D. P., and Banerjee, A. K.**, The 5'-terminal structure of the methylated mRNA synthesized *in vitro* by vesicular stomatitis virus, *Cell*, 5, 51, 1975.
43. **Cartwright, B., Smale, C. J., and Brown, F.**, Dissection of vesicular stomatitis virus into the infective ribonucleoprotein and immunizing components, *J. Gen. Virol.*, 7, 19, 1970.
44. **Butler, P. J. G. and Klug, A.**, Assembly of the particle of tobacco mosaic virus from RNA and disks of protein, *Nature London New Biol.*, 229, 47, 1971.
45. **Ball, L. A.**, Transcriptional mapping of vesicular stomatitis virus *in vivo*, *J. Virol.*, 21, 411, 1977.
46. **Rhodes, D. P. and Banerjee, A. K.**, unpublished observation, cited in Banerjee, A. K., Abraham, G., and Colonno, R. J., Vesicular stomatitis virus: mode of transcription, *J. Gen. Virol.*, 34, 1, 1977.
47. **Schincariol, A. L. and Howatson, A. F.**, Replication of vesicular stomatitis virus. I. Viral specific RNA and nucleoproteins in infected L-cells, *Virology* 42, 732, 1970.
48. **Rhodes, D. P., Moyer, S. A., and Banerjee, A. K.**, *In vitro* synthesis of methylated messenger RNA by the virion-associated polymerase of vesicular stomatitis virus, *Cell*, 3, 327, 1974.
49. **Rose, J. K. and Knipe, D.**, Nucleotide sequence complexities, molecular weights, and poly (A) content of the vesicular stomatitis virus mRNA species, *J. Virol.*, 15, 994, 1975.
50. **Morrison, T., Stampfer, M., Baltimore, D., and Lodish, H. F.**, Translation of vesicular stomatitis messenger RNA by extracts from mammalian and plant cells, *J. Virol.*, 13, 62, 1974.
51. **Knipe, D., Rose, J. K., and Lodish, H. F.**, Translation of individual species of vesicular stomatitis viral mRNA, *J. Virol.*, 15, 1004, 1975.
52. **Both, G. W., Moyer, S. A., and Banerjee, A. K.**, Translation and identification of the viral mRNA species isolated from subcellular fractions of vesicular stomatitis virus-infected cells, *J. Virol.*, 15, 1012, 1975.
53. **Wagner, R. R., Prevec, L., Brown, F., Summers, D. F., Sokol, F., and MacLeod, R.**, Classification of rhabdovirus proteins: a proposal, *J. Virol.*, 10, 1228, 1972.
54. **Repik, P. and Bishop, D. H. L.**, Determination of the molecular weight of animal RNA viral genomes by nuclease digestions. I. Vesicular stomatitis virus and its defective T particle, *J. Virol.*, 12, 969, 1973.

Chapter 5

RNA REPLICATION

Gail Williams Wertz

TABLE OF CONTENTS

I. INTRODUCTION

The replication of the single, negative strand of rhabdoviral RNA can be postulated to occur as a relatively straightforward molecular process. As a first step, replication would involve the synthesis of genome length complementary (plus) strand RNA. Then, using the plus strand for template, progeny negative-strand molecules are synthesized. As with many natural processes, however, the more intense the study of RNA replication becomes, the more complex the process appears. At present, the most accurate statement that can be made concerning the mechanism of replication of rhabdoviral RNA is that it is not understood. Indeed, recent findings concerning the nature and origin of VSV defective interfering (DI) particle RNAs offer unique possibilities for the mechanism of replication of negative strand virus RNAs.

It is the purpose of this chapter to review recent findings bearing on the replication process, to interpret these results, and in light of existing findings, to offer testable models of the replication process. Questions to be considered in detail will be the nature of the template for replication, the existence and nature of intermediate structures in replication, the enzyme(s) for replication, the requirement for concommitant protein synthesis for replication, and the replication of DI particle RNAs.

For comprehensive reviews of work prior to 1975, reference is made to Wagner[1] for a discussion of all aspects of rhabdoviral replication, Emerson[2] for a disussion of structure and function of viral components, and Burke and Russell[3] and Subak-Sharpe and Pringle[4] for a discussion of control of replication process.

II. REPLICATION OF RNA BY STANDARD INFECTIOUS VIRIONS

A. The Question of the Template for Replication

The rhabdovirus genome is a single strand of RNA of negative polarity.[1] No evidence exists for an intermediate role of DNA in the replication process.[5] For these reasons, it seems logical to postulate that replication of this single strand of RNA occurs via a process involving synthesis of a complementary (plus) strand which serves as the template from which progeny negative strands are replicated.

The enzymatic machinery contained in the virion is sufficient for the synthesis and posttranscriptional modification of the complementary mRNA molecules. Five discrete mRNA molecules and a small 48 base-pair leader RNA sequence are synthesized by the virion in vitro or in vivo in the absence of protein synthesis. To date, however, no RNA of the size of virion RNA (4×10^6 daltons or 40S) has been observed in vitro or in vivo in the absence of protein synthesis. The synthesis of virion-length RNA in the infected cell requires protein synthesis. The requirement for protein synthesis is continuous; its inhibition results in a rapid halt of virion-length RNA production. It has been proposed that the individual small mRNAs could serve as templates and their products in turn could be ligated to form virion-size RNA.[6] No positive evidence has been offered for this possibility. Evidence has been presented, based on pulse-chase studies, that small messenger-sized RNAs present in the infected cell are not metabolized to virion-size RNA.[7]

In contrast to virion RNA, which is entirely of negative-strand polarity, intracellular 40S RNAs of both positive and negative polarity have been reported.[8-12]It was observed that 20 to 40% of the total intracellular 40S RNA was complementary to virion RNA (plus-stranded). In addition, Soria et al.[10] reported that the 40S RNA molecules complementary to virion RNA occur in the infected cell cytoplasm in the form of intracellular nucleocapsids. It was found that 20 to 30% of the 40S RNA present in the form of intracellular nucleocapsids was plus-stranded. Additionally, Soria et al.[10]

cited as an unpublished finding the observation that all of the complementary 40S RNA found in the cell was in nucleocapsid form. Smaller complementary mRNAs which do not serve as templates are not found in nucleocapsid form.

The total intracellular 40S virion-sized RNA was analyzed further by Morrison et al.[11] It was determined that total intracellular 40S RNA isolated by velocity gradient separation contained a ratio of plus to minus strands of 40 to 60. This mixture of plus- and minus-strand RNA was shown to migrate in polyacrylamide gels with the same relative electrophoretic mobility as the pure negative-strand RNA extracted from virions. This 40S intracellular RNA was further characterized by measuring its ability to bind to poly(U)-containing glass fiber filters. The results of these experiments showed that 22% of the total intracellular 40S RNA (of which 40% was plus stranded) could bind to poly (U) filters. Unpublished results were quoted that the bound material was greatly enriched for plus strands and these results were interpreted to mean that the 40S plus strand contained poly(A) stretches. It was not shown, however, that the material binding to the poly(U) was completely plus-stranded or that it was 40S in size. Since the RNA applied to the filters had originally been isolated by pooling sucrose fractions, it is possible that some large mRNA could have been present.

In summary, it has been demonstrated that virion-size complementary (+) RNA does exist in VSV-infected cells. This 40S plus strand is present in the cell in the form of a nucleocapsid structure. However, direct proof that this positive strand nucleocapsid structure is the template for replication of progeny virion RNA has yet to be furnished.

B. The Question of Intermediate Structures Involved in RNA Replication

Rapidly labeled RNA containing structures have been implicated by pulse-chase studies as the functional intermediates in the replication of the single-stranded RNAs of bacteria and some animal viruses.[13-15] These replicative intermediate (RI) structures have been described as containing a strand of template RNA with attached nascent progeny strands of opposite polarity. A signal characteristic of these intermediate structures has been that they have some regions of hydrogen-bonded base-pairing between template and product. This results in the molecule being significantly resistant to digestion by ribonuclease A. Since these structures are usually isolated from cells using phenol extraction as a means of deproteinization, it is not possible to determine to what extent base-pairing occurs in the natural state. It has been proposed that the alignment between template and product is maintained solely by the enzyme mediating the reaction, and that extensive hydrogen-bonded base-pairing may occur only as a result of deproteinization and the resultant collapse of molecules against one another. However, if the template and progeny strands are in close association and alignment demanded for base-pairing-mediated replication, then some degree of hydrogen-bonded structure would seem logical. Therefore, the use of partial hydrogen-bonded RNase resistance as a diagnostic characteristic of RI seems valid. Although the normal extent of hydrogen bonding cannot be determined accurately on a deproteinized molecule, the actual relationship between template and progeny strands, because of the demands of the base-pairing replication process, must be one of close alignment and therefore the purified complex may be a close approximation of the in vivo replication complex.

The case of the rhabdovirus, VSV, where the template for replication (either for plus- or minus-strand synthesis) is an RNA-containing nucleocapsid structure, presents a further complexity. How does the enzymatic machinery which catalyzes a base-pairing reaction move along a template tightly covered with N protein? It is most probable in this case that the extent of hydrogen-bonding between template and product is minimal in the natural state due to the presence of nucleocapsid protein. Extensive regions of base-pairing and resultant nuclease resistance are thus probably a function of the association of RNA molecules once the protein has been removed. The nature and number of template and progeny strands, however, should be an accurate reflection of the natural process. It has been shown in the case of VSV that random annealing does not occur between the normally nucleocapsid-encoated virion negative-strand RNA and VSV intracellular mRNA when the two are co-extracted with phenol.[12]

1. Isolation of Intermediate Structures

Two problems have hindered the study of rhabdovirus replication: the difficulty in discriminating between replicative and transcriptive events, and the presence of defective interfering (DI) particles. Many workers have reported the presence of double (DS) or partially double-stranded (pDS) RNA structures that could serve as intermediates.[1] In most of these cases, it was not possible to discriminate among replicative, transcriptive, and DI-particle RNA structures. The most careful study of DS and pDS RNA structures from VSV-infected cells was performed by Schincariol and Howatson.[8] How ever, when the structures were denatured and the size of the RNA molecules subsequently determined, it was found that, at most, 7% of the label present in the denatured structures was of the size of virion RNA, 85% of the material had a sedimentation coefficient less than 28S, and 65% of the material had an S value less than 18S. It is most probable that the structures isolated by Schincariol and Howatson[8] were predominantly transcriptive intermediates.

The problem of discriminating replicative intermediates from transcriptive intermediates has been greatly diminished by the use of temperature-sensitive mutants, as described recently by Wertz,[12] resulting in the isolation and characterization of a probable replicative intermediate structure.

Using the temperature-sensitive mutant *ts*G114, which has been shown to be defective in transcription, but not replication,[9,16] it was possible to establish conditions in which replication, not transcription, was the predominant RNA synthetic event. Cells were infected with *ts*G114 at permissive temperature and at 4 hr postinfection were shifted up to the nonpermissive temperature of 39°C. The data presented in Figure 1 show that under these conditions of infection, transcription, the synthesis of messenger RNA, is inhibited by 80%, whereas replication is increased 100%. RNA was pulse labeled for 10 min in the presence of actinomycin D using these conditions of infection. The cytoplasmic RNA was then extracted and fractionated by LiCl precipitation and chromatography on CF-11 cellulose to yield a radiolabeled molecule that was partially double- and partially single-stranded. Single-stranded or completely double-stranded RNAs were completely removed from the preparation.

The isolated RNA had the following characteristics:

- It was insoluble in high salt, indicating some single-strand components
- It eluted from a cellulose column as RNA with a significant degree of secondary structure
- It was approximately 57% ribonuclease A resistant
- It sedimented in sucrose velocity gradients as a broad heterogeneous peak between 25 and 35S
- It had the electrophoretic mobility of an RNA molecule having a molecular weight of 9 to 12 × 10^6: this is greater than twice the molecular weight of a single strand of virus RNA
- Upon denaturation, the structure resolved to RNA, 60% of which was virion-sized — the remainder, a heterogeneous array of smaller RNAs

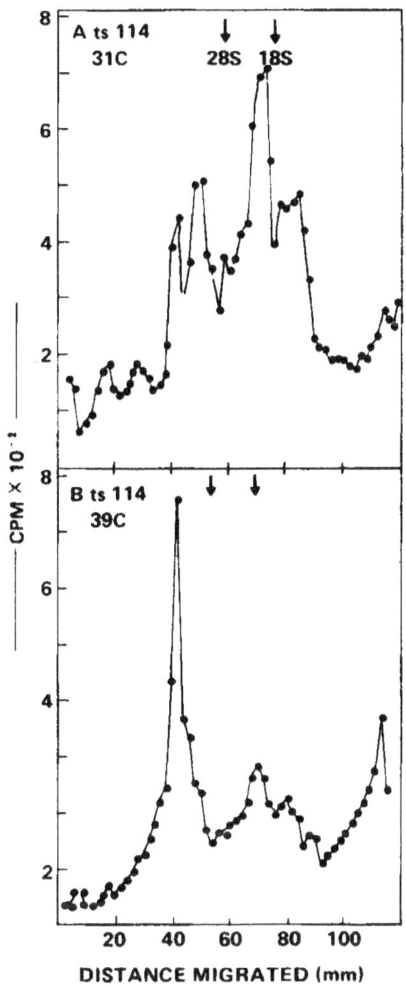

FIGURE 1. Effect of a shift to nonpermissive temperature on synthesis of RNA by tsG114. BHK-21 cells were infected at a multiplicity of one and maintained at 31°C. At 4 hr postinfection, the cultures in (B) were shifted to 39°C for 20 min. At 4 hr and 20 min, [^{3}H] uridine was added to all cultures for 30 min (A at 31°C; B at 39°C). RNA was extracted from cytoplasmic extracts and analyzed by electrophoresis in polyacrylamide gels. (From Wertz, G. W., *Virology*, 85, 271, 1978. With permission.)

- The RNA structure was 60% negative-stranded and 40% positive-stranded, as measured by RNA-RNA hybridization of the RNA from denatured structures with either messenger or virion RNA

Figure 2 shows a profile of the electrophoretic mobility in 2.2% polyacrylamide gels of the partially double-stranded structure before and after denaturation by heating.

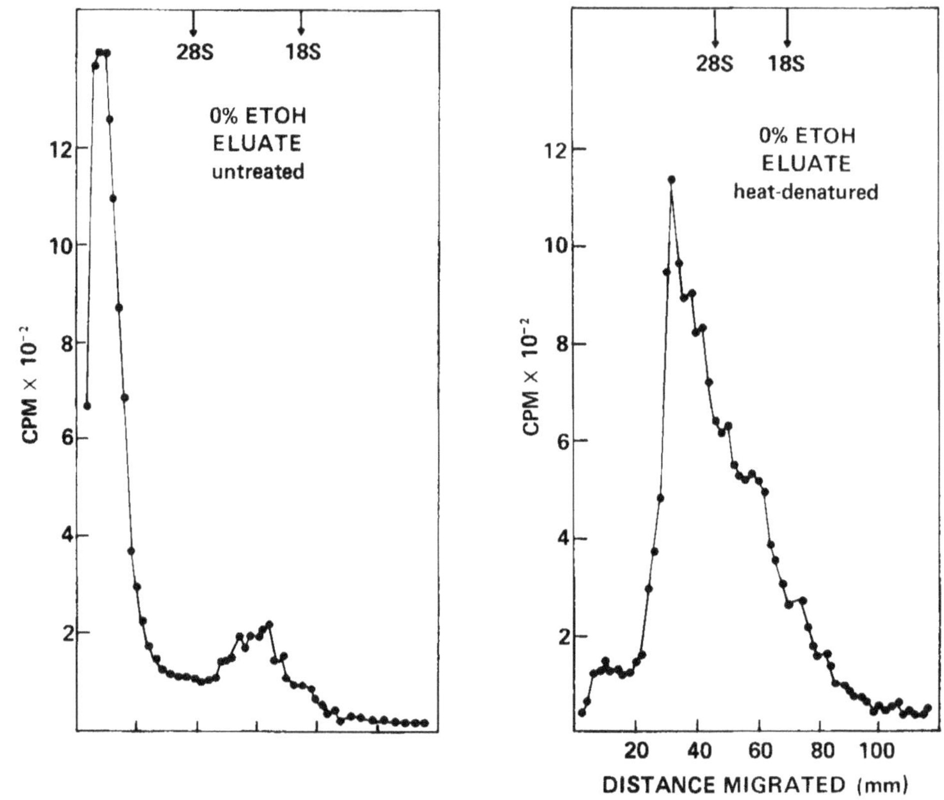

FIGURE 2. Heat denaturation of pDS RNA. RNA from a 10-min pulse of tsG114-infected cells, shifted to 39°C after 4 hr at 31°C, was fractionated by LiCl precipitation and cellulose CF-11 column chromatography. The pDS RNA was denatured by heating. The RNA was analyzed by polyacrylamide gel electrophoresis. (From Wertz, G. W., *Virology*, 85, 271, 1978. With permission.)

The characteristics of the isolated molecule strongly suggest that it is a replicative intermediate structure. The greatest amount of material in a replicative intermediate (either by mass or by labeling) should be in the nontemplate strands; 60% of the material present in the structure was of negative polarity. Also, 60% of the material present was of virion size after denaturation. If, at most, 50% of the material was template, then there exists another 10% that must represent complete or very nearly completed progeny molecules. However, although this structure possesses the characteristics of a replicative intermediate, proof that it is the direct precursor of virion RNA will require pulse-chase studies.

Wertz[12] has also shown that it is possible to drive the intact pDS structure to increased ribonuclease resistance by annealing with RNA of either positive- or negative-strand polarity. This observation indicates that regions of single-strand RNA of both positive and negative polarity exist in the population and suggests that two types of replicative intermediates exist: one to synthesize full-length plus strands (using a 40S negative-strand template) and another to synthesize full-length negative strands (using a 40S plus-strand template). Accordingly, replication can be divided into two phases: synthesis of plus-strand template, which we propose to call virion complementary (vc) RNA replication, and synthesis of progeny negative-strand RNA, which we propose to call virion (v) RNA replication.

C. The Question of the Enzymatic Basis of Replication and the Need for Protein Synthesis

Some of the most complex areas of rhabdoviral biology involve control of RNA replication and transcription, the requirement for protein synthesis in replication, and identification of the protein(s) responsible for replication. The complex and often conflicting findings concerning these areas may be a result of the fact that the processes are closely interrelated and interdependent.

Although the genome of VSV has information for only five gene products and five messenger RNA species are synthesized by the virus, it is still not possible to correlate specific viral proteins with particular viral complementation groups and assign a definite and single function to each protein. This problem may be due to the fact that a defect in the synthesis of one protein or in the activity of that protein may result in pleiotropic effects on the replication process of the virus. The divergent findings concerning attempts to assign a particular gene function to the complementation groups deficient in RNA replication may best illustrate the problem.

1. The Replication-Deficient Complementation Groups

Early work with temperature-sensitive (*ts*) mutants of VSV identified two complementation groups of VSV_{IND}, Groups II and IV, as deficient in replication of RNA.[9,17-20] Also, a Group VI mutant, *ts* M601, and a Group V mutant, *ts*502, recently have been reported to be deficient in RNA replication and in replication and transcription respectively.[21]

Two complementation groups deficient in RNA replication have been identified in the VSV_{NJ} serotype by Lesnaw and Reichmann.[22] It was found that the two groups, E and A, appear to correspond functionally to Groups II and IV of VSV_{IND}, respectively. In addition, Lesnaw and Reichmann[22] reported that the sixth complementation group of VSV_{NJ}, Group F, was unable to synthesize any RNA at nonpermissive temperature.

Conflicting findings have been reported concerning the proteins affected in the replication-deficient complementation groups. Combard et al.,[20] using the complementation Group II mutant *ts*052, found that at nonpermissive temperature, no virion-length RNA was synthesized and no intracytoplasmic nucleocapsids were present. It was also found that the *ts*052 nucleocapsids synthesized at 31°C were less stable than those of wild-type virus when exposed to high temperature and that the RNA was rendered more sensitive to ribonuclease. These and other data led to the conclusion that the *ts*052 defect was in the major nucleocapsid protein, N. However, the data were not sufficient to rule out the possibility that NS protein might be involved in the defect in this group.

In earlier work, Combard et al[19] observed that one major defect in the Group IV mutants was a significant inhibition of N protein synthesis at nonpermissive temperature and at that time, a tentative correlation was drawn between 38S RNA synthesis and N protein.

Another approach to the problem of mapping the proteins of VSV mutants has been reported by Freeman et al.[23] This technique involves the formation of heteroduplexes between virion RNA and the RNAs of *ts* mutants. Heteroduplexes between ^{32}P-labeled *ts*G41 virion RNA and unlabeled mRNA from cells infected with ts^{+} revertants could be distinguished from homoduplexes of ^{32}P-labeled *ts*G41 virion RNA with mRNA from *ts*G41-infected cells by ribonuclease digestion of the hybridized RNA at lowered salt concentrations. It was found that the L, G, and N homo- and heteroduplexes were differentially susceptible to digestion. The N homoduplex was more resistant to digestion than the N heteroduplex. Also, a new duplex, X, was generated in the case of the heteroduplexes. This result was interpreted to suggest that the X duplex resulted from a cleavage within the imperfectly hydrogen-bonded N duplex formed between RNA from *ts* G41 and its revertant. From these results, the *ts* lesion of the group IV mutant *ts* G41 was assigned to the N gene. Oligonucleotide fingerprints of the X and N duplexes will be required to confirm this assignment.

Knipe et al.[21] found that the Group IV mutant, *ts*G41, and a Group IV mutant, *ts* M601, were defective in replication of virion RNA. These mutants also synthesized N protein that was degraded more rapidly than the wild-type N protein.

In the case of the VSV_{NJ} mutants, Wunner and Pringle[24] reported that the *ts* phenotype of the Group D mutant, D1, was associated with the synthesis of N and G proteins with altered electrophoretic migration rates. Reversion of the *ts* phenotype to wild type (WT) was accompanied by a return to normal mobility by N polypeptide. This finding was interpreted to mean that the *ts* phenotype was the result of a defect in N. Lesnaw and Reichmann,[22] however, reported that although VSV_{nj} mutant D1 synthesized only about half as much total RNA at 39°C as at 31°C, it was probably not deficient in replication of RNA since it carried out amplified transcription at 39°C. The VSV_{NJ} Group D mutants were equated with the VSV_{IND} Group V mutants on a functional basis.

In addition to the complementation groups of VSV that are totally defective in replication, there have also been reports of complementation groups that have altered levels of replication at nonpermissive temperature.

Martinet[25] has observed that at 39°C, or after shift up to 39°C, there is a significant decrease in 38S RNA synthesis by mutant *ts* 045 (Group V) or *ts* 023 (Group III); no decrease in messenger RNA synthesis was observed. In a double shift experiment, there was a significant increase in replication of 38S RNA by mutant *ts* 045 if it was first incubated at 32°C, then shifted to 39°C and then returned to 31°C for labeling of RNA. Previous work by Lafay[26] has shown that, in the case of the Group V mutant *ts* 045, the G protein fails to be inserted into the lightest insoluble membrane fraction that can be resolved on discontinuous sucrose gradients. Also, Knipe et al.[21] reported that the Group V mutant *ts* M501 codes for a G protein that does not undergo the final sialylation at the nonpermissive temperature. It was concluded that the defect resides in the G protein itself since the defect cannot be complemeted during coinfection with wild-type virus having G protein that matures normally. On the basis of the findings described above, Martinet[25] postulated that the G and M proteins, tentatively assigned to complementation Groups V and III respectively, while not interacting directly in the replication process, may be important in determining intracellular sites at which the nucleocapsid template apparatus would be in either a replicating or a transcribing mode.

The finding of Martinet[25] that synthesis of virion-size RNA was depressed in the Group V *ts* 045 mutant after shift up to nonpermissive condition has been confirmed by Marnell.[27] Also, the effect of an inhibitor of glycosylation of G protein was used to examine the effect of an altered G protein on replication. Marnell[27] found that the drug glucosamine completely blocks glycosylation of the G protein of wild-type VSV and in the presence of glucosamine, the RNA synthesis phenotype of wild-type VSV was similar to that obtained with mutant *ts* 045 at nonpermissive temperature. RNA replication was decreased and there was an increase in messenger RNA synthesis.

These results suggest that viral membrane proteins may play an indirect role in RNA replication, perhaps by determining sites of replication and/or transcription.

2. The Requirement for Protein Synthesis

Viral RNA replication requires viral protein synthesis. Evidence for this statement comes from the following findings: RNA synthesis in cells infected in the presence of cycloheximide is restricted to primary transcription,[38,39] and no synthesis of virion-size RNA is observed in the in vitro RNA synthesis system (with the exception of a report by Breindl and Holland[28] that synthesis of 40S RNA was observed for a transient period at 6 min after initiation in a linked in vitro transcription-translation system). It has been observed that replication appears to require continuous protein synthesis. Inhibition of protein synthesis at times after onset of replication results in immediate cessation of replication.[29,30] Using the mutant *ts*G114 (Group 1), Perlman and Huang[30] found that if inhibition of protein synthesis was delayed until late in the growth cycle, the shut-off of replication was not complete.

What is the molecular basis for the requirement for protein synthesis in replication? Are one or several proteins synthesized to fulfill this requirement? Experimental approaches using conditional-lethal mutants, metabolic inhibitors, and defective interfering (DI) particles have given rise to numerous possibilities. An early hypothesis was that a new enzyme must be synthesized to serve as a replicase. In light of the fact that 95% of the coding capacity of the virus as we know it (i.e., with no alteration in reading frame) is occupied by the five known structural proteins, it does not seem likely that an entirely new polymerase is made. An alternate possibility is that the transcriptase component of the virus is a core polymerase enzyme which can also function as a replicase either as it is or in an altered form. There are several postulations as to how this could occur. As data to be discussed later will show, the initiation of replication may not in fact require synthesis of a new enzyme or a factor for the alteration of an existing enzyme to serve as replicase. Rather, replication may require alteration of the template or enzyme in such a way that a new recognition site can be utilized or an existing recognition site can be ignored.

To discuss these possibilities, it is easiest to consider the process of replication in two steps. Also, for the sake of discussion, the complete 40S plus strand observed in infected cells in the form of a nucleocapsid will be considered as the template for replication even though the evidence for its function is circumstantial. Replication is divided into at least two components: the synthesis of a 40S plus strand that can serve as template for replication of progeny negative strands, or virion complementary (vc) RNA replication; and the synthesis of progeny negative strands from the full-length plus-strand RNA template, or virion (v) RNA replication.

a. *Synthesis of 40S Virion Complementry (VC) RNA*

It is known that the proteins of the virus nucleocapsid, L and NS, and the template of RNA covered with N are sufficient to synthesize complementary mRNA.[31-33] The synthesis of these mRNAs occurs in an ordered fashion originating at the 3′ end of the virion RNA with synthesis of a leader RNA and progressing toward the 5′ end.[34-36] It is not known whether each message is initiated individually in a sequential fashion or if initiation occurs once at the 3′ end of the genome. In the latter case, the entire length of the genome would be copied and a processing enzyme would cleave the RNA into individual mRNAs on recognition of a specific sequence or a certain secondary structure.

Very recent data suggest that it is possible that VSV mRNAs may arise by a processing mechanism. Wertz and Davis[37] have found that the *E. coli* RNA processing enzyme, ribonuclease III can cleave VSV-40S virion RNA or VSV-40S intracellular RNA which contains both plus and minus strands. After digestion with ribonuclease III, virion RNA that had formerly migrated as a single band with an apparent molecular weight of 4×10^6 is cleaved to yield a set of discrete fragments which range in size from 3.5 to 0.2×10^6 daltons. Similar migration patterns can be obtained after digestion of VSV 40S intracellular RNA, which contains both + and − strand RNA. The extent of the digestion is dependent on the relative concentrations of the enzyme and the RNA. In contrast to the finding with virion length RNA however, the migration of VSV messenger RNAs is unaffected by treatment with RNase III. There results show that specific sites in virion RNA but not mRNA can be recognized by RNase III. This leads us to postulate that primary sequence determined secondary structural features exist in VSV RNAs. These features alone or in combination with the binding of viral proteins may play a role in the regulation of viral RNA synthetic events, one of which may be the synthesis and processing of VSV mRNAs.

If we propose that mRNAs do arise via a processing mechanism, then one explanation for the occurrence of the first step of replication, (vc) RNA replication, may be the absence of processing, resulting in the formation of full-length plus-strand RNA.

It is known that production of vc RNA and mRNA occur at the same time in the infected cell. It is not known if there are separate complexes for transcribing mRNA and for synthesizing vc RNA. Also, it is known that both the 40S negative-strand and the 40S plus-strand intracellular RNAs are found only in nucleocapsid form,[10] that is, a complex of RNA and N protein and the proteins NS and L. RNA covered with N protein, not naked RNA, has been shown to be the active template for transcription.[33] In addition, current evidence[23] favors the assignment of N as the protein of the replication-deficient complementation Group IV. Considered together, these data suggest that one protein-synthetic requirement for replication is for synthesis of N protein in order to form template (RNA with N protein). As a consequence of N binding to RNA, it may protect it from being processed into mRNA.

If this should be the case, the important consideration here is the mechanism by which N discriminates between coating plus-strand RNA to block cleavage and thereby allowing 40S vc RNA to be formed, and not binding so that mRNA is formed. The observed balance between production of mRNA and synthesis of vc RNA may reflect an equilibrium between the cleavage reaction and the blocking of the specific cleavage sites by the binding of the N protein. Another possibility is that the control mechanism may involve specific binding of a protein other than N to a specific sequence or secondary structure in the RNA. Both the phosphoprotein NS and the polymerase L may be candidates for exerting control over the postulated processing activity. The genetic evidence of Martinet[25] also indicates that M and perhaps G protein may be involved in a regulatory function for control of the balance between replication and transcription.

The alternative possibility to consider is that the synthesis of vc RNA may occur as a process entirely separate from mRNA synthesis. Enzyme or template could be altered in such a way that a complex is established which synthesizes exclusively 40S plus-strand RNA. At this time, it is apparent that much more research is needed to elucidate the nature of the transcribing and replicating complexes.

b. Synthesis of 40S Virion (V) RNA

Once the 40S vc RNA is formed, what are the requirements to begin synthesis of virion RNA? First, N protein is required to bind to the vc RNA to form template. Second, replicase must be formed. As discussed above, many factors have suggested that the L protein is the core polymerase that functions both as transcriptase and replicase. The recent work of Johnson et al.,[55] which will be discussed in detail below, indicates that the L protein of the infecting virion may be capable of serving as replicase. Whether or not it functions as is or is altered is unknown. Third, the replicase must bind to the template. Assuming that the polymerase binds to the 3′ end of the minus strand for transcription and the 3′ end of the plus strand for replication, it is of interest to know if these binding sites are the same. Sequencing data have shown that the three bases at the 3′ and the 5′ ends of the virion RNA are potentially complementary.[40] However, after self annealing, the ends of VSV RNA have never been observed in the electron microscope to form pan-handle structures. These structures have been observed with DI particle RNAs.[50,51] This finding may be interpreted to say that extensive complementary terminal sequences have not been detected in VSV RNA by this method. However, recently, the 3′ ends of VSV B and DI particle RNAs have been sequenced.[64] It was observed that 14 of the first 17 bases are the same on the 3′ ends of VSV B and DI particle RNAs. Assuming the loop back model for the generation of defective particles, then the 3′ end of DI is complementary to the 5′ end of B-virion RNA. The implication of this finding is that 14 of the first 17 bases on the 5′ and 3′ ends of VSV RNA are potentially complementary. If this is the case, then small regions of complementarity could exist on the ends of VSV RNA that may not be detectable by techniques of hybridization or electron microscopy. If the binding site for the polymerase includes the first 17 residues of the 3′ ends, this finding indicates that the sites are similar, but not identical, in primary sequence. The effect of the possible three-base difference in sequence on the secondary structure of the molecules and the binding of enzyme remains to be determined.

Finally, an additional factor is required to mediate binding of polymerase to template. Mellon and Emerson[41] have recently shown that the NS protein is required to bind first to the template in order for the L protein to bind. Also of interest at this point is the report of Clinton and Huang[42] that the phosphoprotein NS exists in the cell in two different levels of phosphorylation, only one of which is bound to the nucleocapsid. The different levels of phosphorylation of NS have been suggested as possible controlling factors in the replication-transcription switch.[42] It may be that new NS in a different state of phosphorylation is required to bind to the negative-strand nucleocapsid to get full-length plus strand made and that there is a similar requirement for synthesis of progeny negative strands.

It has not been possible to test the effect of NS on replication because previously *ts* mutants having an altered NS have not been identified. Davis and Wertz,[43] however, have recently described a spontaneous small plaque-size mutant of VSV that has an altered pattern of RNA synthesis in which transcription is increased and replication is decreased. The phosphoprotein, NS, of this mutant has, compared to its parent wild type, an altered migration rate in polyacrylamide gels (Figure 3). The altered NS protein is phosphorylated to the same extent both in its soluble and nucleocapsid-bound state as the NS protein of its wild-type parent virus. This alteration in migration rate then could be accounted for by a 3000 to 4000 dalton increase in the size of the protein. The glycoprotein, G, of this mutant also has an altered electrophoretic migration rate which could be accounted for by a 2000 dalton increase in size. The G protein appears to be glycosylated to the same extent as the wild-type G protein. The isolation of a revertant in the NS and/or G protein will be required to determine whether or not an alteration in NS is associated with altered RNA replication.

Recently, Lesnaw[65] has determined that the VSV New Jersey mutant, El, has an altered NS protein. The El NS protein migrates faster on SDS polyacrylamide gels than the wild type NS, however, it migrates at the same rate as wild type NS in gels containing 4*M* urea. Peptide mapping studies revealed that the El NS protein was missing two low molecular weight peptides as compared to the wild type. Previous work by Lesnaw[22] has shown that the mutant El is defective in replication of viral 40S RNA at nonpermissive temperature. This finding has led Lesnaw to postulate that the NS protein plays a role in replication.

The finding that NS is required for transcription (i.e., NS is required to bind to the RNA template before the transcriptase L can bind[41] and transcription can occur) and

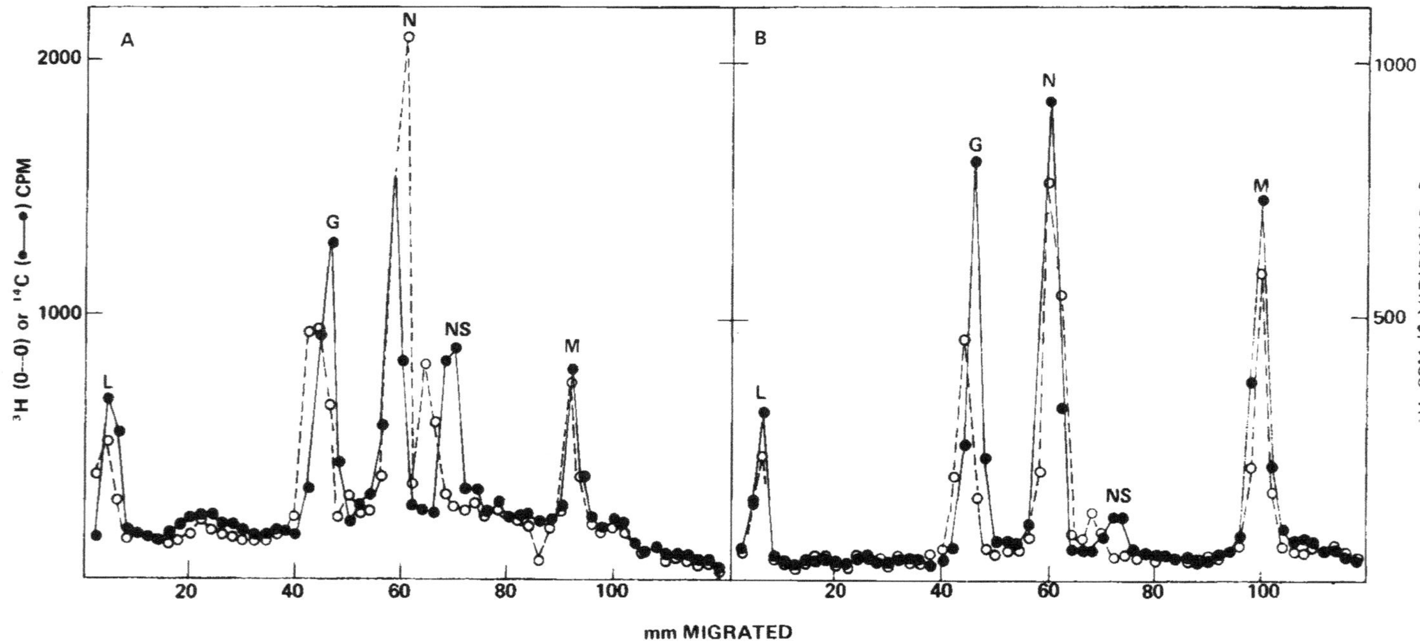

FIGURE 3. Coelectrophoresis of ^{14}C-labeled wild-type VSV proteins and ^3H-labeled S_2VSV proteins synthesized in BHK cells. (A). Cells were labeled between 4 and 5 hr after infection with ^{14}C-amino acid mixture (wild-type VSV, ●——●) or ^3H-amino acid mixture (S_2VSV, O——O), and radioactive cytoplasmic proteins were precipitated with acetone. Samples dissolved in 0.01 *M* sodium phosphate buffer, pH 7.2, 1% SDS, and 1% dithioerythritol were mixed, heated at 100°C for 3 min and electrophoresed in an SDS-containing sodium phosphate-buffered 10% polyacrylamide gel. (B). Parallel cultures to those in (A) were labeled between 1 and 6 hr after infection with ^{14}C-amino acid mixture (wild-type VSV, ●——●) or ^3H-leucine (S_2VSV, O——O). The virus-containing culture supermutants were extracted twice with fluorocarbon, mixed, and centrifuged through a 15 to 45% sucrose gradient. Peak radioactive fractions were pooled, precipitated with two volumes of ethanol, ved, and electrophoresed as described for A.

the finding of Lesnaw[45] that a mutation in NS correlates with a defect in replication but not in transcription suggest that the NS functions both in replication and transcription. It is possible that NS as it exists in the infected cell has two functional domains: one for transcription and one for replication.

D. Summary

Based on the considerations discussed above, the process of replication may be said to involve the following. First, there must be synthesis of a 40S plus strand (vc RNA) to serve as template for the replication of v RNA. This plus-strand template must then be coated with N protein to form the RNA: N protein template. The core enzyme for replication then must bind to the RNA: N protein complex. If we assume that L protein is the core polymerase, then this would require the prior binding of the NS protein. Phosphorylation levels of NS may be important in controlling this step.

Is this plus-strand nucleocapsid complex thus assembled sufficient to replicate progeny RNA? One might assume that the requirements for protein synthesis in v RNA replication would be similar to those in vc RNA replication. The initiation sites and enzyme-binding sites for these two processes, however, may be slightly different. For these reasons, the enzymes and other factors involved may be altered in the two processes both for functional reasons and possibly as a means of controlling the two processes. Currently, no in vitro replicating systems are available. However, one promising area of investigation has been a study of the replication of the DI particle RNAs. This will be discussed in the following section.

III. REPLICATION OF THE RNA OF DEFECTIVE INTERFERING (DI) VIRUS PARTICLES

Defective interfering (DI) particles derive their name from the fact that they contain only a part of the viral genome and they interfere with the replication of a standard B virus.[44] DI particles containing varying amounts of information from the 5′ end (L gene region) or from the 3′ end of the viral genome have been identified.[45-47] The majority of DI particles identified are from the L gene region of the genome.

Early studies indicated that DI particles did not synthesize mRNAs in vitro or replicate themselves except in the presence of co-infecting standard virus.[33,44] Emerson and Wagner[33] showed that the DI particles had an active viral transcriptase, but that the template appeared to be defective.

Subsequent work by Mori and Howatson[48] and by Reichmann et al.,[49] indicated that a very small RNA product that was rich in A residues was synthesized by defective virions. These observations have been confirmed and recent work has shown that DI particles from the two regions of the genome are able to express their genetic information to varying degres. Also, recent results obtained by Perrault and Leavitt[50,51] and Freeman et al.[23] have given new insight into the possible mechanism of origin and replication of these particles. This information, as it relates to replication of the RNA, is discussed below. The subject of the origin of DI particles is discussed in Chapter 12 of this volume.

A. Replication of DI Particles Containing Information From the 3′ End of the Genome

Defective particles generated by the heat resistant (HR) strain of VSV[52] have been shown to contain information from the 3′ half of the viral genome.[45-47] The genetic information contained in these DI particles is capable of being expressed. Chow et al.[53] demonstrated that the information for the G and M proteins could be expressed and these proteins could be synthesized in vivo. Colonno et al.[54] determined that the genomes of HR DI particles contained information for the VSV leader RNA sequence.

Results obtained by this group also demonstrated that in vitro, the HR DI could direct synthesis of VSV leader RNA and four VSV mRNA species that sedimented at 12 to 18S. These RNAs were identical in size, polyadenylation, capping, and methylation when compared with in vitro-synthesized mRNA of *wt* VSV.

Johnson and Lazzarini[55] demonstrated that in cells infected in the presence of cycloheximide, the HR DI-synthesized mRNA is of the 12 to 18S size. Of these RNA species, 76% were complementary to virion RNA and 56% of the RNA could be retained on oligo-d(T) columns, indicating some polyadenylation. No synthesis of mRNA of the size to code for the L gene was observed in the presence of cycloheximide. However, in addition to being able to transcribe, Johnson and Lazzarini[55] found that limited replication of these DI particles occurred in vivo. A 28S RNA was synthesized in addition to the 12 to 18S mRNAs in the infected cells in the presence of protein synthesis. The synthesis of the 28S RNA was inhibited by addition of cycloheximide. The 28S RNA product did not anneal to a DI-particle RNA which contained information for the L gene. Annealing studies demonstrated that ∿37% of the 28S RNA product was plus strand and ∿61% was negative strand. In addition, these 28S RNA species were shown to be present in nucleocapsids. These data were interpreted to mean that the 28S RNA product was not mRNA coding for the L protein, but represented replication of the DI genome RNA.

UV-inactivation experiments have shown that transcription of VSV RNA initiates at the 3′ end of the genome and proceeds in the order 3′ N, NS, M, G, and L with a smaller leader sequence, which is located at the 3′ end, being transcribed first.[34,36] In view of these findings and the information that DI particles contain an active transcriptase and all the structural proteins of the *wt* virus, [33,56] it is not surprising that the HR DI is able to transcribe all of the information in its genome. It contains active enzyme and the initiation site for transcriptase.

The interesting observation was that the RNA was replicated to a limited extent in the absence of any new synthesis of L protein. If the input DI particles contained only negative-strand RNA (see below), then this finding indicates that the input L (and/or template, in a modified form) is competent to produce a complete plus-strand template. Furthermore, this result indicates that if L is the enzyme responsible for replication then, since no new L is being made in the cell, the input L (in its original or a modified form) can associate with the new plus strand and catalyze synthesis of progeny minus strands using it as template. This process is still dependent on synthesis of new protein and Johnson and Lazzarini cited the unpublished observation that G, N, and M proteins were synthesized in cells infected only with DI particles.

These findings, considered with those of Mellon and Emerson,[41] indicate that the need for protein synthesis for replication may be for the synthesis of at least two viral proteins, N and NS. N would be required to bind to RNA to form template and NS would be required to bind to template in order for L to bind. The genetic information for L protein is not present, however. If L is the basic polymerase for the virus, then the possibility exists that, in this case, L can move from one molecule (the input negative strand) to another (the newly synthesized plus-strand template) in order to catalyze the limited replication observed in this system. For unlimited replication, the synthesis of L may be required in addition to the synthesis of N and NS. Further study of this system and of in vitro reconstruction systems will be required to delineate the requirements for replication of this DI particle.

B. Replication of DI Particles Containing Information From The 5′ End of the Genome

Early work of Emerson and Wagner[33] demonstrated that DI particles possessed an active enzyme component and indicated that the template component was responsible for the failure of DI particles to synthesize mRNA in vitro. Subsequent work by Mori and Howatson[48] and Reichmann et al.[49] reported that a small A-rich RNA was synthesized by DI particles.

Emerson et al.[57] have recently reinvestigated the synthesis of RNA by a DI particle containing information from the 5′ end of the genome. The RNA from the DI particle used by Emerson et al.[57] has an RNA that is ~30S. This RNA self-anneals to ~10% over a wide range of concentrations, suggestive of intramolecular base pairing. This DI particle synthesized ~5% as much product RNA per virion as a standard B particle. The DI product was small and distinct from B-particle product. The product RNA annealed to the DI-particle RNA after fragmentation by limited alkali breakage, but it did not anneal to B-particle RNA. The product contained 56% A residues, but the longest run of A was only 5 to 6 bases. These results confirm that the DI particles from the L gene region can serve as template, but that the expression of that genome is limited to the synthesis of a small RNA fragment unique to the DI particle.

In a further attempt to investigate synthesis of RNA by this DI particle, Belle-Isle and Emerson[58] isolated intracellular nucleocapsids containing plus and minus 30S RNA strands. These RNAs self-annealed to 75% in a concentration-dependent fashion, indicating that they were not covalently linked.

The plus and minus RNA-containing nucleocapsids were put in an in vitro system in an attempt to obtain replication, since the plus nucleocapsids were assumed to be the template for replication. The only product of the reaction was a small RNA, which comigrated on polyacrylamide gels with the product synthesized by DI particles in vitro. The polarity of this product was not determined.

Moyer et al.[59] have also isolated DI-particle intracellular nucleocapsid structures. Of the RNA in the nucleocapsids, 30 to 40% is plus-stranded. The plus strands do not have a blocked or methylated 5′ end and they do not bind to oligo-d (T), indicating that no poly-(A) regions are present.

Two types of experimental evidence have suggested that DI particles contain complementary inverted terminal sequences. First, Perrault and Leavitt[50,51] have observed by electron microscopy that DI-particle RNA can form circles (monomers and multimers), some of which contain panhandle-like structures. The ability of the RNA to circularize suggests that the termini of the RNA are complementary.

A second type of evidence comes from oligonucleotide fingerprints of T1 ribonuclease digests of DI- and B-particle RNAs.[23,60] Most of the oligonucleotide spots from the DI RNA were a subset of the spots from B RNA. However, the DI RNA contained two oligonucleotide spots that were not present in similar T1 ribonuclease digests of standard B particle RNA.

These observations have led to the hypothesis that DI may originate by a looping back of the nascent RNA strand to form a hairpin with subsequent realignment of polymerase on the nascent strand and reading of its RNA in the 3′ to 5′ direction toward the 5′ end of the nascent strand. This model could account for the presence of extragenomic sequences and would result in complementary termini of the RNA.

DI particles which contain both plus and minus strands in the same particle have been reported.[50,51,61,62] Perrault and Leavitt[50,51] surveyed DIs originating from various B-particle stocks for the presence of RNA molecules that remained RNase resistant immediately after denaturation. These rapidly annealing RNAs were termed snap-back RNA molecules. The results of this survey indicated that RNA from standard B particles showed little or no self-annealing. However, the RNA from DI particles derived from wild-type virus (both Indiana and New Jersey serotypes), from *ts* mutants, and from persistantly infected carrier cultures showed significant levels of snap-back self-annealing. Some DI particles were found to have complementary RNA that did self-anneal, but not snapback. The snap-back molecules are postulated to contain a covalent linkage between complementary terminal sequences.

Perrault and Leavitt[51] further reported the isolation of RNAse-resistant molecules 60 to 70 nucleotides in length from the melted and quick-cooled DI RNAs. Preliminary data indicate that these small components are the complementary terminal sequences. Recent results have estabished that the 5′-end and not the 3′-end terminal sequence of ~70 nucleotides of the DI RNA is found in the full-size B-virion genome RNA.

C. Summary

Previous work has shown that the VSV B-virion RNA contains a 3-base sequence on the 3′ and the 5′ end which is potentially complementary. Other than this, there have been no indications of significant regions of homology in this molecule, as indicated by ability of molecules to self-anneal or circularize.[50,51] On the other hand, DI-particle RNAs do appear to contain inverted complementary terminal sequences, which can give rise to circular molecules after melting and quick cooling.[50,51] These findings, however, do not rule out the possibility that B-particle ends could have regions of homology great enough to allow hybridization and formation of circles and/or multimers. The failure to observe these structures may be due to the size of the B RNA molecules. Kolakofsky[63] observed that although DI-particle RNAs of Sendai virus could circularize, the frequency of circles observed was lower in RNA preparations of large DI RNAs compared to the frequency in smaller DI RNAs. Recently, however, the 3′ ends of VSV B and DI RNA have been sequenced and 14 out of the first 17 bases were reported to be the same.[64]

Assuming the loop back model for generation of defective RNA described above, then the 3′ end of DI RNA is complementary to the 5′ end of B-virion RNA. This implies that all but three of the first 17 bases on the 5′ and 3′ ends of VSV RNA are potentially complementary.

If this is so, then the indication is that the 3′-end initiation site for transcription on the minus strand is similar, but not identical, in primary sequence to the 3′-end initiation site for vRNA replication on the plus strand of 40S RNA. While this difference in primary sequence is not great, it may be significant in terms of secondary structure and/or enzyme binding.

Several DI particles containing information from the 5′ end of the genome appear to have inverted terminal sequences. If these partial genomes do arise by a doubling back of the enzyme to read the nascent RNA, then the 5′ end of the DI-particle minus strand is the same as the 5′ end of the B-particle minus strand, but the 3′ end is not. Rather, the 3′ end is complementary to the 5′ end of the DI RNA. The plus-strand template for replication of this DI would have a 3′ end the same as that of the B particle for initiation of replication. In addition, it would also have a 5′ end complementary to that 3′ end. This fact, coupled with the absence of a 3′ initiation site for transcription on the DI minus strand, may confer a strong replicative advantage on the DI particle over the B particle. These findings may indicate that primary sequence and secondary structure may play an important role in recognition and discrimination of enzyme-binding sites for replication vs. transcription.

Obviously, factors other than sequence and initiation site must be involved in control of the replication of DI RNA, as evidenced by the finding of Emerson et al.[57] that the DI-particle RNAs synthesize only a small RNA sequence in in vitro systems. The other requirements for replication may be analogous to those discussed for B particles, which can only transcribe in vitro and require concomitant synthesis of proteins for replication as described above.

Further, the finding of Perrault[61] of covalently linked + and − RNA in DI particles has led to the suggestion that a self-priming type of mechanism may occur in the replication of DI RNA, as well as in B-particle RNA replication. In such a reaction, the 3′ end of the replicating RNA, folded back upon a nearby complementary region, would become the primer for RNA synthesis toward the 5′ end of the template molecule.

The problem of rhabdoviral RNA replication is obviously complex. This problem is interesting not only as a means to understand how viruses replicate so that we may further our knowledge of how to control viral diseases, but it is also an interesting system in which to study a unique biological gene system and the control of the expression of that genetic information. Further investigation of this problem using molecular, biochemical, and electron microscopic techniques is required.

ACKNOWLEDGMENTS

I thank Drs. S. Bachenheimer and C. Y. Kang for many helpful discussions; P. Printz, C. Martinet, J. Lesnaw, A. Huang, S. Emerson, and S. Moyer for making available data prior to publication; and N. Davis, K. Wertz, E. Dubovi, and L. Marnell for critical reading of the manuscript. This work was supported by grant BMS75-03544 from the National Science Foundation and by Public Health Service grant AI-12464 from the National Institute of Allergy and Infectious Disease.

REFERENCES

1. **Wagner, R. R.**, Reproduction of Rhabdoviruses, in *Comprehensive Virology*, Vol. 4, Fraenkel-Conrat, H. and Wagner, R. R., Eds., Plenum Press, New York, 1975, 1.
2. **Emerson, S.U.**, Vesicular stomatitis virus: structure and function of virion components, *Curr. Top. Microbiol. Immunol.*, 73, 1, 1976.
3. **Burke, D. C. and Russell, W.C.**, Eds., Control of nucleic acid replication in cells infected with animal viruses, in *Control Processes in Viral Multiplication*, Cambridge University Press, New York, 1975, 253.
4. **Subak-Sharpe, J. H. and Pringle, C. R.**, Viral information contolling DNA and RNA animal virus replication, in *Control Processes in Viral Multiplication*, Burke, D. and Russell, W., Eds., Cambridge University Press, New York, 1975, 363.
5. **Bader, J. P.**, The requirement for DNA synthesis in the growth of Rous sarcoma and Rous-associated viruses, *Virology*, 26, 253, 1965.
6. **Kiley, M. P. and Wagner, R. R.**, Ribonucleic acid species of intracellular nucleocapsids and released vions of vesicular stomatitus virus, *J. Virol.*, 10, 224, 1972.
7. **Wertz, G. W.**, A method to examine viral RNA metabolism in cells in culture: metabolism of vesicular stomatitis virus RNA, *J. Virol.*, 16, 1340, 1975.
8. **Schincariol, A. L. and Howatson, A. F.**, Replication of vesicular stomatitis virus. I. Viral specific RNA and nucleoprotein in infected L cells, *Virology*, 42, 732, 1970.
9. **Perlman, S. M. and Huang, A. S.**, Virus specific RNA specified by group I and IV temperature sensitive mutants of vesicular stomatitis virus, *Intervirology*, 2, 312, 1973.
10. **Soria, M., Little, S. P., and Huang, A. S.**, Characterization of vesicular stomatitis virus nucleocapsids. I. Complementary 40S RNA molecules in nucleocapsids, *Virology*, 61, 270, 1974.
11. **Morrison, T. G., Stampfer, M. Lodish, H. F., and Baltimore, D.**, *In vitro* translation of vesicular stomatitis virus messenger RNAs and the existence of a 40S 'plus' strand, in *Negative Strand Viruses*, Vol. 1, Mahy, B. and Barry, D., Eds., Academic Press, New York, 1975, 293.
12. **Wertz, G. W.**, Isolation of possible replicative intermediate structures from vesicular stomatitis virus infected cells, *Virology*, 85, 271, 1978.
13. **Bishop, J. M. and Koch, G.**,Infectious replicative intermediate of poliovirus: purification and characterization, *Virology*, 37, 521, 1969.

14. **Noble, J. and Levintow, L.**, Dynamics of poliovirus-specific RNA synthesis and the effects of inhibitors of virus replication, *Virology*, 40, 634, 1970.
15. **Spiegelman, S., Haruna, I., Pace, N. R., Mills, D. R., Bishop, D. H. L., Claybrook, J. R., and Peterson, R.**, Studies in the replication of viral RNA, *J. Cell. Physiol.*, 70, 35, 1967.
16. **Hunt, D. M., Emerson, S. U., and Wagner, R. R.**, RNA-temperature sensitive mutants of vesicular stomatitis virus: L protein thermosensitivity accounts for transcriptase restriction of group I mutants, *J. Virol.*, 18, 596, 1976.
17. **Printz-Ane, C., Combard, A., and Martinct, C.**, Study of the transcription and the replication of vesicular stomatitis virus by using temperature sensitive mutants, *J. Virol.*, 10, 889, 1972.
18. **Unger, J. T. and Reichmann, M. E.**, RNA synthesis in temperature sensitive mutants of vesicular stomatitis virus, *J. Virol.*, 12, 570, 1973.
19. **Combard, A., Martinet, C., Printz-Ane, C., Friedman, A., and Printz, P.**, Transcription and replication of VSV: effects of temperature sensitive mutations in complementation group IV, *J. Virol.*, 13, 922, 1974.
20. **Combard, A., Printz-Ane, C., Martinet, C., and Printz, P.**, Temperature sensitive defect of vesiclar stomatitis virus in complementation group II, *J. Virol.*, 21, 913, 1977.
21. **Knipe, D., Lodish, H. F., and Baltimore, D.**, Analysis of the defects of temperature-sensitive mutants of vesicular stomatitis virus: intracellular degradation of specific viral proteins, *J. Virol.*, 21, 1140, 1977.
22. **Lesnaw, J. and Reichman, M. E.**, RNA synthesis by temperature sensitive mutants of vesicular and stomatitis virus, New Jersey serotype, *Virology*, 63, 492, 1975.
23. **Freeman, G. J., Rao, D., and Huang, A.**, Genome organization of vesicular stomatitis virus: mapping of *ts* G41 and the defective interfering T particle, presented at Negative Strand Virus meeting, Cambridge, England, 1977.
24. **Wunner, W. H. and Pringle, C. R.**, A temperature-sensitive mutant of vesicular stomatitis virus with two abnormal virion proteins, *J. Gen. Virol.*, 23, 97, 1974.
25. **Martinet, C.**, personal communication, 1978.
26. **Lafay, F.**, Envelope proteins of vesicular stomatitis virus: effect of temperature-sensitive mutations in complementation groups III and V, *J. Virol.*, 14, 1220, 1974.
27. **Marnell, L.**, unpublished data, 1978.
28. **Breindl, M. and Holland, J.**, Studies on the *in vitro* transcription and translation of vesicular stomatitis virus mRNA, *Virology*, 73, 106, 1976.
29. **Wertz, G. W. and Levine, M.**, RNA synthesis by vesicular stomatitis virus and a small plaque mutant: effects of cycloheximide, *J. Virol.*, 12, 253, 1973.
30. **Perlman, S. M. and Huang, A. S.**, RNA synthesis of VSV V. Interaction between transciption and replication, *J. Virol.*, 12, 1395, 1973.
31. **Bishop, D. H. L. and Roy, P.**, Dissociation of vesicular stomatitis virus and relation of the virion proteins to the viral transcriptase, *J. Virol.*, 10, 234, 1972.
32. **Szilágyi, J. and Uryvayev, L.**, Isolation of an infectious ribonucleo-protein from vesicular stomatitis virus containing an active RNA transcriptase, . *Virol.*, 11, 278, 1973.
33. **Emerson, S. and Wagner, R. R.**, Dissociation and reconstitution of the transcriptase and template activities of vesicular stomatitis B and T virions, *J. Virol.*, 10, 297, 1972.
34. **Ball, L. A. and White, C. N.**, Order of transcription of genes of vesicular stomatitis virus, *Proc. Natl. Acad. Sci. U.S.A.*, 73, 442, 1976.
35. **Abraham, G. and Banerjee, A. K.**, Sequential transcription of the genes of vesicular stomatitis virus, *Proc. Natl. Acad. Sci. U.S.A.*, 73, 1504, 1976.
36. **Colonno, R. J. and Banerjee, A. K.**, A unique RNA species involved in initiation of vesicular stomatitis virus RNA transcription *in vitro*, *Cell*, 8, 197, 1976.
37. **Wertz, G. W. and Davis, N. L.**, unpublished data, 1978.
38. **Marcus, P. I., Engelhardt, D. L., Hunt, J. M., and Sekellick, M. J.**, Interferon action: inhibition of vesicular stomatitis virus RNA synthesis induced by virion bound polymerase, *Science*, 174, 593, 1971.
39. **Huang, A. S. and Manders, E. K.**, Ribonucleic acid synthesis of vesicular stomatitis virus. IV. Transcription by standard virus in the presence of defective interfering particles, *J. Virol.*, 9, 909, 1972.
40. **Banerjee, A. K., Abraham, G., and Colonno, R. J.**, Vesicular stomatitis virus: mode of transcription, *J. Gen. Virol.*, 34, 1, 1977.
41. **Mellon, M. and Emerson, S.**, The binding of L and NS proteins to the ribonucleocapsid of vesicular stomatitis virus presented at Negative Strand Virus meeting, Cambridge, England, 1977.
42. **Clinton, C. M. and Huang, A. S.**, Heterogeneity in the phosphorylation of vesicular stomatitis virus NS protein, Paper S777, Abstracts of the Annual Meeting of the American Society of Microbiology, New Orleans, 1977.

43. **Davis, N. L. and Wertz, G. W.**, Altered proteins synthesized by a replication deficient mutant of vesicular stomatitis virus, Paper S343, Abstracts of the Annual Meeting of the American Society for Microbiology, New Orleans, 1977.
44. **Huang, A. S.**, Defective interfering viruses, *Annu. Rev. Microbiol.*, 27, 101, 1973.
45. **Leamnson, R. N. and Reichmann, M. E.**, The RNA of defective vesicular stomatitis virus particles in relation to viral cistrons, *J. Mol. Biol.*, 85, 551, 1974.
46. **Schnitzlein, W. and Reichmann, M. E.**, The size and cistronic origin of defective vesicular stomatitis virus particle RNAs in relation to homotypic and heterotypic interference, *J. Mol. Biol.*, 101, 307, 1976.
47. **Stamminger, G. and Lazzarini, R.**, Analysis of the RNA of defective VSV particles, *Cell*, 3, 85,1974.
48. **Mori, H. and Howatson, A. F.**, *In vitro* transcriptase activity of vesicular stomatitis virus B and T particles: analysis of product, *Intervirology*, 1, 168, 1973.
49. **Reichmann, M. E., Villarreal, L. P., Kohne, D., Lesnaw, J., and Holland, J. J.**, RNA polymerase activity and poly(A) synthesizing activity in defective T particles of vesicular stomatitis virus, *Virology*, 58, 520, 1974.
50. **Perrault, J.and Leavitt, R. W.**, Characterization of snap-back RNAs in vesicular stomatitis defective interfering virus particles, *J. Gen. Virol.*, 38, 21, 1977.
51. **Perrault, J. and Leavitt, R. W.**, Complementary terminal repeat sequences in single-stranded RNAs and snap-back RNAs from vesicular stomatitis virus defective interfering paticles, *J. Gen. Virol.*, 38, 35, 1977.
52. **Petric, M. and Prevec, L.**, Vesicular stomatitis virus — a new interfering particle, intracellular structures and specific RNA, *Virology*, 41, 615, 1970.
53. **Chow, J. M., Schnitzlein, W. M., and Reichmann, M. E.**, Expression of genetic information contained in the RNA of a defective interfering particle of vesicular stomatitis virus, *Proc. Natl. Acad. Sci. U.S.A.*, 77, 579, 1977.
54. **Colonno, R. J., Lazzarini, R. A., Keene, J. D., and Banerjee, A. K.**, *In vitro* synthesis of messenger RNA by a defective interfering particle of vesicular stomatitis virus, *Proc. Natl. Acad. Sci. U.S.A.*, 74, 1884, 1977.
55. **Johnson, L. D. and Lazzarini, R. A.**, Replication of viral RNA by a defective interfering vesicular stomatitis virus particle in the absence of helper virus, *Proc. Natl. Acad. Sci. U.S.A.*, 74, 4387, 1977.
56. **Wagner, R. R., Schnaitman, T. C., and Snyder, R. M.**, Structural proteins of vesicular stomatitis viruses, *J. Virol.*, 3, 395, 1969.
57. **Emerson, S. U., Dierks, P. M., and Parsons, J. T.**, RNA synthesis by defective particles of vesicular stomatitis virus, presented at Negative Strand Virus Meeting, Cambridge, England, 1977.
58. **Belle Isle, H. D. and Emerson, S.U.**, Defective interfering particles as a model system for vesicular stomatitis virus replication, Paper S108, Abstracts of the Annual Meeting of the American Society for Microbiology, New Oleans, 1977.
59. **Moyer, S., Holmes, K., Hamilton, P., and Moyer, R.**, Characterization of the positive strand (+) RNA species isolated from the nucleocapsids of defective interfering (DI) particles of vesicular stomatitis virus (VSV), Paper S109, Abstracts of the Annual Meeting of the American Society for Microbiology, New Orleans, 1977.
60. **Huang, A. S.**, Viral pathogenesis and molecular biology, *Bacteriol. Rev.*, 41, 811, 1977.
61. **Perrault, J.**, Cross-linked double-stranded RNA from a defective vesicular stomatitis virus particle, *Virology*, 70, 360, 1976.
62. **Lazzarini, R. A., Weber, G. H., Johnson, L. P., and Stamminger, G. M.**, Covalently linked message and anti-message (genomic RNA) from a defective vesicular stomatitis virus particle, *J. Mol. Biol.*, 97, 289, 1975.
63. **Kolakofsky, D.**, Isolation and characterization of Sendai Virus DI-RNAs, *Cell*, 8, 547, 1976.
64. **Lazzarini, R. A.**, personal communication, 1978.
65. **Lesnaw, J. A.**, personal communication, 1978.

Chapter 6

RHABDOVIRAL ASSEMBLY AND INTRACELLULAR PROCESSING OF VIRAL COMPONENTS

Trudy G. Morrison

TABLE OF CONTENTS

I. INTRODUCTION

The morphogenesis of rhabdoviruses is a complex process which encompasses such diverse events as protein synthesis, glycosylation reactions, protein-protein interactions, and protein-nucleic acid interactions, as well as membrane assembly. Many of the details of rhabdovirus morphogenesis are not yet fully understood; however, results obtained with a wide variety of enveloped viruses suggest a general description of rhabdovirus assembly where details are absent. Furthermore, principles elucidated in studies of assembly of cellular membranes have been useful in understanding rhabdovirus morphogenesis.

Studies of the morphogenesis of rhabdoviruses have concentrated primarily on vesicular stomatitis virus (VSV). As is discussed extensively in Chapters 5 to 9 in Volume I, the VSV is composed of an inner core and the outer envelope.[1] The core of virus particles contains the genome RNA as well as the nucleocapsid (N) protein, the NS protein, and the transcriptase (L) protein. The envelope of virus particles is a membrane, derived from host cell membranes,[2] which contains two viral-encoded proteins, the glycoprotein (G) and the membrane or matrix (M) protein.[3] The glycoprotein forms the spikes or external projections of the virus particle[4] and is located primarily on the external surface of the virion membrane.[5-7] The M protein is located on the inside surface or within the virion membrane.[5-7]

In most host cell systems which have been extensively studied, the immediate precursor of mature virus particles are regions of infected cell plasma membrane which contain these virion components.[2] These modified regions of host cell plasma membrane bud out to form mature virions, the inner side of the cellular membrane forming the inner side of the virion membrane and the external surface of the cell becoming the external surface of the virion.[8]

The components of virus particles arrive at the plasma membrane by one of three apparently different pathways.[9,10] The most extensively studied pathway is that followed by the glycoprotein. The glycoprotein is incorporated into plasma membranes by a series of steps which seem to parallel the path of host cell membrane glycoproteins to the plasma membrane. This process begins with the initiation of the synthesis of the glycoprotein and involves attachment of the protein to intracellular membranes, transport of the protein across the membranes, migration of the protein from intracellular membranes to the plasma membrane, and the stepwise glycosylation of the protein.

The pathway of the membrane protein to the plasma membrane is much less well understood, but it seems clear that this protein does not follow the same sequence of events involved in the glycoprotein maturation. The third pathway in VSV morphogenesis is followed by the ribonucleoprotein core of the virus. The core structure is first assembled from its component parts and then migrates to regions of host cell plasma membranes which also contain viral membrane proteins.

Below, each of these three pathways in viral assembly will be considered in detail as well as the events at the plasma membrane which lead to the budding of mature virus particles.

II. GLYCOPROTEIN MATURATION

A. Synthesis and Membrane Attachment of the Glycoprotein

The first step in the maturation of the VSV glycoprotein is the attachment of the protein to intracellular membranes. There is considerable evidence that this association is due to the properties of the nascent polypeptide chain itself.

The glycoprotein is synthesized exclusively in association with membranes on membrane-bound polyribosomes. This fact has been shown by cell-fractionation studies in which the location of the messenger RNA (mRNA) engaged in the synthesis of the glycoprotein was determined. There is general agreement that close to 100% of the polyribosome-associated glycoprotein mRNA is found in membrane-associated polyribosomes.[4,12,13]

Two groups have independently presented evidence that strongly suggests that the localization of the Vesicular Stomatitis glycoprotein mRNA in membrane-bound polyribosomes is because of an affinity of the nascent polypeptide chain for membranes. Grubman et al.[14] have shown that newly made G mRNA can be found in association with nonmembrane-bound 40S ribosomal subunits and subsequently nonmembrane-bound 80S monosomes. However, polyribosome-associated G mRNA is found exclusively in membrane-associated polyribosomes. The entry of the G mRNA into membrane-bound polyribosomes can be prevented by treating infected cells with pactamycin, an inhibitor of polypeptide chain initiation. Cycloheximide, an inhibitor of polypeptide chain elongation, does not inhibit formation of membrane-bound polyribosomes. Both Grubman et al.[14] and Lodish and Froshauer[15] have shown that puromycin treatment of infected cells releases 80% of the glycoprotein mRNA from the membrane fraction of the cell. These results taken together suggest that the association of the glycoprotein mRNA with membranes is dependent upon polysome formation and the initiation of protein synthesis, and that the nascent glycoprotein chain is responsible for the attachment of polyribosomes to membranes.

An observation complementary to these results is the finding that the VSV glycoprotein is always found in association with the membrane fraction of the cells: no free glycoprotein is even found even in very short labels with radioactive amino acids.[9,16] In addition, puromycin treatment of pulse labeled infected cells does not release glycoprotein from membranes arguing for an immediate and direct attachment of the protein to membranes.[16]

Blobel and Dobberstein[17] have proposed that secreted proteins which ultimately are found on the external surface of membranes are synthesized with an animo terminal sequence, a signal sequence, which is very hydrophobic, and therefore, responsible for the attachment of the protein to membranes. The results summarized above have led to the proposal that the VSV glycoprotein has such a signal sequence.[16] In further support of this idea, Rose[18] has sequenced the ribosome binding site on the glycoprotein messenger RNA and has found that the codons immediately adjacent to the initiator codon predict an NH_2 terminal amino acid sequence of lys–cys–leu–leu–tyr–leu, which contains many hydrophobic residues.

Katz et al.[19] and Toneguzzo and Ghosh[20] have reported that if purified membranes are added to a wheat germ cell-free protein synthesizing system directed by VSV mRNA, the VSV glycoprotein is inserted into the membrane and glycosylated. Using this system, Rothman and Lodish[21] have varied the time of addition of membranes to the reaction and found evidence that the glycoprotein must be membrane associated by the time approximately the first 80 amino acids are polymerized or the molecule does not properly insert into membranes as assayed by subsequent glycosylation. Given the assumption that the cell-free system reflects requirements in vivo, these results imply that in vivo the glycoprotein associates with membranes very soon after the initiation of synthesis of the protein.

Thus, the results summarized above suggest that the VSV glycoprotein binds to membranes even before its synthesis is completed.

B. Transport of the Glycoprotein Across the Membrane

Membranes are thought to be asymmetric with little random movement of membrane components between the two surfaces of the membrane.[22] Thus, proteins which exist primarily on the external surface of cells or which are transmembranal require a special mechanism for their transport across the membrane barrier of the cell. The VSV glycoprotein is such a protein; synthesized in the cytoplasm, it ultimately must be transported to the external surface of infected cells.[23]

The signal sequence hypothesis, originally proposed by Blobel and Dobberstein[17] to account for the transport of secreted proteins across membranes, could also account for the membrane assembly of membrane proteins which are found on the external surface of cells. Blobel and Dobberstein proposed that proteins destined for the external surface of cells are synthesized with a hydrophobic NH_2-terminal amino acid sequence, which is responsible not only for the attachment of the protein to membranes as discussed above, but the formation of a protein channel through which the entire protein can traverse the membrane from the cytoplasmic side to the lumen of the endoplasmic reticulum.

As discussed in the previous section, there is evidence that the VSV glycoprotein is synthesized with a very hydrophobic NH_2-terminal amino sequence.[18] Certainly, the nascent protein immediately binds to intracellular membranes. Evidence has been presented that the glycoprotein is indeed transferred across the membrane in intracellular membranes.[24] Membranes have been isolated from pulse-labeled VSV infected cells. Membrane preparations derived from intracellular membranes are largely in the form of sealed cytoplasmic side out vesicles[25] and in fact, the membrane preparations are in a conformation such that ribosomes are on the outside surface of the vesicles. Proteins or portions of protein which are exposed on the surface of membranes should be susceptible to proteolytic digestion, while proteins which are embedded in the membrane bilayer or which are inside a sealed vesicle should be resistant to digestion. Upon trypsin digestion of membranes prepared from VSV-infected cells, no full-sized glycoprotein (69,000 daltons) was detected. However, the majority of the glycoprotein was found in nearly full-sized fragments with a molecular weight of 66,000 daltons.[24] Fragments this large are not likely to be protected from digestion because they are embedded in the membrane bilayer, but rather because they are inside a sealed vesicle. Kriebick and Sabatini[26] have shown that material inside membrane vesicles may be rendered susceptible to proteolytic digestion if the membrane vesicles are made porous with very low concentrations of the detergent deoxycholate. It was found that at low concentrations of deoxycholate (0.025%), trypsin digests the glycoprotein to very small molecular weight material.[24] These results are consistent with the idea that newly made glycoprotein is transported cross the membrane of the cell from the cytoplasmic side of the membrane into the lumen of the endoplasmic reticulum.

It is significant that no full-sized glycoprotein is found after trypsin digestion of membranes, but that the largest protected fragment obtained is approximately 3000 daltons smaller than intact glycoprotein. This result suggests that intracellular glycoprotein is transmembranal and that a small portion of one end of the glycoprotein extrudes from the cytoplasmic side of the membrane and is, therefore, susceptible to trypsin.

Similarly, Katz et al.[19] have reported that if membranes are added to a cell-free protein synthesizing system directed by VSV mRNA, the newly made VSV glycoprotein becomes largely resistant to trypsin. Full-size glycoprotein is reduced to a polypeptide with a molecular weight of 66,000 daltons. However, glycoprotein made in the absence of membranes is fully sensitive to trypsin digestion. Katz et al.[19] interpreted this result to mean that the glycoprotein is inserted into membranes in vitro and that the glycoprotein is transmembranal. Katz et al.[19] state that it is the carboxyl-terminal end of the molecule which is exposed on the cytoplasmic face of the membrane. This result is consistent with the finding that most plasma-membrane proteins studied are anchored to the plasma membrane at their carboxyl-terminal end.[27]

The precise mechanism for the transport of the glycoprotein across the membrane is unclear. It is possible that the synthesis of the molecule, in fact, drives the molecule across the membrane. Alternatively, glycosylation of the molecule may play a role in the transport. As will be discussed below, there is evidence that the initial glycosylation of the molecule occurs as the protein is being transported across the membrane.[21] Therefore, the role of glycosylation in the transport of the glycoprotein across the membrane was investigated.

Tunicamycin is an antibiotic which blocks the formation of lipid sugar intermediates required for the addition of carbohydrate to proteins.[28,29] Leavitt et al.[30] have shown that tunicamycin-treated VSV-infected cells synthesize the unglycosylated form of the VSV glycoprotein (molecular weight 63,000 daltons). It was found that this unglycosylated glycoprotein synthesized in tunicamycin-treated cells becomes membrane bound and appears to be transported across the membrane normally.[100] Trypsin digestion of membranes prepared from tunicamycin-treated infected cells cleaves a 3000 dalton fragment from the molecule. This result argues that the unglycosylated G protein is also transported to the lumen of the endoplasmic reticulum in much the same manner as the fully glycosylated G protein: a small portion of the molecule remains on the cytoplasmic face of the membrane. Thus the initial glycosylation reaction seems to play no role in the transport of the protein across the membrane.

C. Proteolytic Cleavage of the Glycoprotein Molecule

The signal sequence hypothesis of Blobel and Dobberstein[17] includes the idea that the amino terminal amino acid sequence, or signal sequence, is proteolytically cleaved from the protein soon after the amino terminus emerges on the lumenal side of the membrane. Thus far there has been no direct evidence for any proteolytic cleavage of the VSV glycoprotein. However, indirect evidence suggests such a cleavage. Schloemer and Wagner[31] have reported that the amino-terminal amino acid on the virion-associated glycoprotein is alanine. As described above, by sequencing the ribosome binding site, Rose[18] has found that the codons immediately adjacent to initiation codon predict an amino acid sequence of lys−cys−leu−leu−tyr−leu. Thus, in order to reconcile these two results, it is necessary to propose that at least six amino acids are removed from the amino terminus of the glycoprotein.

Thus far it has been difficult to determine directly if the glycoprotein is subjected to a minor cleavage. The obvious way is to compare the glycoprotein made in a cell-free system devoid membrane with the glycoprotein made in vivo. However, the addition of carbohydrate to the in vivo protein makes any difference seen between the two proteins difficult to interpret. The protein made in a cell-free system was compared with that made in infected cells in the presence of tunicamycin.[100] The glycoprotein made in the presence of tunicamycin is unglycosylated,[30] but as discussed above, is membrane associated and is apparently transported across the membrane normally. The glycoprotein made in a cell-free system devoid of membranes migrates on polyacrylamide gels slightly slower than the protein made in tunicamycin-treated cells. Furthermore, minor differences were found in the two proteins by the Cleveland[32] procedure for peptide mapping.

Cleveland et al.[32] have devised a procedure whereby polypeptides can be partially cleaved in the presence of detergent, sodium dodecyl sulfate, and the polypeptide fragments obtained are resolved on polyacrylamide gels. Using this procedure, Cleveland et al. have compared P23, the major head protein of bacteriophage T4, with the precursor protein P23*. P23 is known to be a result of a proteolytic cleavage of a 10,000 dalton amino acid sequence from the NH_2-terminal end of P23* protein.[33] At low-enzyme concentrations, several differences in the fragment patterns obtained from the product and the precursor were observed, while at higher concentrations of enzyme, only two differences were seen.

A similar comparison after two unglycosylated VSV G proteins at low enzyme concentration resulted in two differences in the polypeptide patterns obtained; while at high-enzyme concentrations, only one difference was seen.[100] These results are suggestive of a minor proteolytic cleavage of the glycoprotein in vivo.

D. Migration to the Plasma Membrane

There is a considerable delay between the completion of synthesis of the glycoprotein and its appearance at the surface of infected cells and in completed virions. Estimates of the time between synthesis and incorporation into virions have ranged from 20 to 30 min.[9,10,34]

It has been proposed by several investigators that the glycoprotein arrives at the plasma membrane by way of cellular-smooth membranes: that the glycoprotein migrates from rough endoplasmic reticulum, to smooth endoplasmic reticulum, to the Golgi membranes, and then to the plasma membrane.[9,10,34-36] The results presented in support of this model are not clear cut, but they do show a general trend which suggests such a pathway.

The type of experiment which has been done to show this migration is to pulse label infected cells with radioactive amino acids and then fractionate cell membranes according to their density in sucrose gradients. The heaviest density membranes are thought to be derived from rough endoplasmic reticulum because polyribosomes are found in this fraction. Lighter density membranes are presumed to be derived from smooth endoplasmic reticulum and the Golgi organelle. There has been no reported attempt to separate these two classes of smooth membranes. The location of the plasma membrane in these gradients has been determined by labeling the surface of cells prior to their disruption with either radioactive wheat germ agglutinin[10] or by iodination[9] and assaying the gradient for the location of the cell-surface label.

It seems clear that this sort of procedure does not result in clean preparations of the various classes of membranes: there is undoubtedly a great deal of cross contamination. This contamination may be responsible for some of the ambiguities in the results (see below).

By using long periods of radioactive labeling (3 hr), Lafay[36] and Wagner et al.[35] observed that the VSV glycoprotein could be found in all classes of cellular membranes, although in these long labels, the glycoprotein was found preferentially in lighter density membranes.

Many groups have pulse-labeled infected cells with radioactive amino acids and then chased with nonradioactive amino acids in order to locate newly made glycoprotein and to follow its fate with time.[9,10,34,36] Newly made glycoprotein is preferentially located in the rough endoplasmic reticulum. With increasing chase times, G protein concentration in the rough endoplasmic reticulum decreases and increases in lighter density membrane fractions.

There are a number of problems with these experiments which complicate interpretation. First, not all of the labeled glycoprotein can be chased into mature virions.[9] In another study, much of the labeled G protein found in light density membranes could not be chased into plasma membrane.[10] Third, significant amounts of newly made glycoprotein are located in light-density membranes or in the plasma membrane fraction.[10,37] In light of these observations, there is no assurance that the major pathway of the glycoprotein to virions is through the smooth membrane.[10] However, it seems quite possible that these observations could be due to incomplete separation of various membrane fractions and perhaps sticking of virus particles to membranes.[9]

Additional evidence for the obligatory migration of VSV glycoprotein through intracellular membranes to the plasma membrane comes from mutant studies. There exist several temperature-sensitive mutants which have defective glycoproteins, for example *ts*045[38] and *ts*M501.[39] Lafay[36] has shown that at nonpermissive temperature the *ts*045 glycoprotein is found primarily in heavier density membranes. Similar results have been obtained with *ts*M501.[40] Neither mutant glycoprotein can be found at the cell surface at nonpermissive temperature.[36,40] In addition, there is evidence that these mutant proteins do not contain sialic acid at nonpermissive temperature.[40] Since the enzymes which add sialic acid to glycoproteins are thought to be located in the smooth membrane fraction of the cell,[41] it seems likely that these defective glycoproteins never reach the smooth membranes of cells at nonpermissive temperature.

The mechanism responsible for the migration of a protein through intracellular membranes is completely unknown. However, the failure of the glycoprotein made by mutant *ts*045 and *ts*M501 to migrate to the plasma membrane at nonpermissive temperature suggests that the glycoprotein itself plays an active role in this process.[40]

As discussed below, the carbohydrate moieties of the glycoprotein have two components: a core structure which is located proximately to the polypeptide chain and more peripherally located sugars which are, for the most part, attached to the core.[42,43]

Leavitt et al.[30,44] have reported that failure to add the core oligosaccharide to the glycoprotein inhibits the appearance of the glycoprotein at the surface of infected cells and the formation of virus particles.[30] In addition, failure to add core oligosaccharides inhibits the migration of the glycoprotein to smooth membranes.[100] As described above, tunicamycin-treated infected cells synthesize an unglycosylated glycoprotein.[30] The glycoprotein synthesized in tunicamycin-treated cells is membrane associated, but it is not transported to light-density membrane fractions.[100] This result could indicate that correct glycosylation is required for the migration of the protein. Alternatively, the absence of carbohydrate could alter the conformation of the molecule such that it is incapable of participating in the normal mechanisms responsible for the migration.

In contrast, the failure to add the more terminal sugars of the carbohydrate chain present on the glycoprotein does not appear to affect the maturation of the glycoprotein. Gottlieb et al.[45] have reported a mutant of Chinese hamster ovary cells is defective in Nacetylglucosaminyltransferase activity. Infection of these cells with VSV results in a normal yield of infectious virus particles[46] even though the carbohydrate on the glycoproteins of the released virus lacked some terminal sugar.[46]

The last step of the migration of the glycoprotein is the insertion of the glycoprotein into plasma membranes. There is not direct evidence which bears on the mechanism of insertion, but a likely model has been proposed. Knipe et al.[40] have suggested that the glycoprotein which has migrated to the Golgi organelle from the endoplasmic reticulum is incorporated into vesicles. These vesicles have the same orientation as the endoplasmic reticulum: the majority of the glycoprotein molecule is inside the vesicle and the glycoprotein is attached to the membrane of the vesicle at its carboxyl terminal end. The vesicle migrates to the plasma membrane, fuses with the plasma membrane and the inside surface of the vesicle becomes contiguous with the outside surface of the cell. Thus, the glycoprotein is deposited such that it is located primarily on the external surface of the plasma membrane.

This model predicts that the orientation of the glycoprotein is the same as in the endoplasmic reticulum: the carboxyl terminus is embedded in the membrane. However, Schloemer and Wagner[31] have reported that in the virion, the amino terminus is embedded in the virion membrane. Resolution of this problem awaits future work.

The migration of the glycoprotein to the plasma membrane is apparently independent of all other viral proteins. The only mutations which reportedly affect the migration of the glycoprotein are in the glycoprotein itself.[36,40]

E. Glycosylation of the Glycoprotein

The carbohydrate moieties of viral glycoproteins are host specified; that is, the enzymes required for the addition of the oligosaccharides are host enzymes. Thus the composition and the sequence possibly vary somewhat with the host cell.[47,48]

The structure of the carbohydrate on the VSV glycoprotein made in Hela cells has been characterized in some detail. The carbohydrate is primarily in the form of two oligosaccharide chains per glycoprotein molecule.[42] Moyer et al.[49] have presented evidence which suggests that each chain is linked to the polypeptide backbone by an asparagine N acetylglucosamine linkage. Each oligosaccharide chain is composed of a core structure of N acetylglucosamine and repeating mannose residues, which are located proximally to the polypeptide, and distal sugars linked, for the most part, to the mannose core. These distal sugars are galactose, N acetylglucosasmine, sialic acid, and fucose.[42,43]

Extensive studies of carbohydrate associated with many different glycoproteins has revealed that there is a large class of glycoproteins which have oligosaccharide chains with structures very similar to that found on the VSV glycoprotein: a core structure, composed of N acetylglucosamine and mannose residues, which is linked to the polypeptide through an asparagine residue, and variable peripheral sugars.[50] Evidence is beginning to accumulate that the core oligosaccharides are assembled on lipid-carrier molecules and once assembled, are transferred in bloc to the polypeptide. Then additional peripheral sugars are added in a stepwise fashion from their respective sugar nucleotides to the core oligosaccharide.[51] Thus far, evidence suggests that the glycosylation of the VSV glycoprotein proceeds in a similar stepwise manner.

Hunt and Summers[41] have combined cell fractionation with pulse labeling with radioactive sugars in order to examine different steps in the glycosylation of the VSV glycoprotein in Hela cells. Three classes of cellular membranes were resolved in discontinuous sucrose gradients: rough endoplasmic reticulum, smooth membranes, and plasma membranes. Mannose, which is located in the core region of the oligosaccharide chain, is incorporated first into the rough endoplasmic reticulum and after longer labeling periods, can be found in light density and plasma membranes. Fucose and galactose, both peripheral sugars, are first incorporated into light density membranes and after longer labeling periods, can be found in plasma membrane. Glucosamine, which is found in both the core and peripheral regions, seems to be incorporated in both heavy- and light-density membranes. Thus, components of the core seem to be added in the rough endoplasmic reticulum, while peripheral sugars are added in smooth membranes.

Hunt and Summers[41] also present evidence which suggests that glucosamine and mannose residues are added as a block to the G protein. Gel filtration of glycopeptides derived from pronase-treated pulse-labeled material yield a single major peak of material with a molecular weight of 1600 to 1800, rather than a series of smaller molecular weight glycopeptides. The latter result would have been expected if the core sugars were added sequentially.

In experiments of a completely different nature, Knipe et al.[23] and Rothman and Lodish[21] have also presented evidence that glycosylation of the VSV glycoprotein proceeds in steps. The addition of carbohydrate to polypeptides alters the migration of the protein on polyacrylamide gels: the higher percentage of carbohydrate, the slower the glycoprotein migrates.[52] Virion-associated, and therefore fully glycosylated, VSV

G protein migrates with an apparent molecular weight of 69,000 daltons. This form of the glycoprotein has been termed G2 by Knipe et al.[23] The VSV glycoprotein synthesized in membrane-free, cell-free extracts directed by VSV mRNA is completely unglycosylated (Go) and migrates with an apparent molecular weight of 63,000 daltons.[11,53] Knipe et al.[23] have pulse-labeled VSV-infected cells and have found two intracellular forms of the G protein: G2 and a species of glycoprotein with an apparent molecular weight between Go and G2. This form of glycoprotein was termed G1 (apparent molecular weight of 65,000 daltons). Treatment of the G2 form of the protein with neuraminidase converts it to G1. This result suggests that the difference between the two forms is at least the addition of sialic acid. There is no evidence that the G1 form of the glycoprotein is a homogeneous molecule: it may be composed of glycoprotein molecules with only the core oligosaccharides and glycoprotein molecules containing the core and peripheral sugars except sialic acid.

The G1 form of the glycoprotein is found in the rough endoplasmic reticulum, as well as light-density membranes. In contrast, the G2 form is found only in light-density membranes.[23] In a short 5 min pulse of radioactive methionine, only G1 is detected in infected cells. After a chase period of 20 min, G2 can be detected.[23] These results suggest that the addition of sialic acid takes place long after the addition of more proximal sugars and in a different cell location.

Katz et al.[19] have reported a cell-free protein synthesizing system which is capable of glycosylating VSV G protein in vitro. Membranes purified from dog pancreas are added to a wheat germ cell-free system directed by VSV RNA. Using this cell-free system, Rothman and Lodish[21] have seen another form of G, which they have called G1/2. To reactions containing membranes, they have added triton X100 at various times after the initiation of synthesis of the glycoprotein in order to disrupt the membranes. Triton has no affect on protein synthesis. If triton is added at the appropriate time, G1/2 is seen. They propose that this is the glycoprotein containing one of the two oligosaccharide chains.

Go is never seen in infected cells even after very short radioactive pulse labels.[23] This result argues that the glycoprotein is glycosylated very soon after or during the synthesis of the protein. In the cell-free system, Rothman and Lodish[21] have shown that glycosylation occurs before the completion of the nascent polypeptide chain. In this system, one polypeptide chain is completed in about 45 min. However, glycosylation is complete by 29 min after the initiation of synthesis of the protein. If the in vitro system is an accurate representation of events in vivo, these results argue that the initial glycosylation takes place on the nascent polypeptide chain.

The results taken together suggest that the glycosylation of the VSV glycoprotein proceeds in a stepwise fashion and is not complete until approximately 20 min after the initiation of synthesis of the molecule. The first sugars, presumably the two core structures, are added sequentially to the nascent polypeptide chain in the rough endoplasmic reticulum. The protein then migrates to the smooth membranes of the cell where peripheral sugars are added.

F. Summary of Glycoprotein Maturation

Available evidence suggests that the VSV glycoprotein follows a rather complex series of steps which lead it ultimately to the outside surface of the host cell plasma membrane. The nascent protein first binds to intracellular membranes at its site of synthesis and evidence suggests that this binding occurs as the protein is being synthesized. The protein is transported across the membrane of the endoplasmic reticulum such that most of the molecule is on the lumenal surface of the membrane. However, there is evidence that the protein is transmembranal and that a segment of at least

3000 daltons is on the cytoplasmic surface of the membrane. The protein then migrates from the rough endoplasmic reticulum to smooth endoplasmic reticulum and perhaps the Golgi organelle. During this transit, the protein is progressively glycosylated: core oligosaccharides are added in the rough endoplasmic reticulum while peripheral sugars are added in smooth membranes and/or the Golgi organelle. The glycoprotein is then incorporated into the host-cell plasma membrane.

III. MEMBRANE PROTEIN MATURATION

The maturation of the M protein is not well understood. This protein is ultimately found on the inner surface or within the virion membrane: its precise conformation with respect to the virion membrane is not clear.[5-7] However, it is fairly certain that the M protein is not exposed to the outer surface of virion membrane or plasma membrane.[6,7] Thus, in contrast to the glycoprotein, the M protein need not be transported across the cell membrane.

Also, in contrast to the glycoprotein, the M protein can be incorporated into virions very soon after its synthesis. Estimates of the lag between synthesis and the appearance of the M protein in virions has ranged from 8 to 10 min.[9,10,34] Furthermore, in experiments in which viral proteins were pulse labeled with radioactive amino acids followed by a chase with nonradioactive amino acids, the M protein was seen to accumulate in plasma membranes within 5 min after the onset of the chase.[37] These results clearly show that the M protein can be incorporated into the plasma membranes and virions much more rapidly than the G protein. Thus it seems likely that the M protein is inserted into the host-cell plasma membrane by a process distinct from glycoprotein maturation.[10,40]

The M protein appears to be synthesized primarily on nonmembrane-associated polyribosomes.[11-13] There are some reports that a fraction of the polyribosome-associated M protein mRNA can be found in the membrane fraction of cells,[11-15] but this association can be reduced upon subsequent repurification of the membranes.[15] In any event, the M protein mRNA does not show the exclusive association with membranes characteristic of glycoprotein mRNA.

In light of the observations concerning the nature of the attachment of glycoprotein mRNA to membrane-bound polyribosomes, these results suggest that the M protein amino terminus does not have an affinity for host-endoplasmic reticulum, and in fact, significant amount of newly made M protein can be found in the soluble fraction of infected clls.[9,16,54] After a very short pulse label with radioactive methionine, Knipe et al.[9] and David[16] report 50 and 56%, respectively, of the M protein in the soluble fraction of infected cells. Furthermore, if 0.1 *M* NaCl is included in the buffers used in cell fractionation up to 70% of the nascent, M protein can be found in the soluble fraction of infected cells.[9]

Although initially a large fraction of nascent M protein is nonmembrane associated, with time, the protein appears to accumulate in membrane fractions of infected cells. Following a 2-min pulse label and a 60-min chase, David[16] reports that 100% of intracellular M protein is membrane associated. Under similar conditions, Knipe et al.[9] find that 20% of the intracellular M protein is still in the soluble fraction of the cell.[9] However, both reports agree that the M protein becomes membrane associated with time.

Hunt and Summers[10] report that the M protein appears to accumulate in all classes of cellular membrane: the rough endoplasmic reticulum, smooth membranes, and plasma membranes. Furthermore, no apparent movement of the M protein from one class of membranes to another was detected. After a short pulse of radioactive amino acids followed by a chase with nonradioactive precursor, the amount of radioactive M protein did not decrease in any class of membranes. However, in similar experiments, Knipe et al.[9] report a transient association with membranes characterized as rough endoplasmic reticulum followed by a transient association with light-density membranes characterized as plasma membranes. These results have been interpreted to indicate that, following its synthesis, the M protein is released from polyribosomes into the soluble fraction of infected cells and is then directly incorporated into plasma membranes.[9,10,40]

There are a number of observations concerning the M protein which are difficult to understand in terms of this model of M protein maturation. It has been shown that the M protein has a clear-cut affinity for membranes in vitro. Cohen et al.[54] isolated the M protein from the soluble fraction of infected cells and showed that it could bind to isolated membranes in vitro. The M protein has been synthesized in a cell-free system directed by VSV mRNA and this protein also was found to have an affinity for uninfected membranes.[55] Futhermore, the M protein-membrane complex formed in vitro was found to be very stable. No condition short of disrupting the membrane will release the M protein from membranes.[100] Given this affinity for membranes, it is unclear why newly made M protein is found initially in the soluble fraction of cells, rather than immediately attaching to membranes upon completion of the molecule.

A second puzzling problem is the observation that the M protein accumulates in all classes of intracellular membranes, rather than just in fractions enriched in plasma membrane.[9,10] It is possible that this finding is a result of contamination of various membrane fractions with plasma membranes and/or budding virus particles. Alternatively, the association of M protein with intracellular membranes may not be directly related to virion formation.

Knipe et al.[40] argue that the M protein is incorporated into a region of the plasma membrane which already contains the glycoprotein. It is also argued that M protein binding is stabilized by the binding of the N protein. These proposals have been made because at nonpermissive temperatures, cells infected with temperature-sensitive mutants defective in the G protein or the N protein seem to have less membrane-associated M protein than wild-type infected cells. This observation is puzzling in light of the observed affinity of the M protein to uninfected cell membranes.

It seems clear that many of the questions concerning M protein maturation will be resolved when more is known about the chemistry of the M protein molecule and the sorts of interactions it undergoes in infected cells.

IV. THE RIBONUCLEOPROTEIN CORE

The components of the ribonucleoprotein core of virus particles — the genome RNA, the nucleocapsid protein, the NS protein, and the transcriptase — arrive at the plasma membrane by a third process which seems to be distinct from pathways followed by the glycoprotein and the M protein.[34,35,54] Much of the published data concerning this pathway deals only with the VSV N protein, but because of the localization of the NS and L proteins in the core of the virus,[55] these proteins may follow the same general series of steps as the N protein. However, verification of this possibility awaits further work which focuses on these proteins.

As observed with the M protein, the protein components of the core are synthesized largely on nonmembrane-associated polyribosomes[11-13,15] and the nascent polypeptides are released into the soluble fraction of infected cells.[9,16,35] After a short pulse label with a radioactive amino acid, up to 50% of the N protein is found in the soluble fraction of cells.[9] With time, however, the N protein is found in a rapidly sedimenting structure with a sedimentation coefficient of 120.[56-58] This structure has been characterized as the ribonucleoprotein core which contains genome-sized RNA, N protein, NS protein, and L protein.[56-58] This structure seems to be assembled in the soluble fraction of infected cells:[35] there is no evidence that membranes are involved.

The ribonucleoprotein core has been partially assembled in a cell-free system. Emerson and co-workers[59,60] have shown that high salt treatment of the core solubilizes the VSV NS and L proteins and leaves the RNA complexed with N protein.[57] The RNA-N protein complex was mixed with purified L and NS proteins and the complete ribonucleoprotein complex reformed properly as assayed by its capacity for in vitro transcription.[59,60]

There is no report of in vitro assembly of the RNA-N protein complex from its component parts. However, Soria[61] has reported that the N protein will detach from the RNA at high temperatures. Thus, at this time, there is very little detailed information about the steps in the assembly of the ribonucleoprotein core. The assembly process may involve protein-protein and protein-nucleic acid interactions which depend upon the properties of the individual core proteins themselves.

There is, however, one puzzling observation which , if verified, may indicate that nucleocapsid assembly is somewhat more complicated than this simple model. Knipe et al.[40] have reported that at nonpermissive temperature, cells infected with *ts*502 have a defect in the assembly of nucleocapsids.[40] In cells labeled at 39°C, nearly all the N protein was found in the soluble fraction. The mutation in this virus has been assigned to complementation group V[62] or the VSV glycoprotein.[1]

Once assembled, the core structure is thought to attach to regions of the plasma membrane which already contain viral proteins. This model has been proposed based on electron microscopic observations of cells infected with paramyxoviruses and myxoviruses. These viruses are constructed in a fashion very similar to rhabdoviruses — they are composed of an outer envelope which is modified with viral glycoproteins and membrane proteins and an inner core structure.[63] Furthermore, these viruses bud from the surface of infected cells.[63] Numerous studies have presented electron micrographs which show the viral nucleocapsid attached to the inner side of a region of plasma membrane which also contains viral glycoprotein on the outside surface.[63-68]

Wagner et al.[69] have presented electron micrographs of the plasma membranes of VSV-infected L cells. While the location of the glycoprotein on the outer surface is visualized by ferritin-labeled antibodies, the presence of nucleocapsid structures under these regions of host cell membrane is not obvious.[69]

Cohen et al.[54] and Atkinson et al.[37] have purified infected cell-plasma membranes and have shown that the N protein can be found in association with these membranes. It is likely that this association reflects the binding of the core structure; however, the core itself was not directly shown to be present in these membranes. Atkinson et al.[37] showed that there was a 15-min lag between the synthesis of the N protein and its incorporation into plasma membranes. Perhaps this lag reflects the time required for the assembly of the core structure.

Thus, the migration of the assembled core structure to plasma membranes and its association with plasma membranes has not been studied in detail. However, circumstantial evidence suggests that it does bind to host-cell membranes.

What causes the structure to migrate to the plasma membrane is completely unknown. In addition, the interactions responsible for the binding of the core to plasma membranes are incompletely understood. However, it seems clear that the core structure must interact with other viral proteins present in the plasma membrane. Studies with temperature-sensitive mutants and studies of phenotypically mixed particles which result from coinfection of cells with VSV and other enveloped viruses have suggested that the M protein and perhaps the G protein are required for core binding.

Knipe et al.[40] explore the effect of mutations in M and G proteins on the incorporation of N protein into membranous structures and virus particles. At nonpermissive temperatures, cells infected with *ts*M301 (a Group III or M protein mutant)[1,62] did not produce virus particles and the N protein remained in core structures or in the soluble fraction of the cell. Similar results were obtained with virus containing mutations in the glycoprotein. However, in these cases, the M protein assembly into membranes also appeared defective. These results suggest that at least the M protein is required for core binding.

Compans and Choppin[70] and McSharry et al.[71] have presented results of phenotypic-mixing experiments which also suggest that the M protein is important. Cells coinfected with SV5, a paramyxovirus, and VSV yield VSV, SV5, and phenotypically mixed virus (pseudotypes). Pseudotypes which contained the VSV genome always contained VSV M protein — never the SV5 M protein. From this result, it has been argued that the M protein-core interactions are necessary for assembly of virus particles.

Pseudotypes can also be formed between VSV and RNA tumor viruses:[72-74] these phenotypically mixed virus contain the VSV genome and tumor virus surface antigens.[72-74] Witte and Baltimore[75] have used VSV virus which has a temperature-sensitive mutation in the M protein and have shown that at nonpermissive temperature, pseudotypes do not form. This result again argues that the VSV M protein is required for the assembly of the VSV core into virus particles: the tumor virus envelope protein will not substitute.

The direct requirement of the glycoprotein in the attachment of nucleocapsids to plasma membrane is less clear. After coinfection with VSV and SV5, the majority of the pseudotypes which contain the VSV genome have both the VSV glycoprotein and the SV5 glycoproteins.[70,71] This finding led to the suggestion that there are no stringent requirements on the type of glycoprotein which may be incorporated into virions and that the viral spikes do not recognize specific components on the inner surface of plasma membranes.[70,71]

Pseudotypes formed between VSV and RNA tumor virus have also been shown to contain glycoprotein from both viruses.[75,76] However, Witte and Baltimore[75] have argued that this result indicates that at least some VSV glycoprotein is required to initiate the binding of other internal components of the VSV virion to membranes. Even if this is so, it is not clear whether or not the glycoprotein plays a direct role in the binding of the core to the plasma membrane. An absolute requirement for some VSV glycoprotein may reflect the necessity for a specific interaction between VSV glycoprotein and VSV M protein.

There are a number of other unanswered questions concerning the assembly of the ribonucleoprotein complex into virions. For example, infected cells contain ribonucleoprotein complexes which contain RNA complementary to virion RNA.[77] These complexes are undoubtedly intermediates in the replication of viral RNA.[77,78] These structures, however, are apparently not incorporated into virions since RNA isolated from virions will not anneal to form double-stranded RNA structures.[79] The mechanism responsible for the selection of core structures containing only the negative-stranded RNA for assembly into virions is unknown.

Another interesting observation which may be related to morphogenesis is the fact that the NS protein is phosphorylated.[80-83] The protein is phosphorylated early in infection and the phosphorylated protein is incorporated into the ribonucleoprotein complex.[80-82] However, the role of this modification in morphogenesis is not clear at this time.

In summary, evidence to date suggests that the ribonucleoprotein core of VSV is assembled in the cytoplasm from components which exist in the soluble fraction of infected cells. The ribonucleocapsid may spontaneously come together as a result of the properties of the individual proteins themselves. Once assembled, the core migrates to regions of the plasma membrane, which already contain viral membrane components. Evidence strongly suggests that the M protein is required for the binding of the core structure to plasma membranes. Whether the glycoprotein directly participates in this binding is not clear.

V. EVENTS AT THE PLASMA MEMBRANE

A. Introduction

The final stages in the formation of mature virus particles occur at the plasma membrane.[63] Extremely little information is available concerning the detailed events at this stage. However, by analogy with other enveloped viruses, one may speculate that there are at least three components to this part of morphogenesis. First, virion components must be inserted into plasma membranes. Second is the formation of patches or regions of plasma membrane which contain high concentrations of virion components. Finally, these patches bud out to form mature virions. Insertion, patch formation, and budding are quite likely dependent upon each other. However, for convenience, each will be discussed separately. Insertion of viral components into the plasma membrane has been discussed extensively above and will not be considered further.

B. Patch Formation

1. Properties of the Patches

Patches of viral components on the plasma membrane of infected cells have been most clearly demonstrated following infection with paramyxoviruses, myxoviruses, and togaviruses by adding ferritin-tagged antiviral antibodies to infected cells.[64-67,84] The ferritin can be visualized adhering to localized regions of the surfaces of infected cells. In the case of VSV-infected cells, hybrid antibodies labeled with ferritin bound to localized regions early in infection, but seemed to label the entire surface of the plasma membrane late in infection.[69] As discussed above, ribonucleocapsids can be clearly seen on the cytoplasmic side of these patches following paramyxovirus and togavirus[67,84] infection while ribonucleocapsids are not obvious under the patches formed following VSV infection.[69] Either the VSV ribonucleocapsids are not easily seen or they do not bind until, or concomitant with, budding. The M protein is presumably located in these patches, although it has not been directly demonstrated.

Electron micrographs reveal that the patches of membrane which contain virion components are continuous with host cell plasma membrane.[67] However, several lines of evidence suggest that most of the host-cell plasma membrane proteins are excluded from these patches. First, in the case of paramyxovirus infection, ferritin-labeled antibodies raised against the surface proteins of uninfected cells do not bind to the patches, but do bind to adjacent regions of the plasma membrane.[64] Second, freeze-fracture studies of VSV infected cell-plasma membranes have shown that host-membrane structural proteins which extend into or across the nonpolar region of the membrane bilayer are lost from the regions of the plasma membrane as they become viral envelopes.[85]

That these patches contain very little host cell protein is also suggested by the fact that mature virions contain very litle host protein. If purified virions are disrupted and the protein components resolved on polyacrylamide gels, the major proteins present, as determined by coomassie blue staining, are viral proteins.[82] The major protein species seen in preparations of uninfected cell plasma membranes are largely absent in virion preparations.[82]

However, there are indications that some specific host-cell membrane proteins are present in virions. Hecht and Summers[86] have presented evidence which suggests that the H-2 histocompatibility antigen found on murine cells can be detected in purified virus grown in these cells. In addition, there are reports of enzymatic activities in VSV virions which are thought to be derived from the host cell. For example, Strand and August[87] reported a protein kinase, and Roy and Bishop[88] reported nucleoside triphosphate phosphotransferase activity. At this time, it is unknown if these host-cell proteins are passively incorporated into virions during budding or whether there is a mechanism for the incorporation of specific host components into patches and, therefore, virions.

In contrast, it seems quite clear that host cell plasma membrane lipid components are present in these patches and, therefore, the completed virions.[89,90]

2. Possible Mechanisms for Patch Formation

How these patches are formed is completely unknown. Several alternate modes of assembly seem possible. These speculations are modifications of the alternatives proposed by Compans and Caliguri.[91] It is possible that the glycoprotein is randomly inserted into the host-cell plasma membrane either by the fusion of a vesicle containing the glycoprotein or by some other mechanism (see Section II). The fluid mosaic model of cell membranes includes the idea that components of membranes are mobile in the lateral plane of the membrane.[92] Thus, glycoprotein molecules, once inserted, may aggregate by moving together. Aggregation may be promoted by glycoprotein-glycoprotein interactions within the hydrophobic domain of the membrane and/or at either side of the plasma membrane.[22] Alternatively, glycoprotein aggregation may be promoted by the interaction of the glycoprotein molecules with viral components on the cytoplasmic side of the plasma membrane.[22] For example, the M protein may interact with the glycoprotein as well as with itself. Thus, M protein aggregation would favor glycoprotein aggregation. Another possibility is that the glycoprotein molecules and/or the M protein molecules are aggregated by interactions with the ribonucleocapsid structure which binds to the inner surface of the plasma membrane. In the latter case, the large core structure would provide a nucleation site for the formation of a patch.

Whatever the mechanism for the aggregation of viral components, the aggregation could exclude host cell protein except perhaps those proteins which may favorably interact with viral components.

An alternative mechanism for the formation of patches is the formation intracellularly of membrane vesicles synthesized *de novo* in the Golgi organelle. High concentrations of glycoprotein may be incorporated into these vesicles. These vesicles could then fuse with the host plasma membrane and the fusion event itself would generate patches of glycoprotein on the cell surface.

There is not enough information available to distinguish between these or any other models.

C. Budding of Virus from Cells

The last stage in morphogenesis is the budding of the virus particle from host-cell plasma membrane. The budding process is the outfolding of a patch of plasma membrane which contains viral components. The emerging virus particle is surrounded by a unit membrane which is still contiguous with the host-cell plasma membrane.[69,93-96] However, the emerging virus particle finally separates itself from the cell, the outside surface of the cell becoming the outside surface of the virion, and the inner side of the plasma membrane becoming the inner side of the virion.[8,97]

What initiates the outfolding of the patch is completely unknown. It is possible that when the core structure binds to the plasma membrane, its interaction with viral-membrane proteins causes the entire patch to assume a conformation which causes the outfolding to the patch.

Alternatively, one of the virion components may change its conformation with respect to the membrane and thus initiate the budding. Work with toga viruses has suggested that the cleavage of one of the glycoprotein spikes, PE2, is required to initiate budding of the virus particle. A temperature-sensitive mutant which is defective in this cleavage at high temperature accumulates patches in the host-cell plasma membrane which contain the viral proteins, but no budding of the patches occurs.[98,99]

There is, however, no evidence at this time that the VSV glycoprotein is cleaved at the cell surface. It is interesting that freeze-fracture studies of budding virus reveal no obvious transmembranal structures.[85] However, studies of the glycoprotein in intracellular membranes suggest that intracellular glycoprotein is transmembranal.[19,24] These two observations could be reconciled if there was a cleavage of the glycoprotein at the plasma membrane.

As the viral patch buds from the cell, the cylindrical-shaped structure of the mature virus particle becomes evident. What determines this precise shape, which is unique among enveloped viruses, is unclear. That the external regions of the glycoprotein spikes are not responsible for the shape is shown by the fact that the following protease treatment which removes the spikes, the shaved virions still have a bullet shape.[3] In addition, phospholipase C treatment of virions, which removes much of the lipid, yields a protein scafold which is still bullet shaped.[3] That the ribonucleoprotein core plays a role in determining the shape of the virion is suggested by the fact that defective interfering particles which have a shortened genome and, therefore, core structure, are not full-sized bullet-shaped particles, but are shortened particles.[2] However, it seems likely that the structure is determined by the specific interactions of the core structure and the membrane of the virus, since the core structures isolated from virus particles do not seem to be in a helical or bullet shape.[2]

The outfolding of the virus patch may occur in several different orientations. Most electron micrographs reveal budding virus emerging rounded end first.[93-96] However, there are several reports of budding virus emerging from the cell sideways.[85,94] This observation may be due to mature virus which has adsorbed to the cell in this orientation, however, the fact that freeze-fracture studies also show virus budding sideways argues against this interpretation.[85] Until more is understood about the budding process and what determines the bullet-shaped structure of the virion, this observation will remain enigmatic.

Finally, the emerging virus particle separates from the infected cell. The mechanism whereby this separation occurs is completely unknown. Perhaps this final separation occurs by a fusion of the plasma membranes which are adjacent to the viral patch. Some electron micrographs of mature virions show amorphous membrane bag at the bottom, or blunt end, of the bullet.[96] This structure has been called the terminal hemisphere by Orenstein et al.[96] Glycoprotein spikes are apparently absent from this membrane. Other micrographs show a blunt end which appears to have membrane drawn tightly across, and glycoprotein spikes cover the blunt end.[2,3,93-95] Which picture reflects the true form of the virion is not clear at this time, but resolution of this problem is important in deciding upon the mechanism of separation of the virus from the cell.

REFERENCES

1. **Wagner, R.**, Reproduction of rhabdoviruses, in *Comprehensive Virology*, Vol. 4, Fraenkel-Conrat, H. and Wagner, R. Eds., Plenum Press, New York, 1975, Chap. 1
2. **Nakai, T. and Howatson, A. F.**, The fine structure of vesicular stomatitis virus, *Virology*, 35, 208, 1968.
3. **Cartwright, B., Smale, C. J., and Brown, F.**, Surface structure of vesicular stomatitis virus, *J. Gen. Virol.*, 5, 1, 1969.
4. **Cartwright, B., Talbot, P., and Brown, F.**, The proteins of biologically active subunits of vesicular stomatitis virus, *J. Gen. Virol.*, 7, 267, 1970.
5. **Mudd, J.**, Glycoprotein fragment associated with vesicular stomatitis virus after proteolytic digestion, *Virology*, 62, 573, 1974.
6. **Moore, N., Kelley, J., and Wagner, R.**, Envelope proteins of vesicular stomatitis virions: accessibility to iodination, *Virology*, 61, 292, 1974.
7. **Eger, R., Compans, R., and Rifkin, P.**, The organization of the proteins of vesicular stomatitis virions: labeling with pyridoxal phosphate, *Virology*, 66, 610, 1975.
8. **Patzer, E. J. and Wagner, R. R.**, Phospholipid asymmetry of the membrane of vesicular stomatitis virus, *Am. Soc. Microbiol. Abstr.*, 5, 114, 1977.
9. **Knipe, D., Baltimore, D., and Lodish, H.**, Separate pathways of maturation of the major structural proteins of vesicular stomatitis virus, *J. Virol.*, 21, 1128, 1977.
10. **Hunt, L. and Summers, D.**, Association of vesicular stomatitis virus proteins with Hela cell membranes and released virus, *J. Virol.*, 20, 637, 1976.
11. **Morrison, T. and Lodish, H.**, Site of synthesis of membrane and nonmembrane proteins of vesicular stomatitis virus, *J. Biol. Chem.*, 250, 6955, 1975.
12. **Grubman, M., Moyer, S., Bannerjee, A., and Ehrenfeld, E.**, Subcellular localization of vesicular stomatitis virus messenger RNAs, *Biochem. Biophys. Res. Comm.*, 62, 531, 1975.
13. **Both, G., Moyer, S., and Bannerjee, A.**, Translation and identification of the viral mRNA species isolated from subcellular fractions of vesicular stomatitis virus infected cells, *J. Virol.*, 15, 1012, 1975.
14. **Grubman, J. J., Weinstein, J. A., and Shafritz, D. A.**, Studies on the mechanism for entry of vesicular stomatitis virus glycoprotein mRNA into membrane bound polyribosome complexes, *J. Cell Biol.*, 74, 43, 1977.
15. **Lodish, H. F. and Froshauer, S.**, Binding of viral glycoprotein mRNA to endoplasmic reticulum membranes is disrupted by puromycin, *J. Cell Biol.*, 74, 74, 1977.
16. **David, A.**, Assembly of the vesicular stomatitis virus envelope: transfer of viral polypeptides fom polysomes to cellular membranes, *Virology*, 78, 98, 1977.
17. **Blobel, G. and Dobberstein, A.**, Transfer of proteins across membranes. II. Reconstitution of functional rough microsomes from heterologous components, *J. Cell Biol.*, 67, 82, 1975.
18. **Rose, J. K.**, Ribosome recognition sites in vesicular stomatitis messenger RNA in *Negative Strand Viruses and the Host Cell*, Mahy, B. W. J. and Barry, R. D., Eds., Academic Press, London, 47, 1978.
19. **Katz, F., Rothman, J. E., Lingappa, V. R., Blobel, G., and Lodish, H. F.**, Membrane assembly *in vitro*: synthesis, glycosylation, and asymmetric insertion of a transmembrane protein, *Proc. Natl. Acad. Sci. U.S.A.*, 74, 3278, 1977.
20. **Toneguzzo, F. and Ghosh, H. P.**, Synthesis and glycosylation *in vitro* of glycoprotein of vesicular stomatitis virus, *Proc. Natl. Acad. Sci. U.S.A.*, 74, 1516, 1977.
21. **Rothman, J. E. and Lodish, H. F.**, Synchronised transmembrane insertion and glycosylation of a nascent membrane protein, *Nature (London)*, 269, 775, 1977.
22. **Nicholson, G. L.**, Transmembrane control of the receptors on normal and tumor cells. I. Cytoplasmic influence over cell surface components, *Biochim. Biophys. Acta*, 457, 57, 1976.
23. **Knipe, D., Lodish, H., and Baltimore, D.**, Localization of two cellular forms of the vesicular stomatitis viral glycoprotein, *J. Virol.*, 21, 1121, 1977.
24. **Morrison, T. G. and McQuain, C. O.**, Assembly of viral membranes: nature of the association of vesicular stomatitis virus proteins to membranes, *J. Virol.*, 26, 115, 1978.
25. **Depierre, J. W. and Ernster, L.**, Enzyme topology of intracellular membranes, *Ann. Rev. Biochem.*, 46, 46, 1977.
26. **Kreibick, G. and Sabatini, D.**, Microsomal membranes and the translational apparatus of eukaryotic cells, *Fed. Proc. Fed. Am. Soc. Exp. Biol.*, 32, 2133, 1973.
27. **Marchesi, V. T., Furthmayr, H., and Tomita, M.**, The red cell membrane, *Ann. Rev. Biochem.*, 45, 1976, 667.

28. **Takatsuki, A., Arima, K., and Tamura, G.,** Tunicamycin, a new antibiotic. I. Isolation and characterization of tunicamycin, *J. Antibiot.*, 24, 215, 1971.
29. **Takatsuki, A., Kohro, K., and Tamura, G.,** Inhibition of biosynthesis of polyisoprenol sugars in chick embryo microsomes by tunicamycin, *Agric. Biol. Chem.*, 39, 2089, 1975.
30. **Leavitt, R., Schlesinger, S., and Kornfeld, S.,** Tunicamycin inhibits glycosylation and multiplication of sindbis and vesicular stomatitis viruses, *J. Virol.*, 21, 375, 1977.
31. **Schloemer, R. and Wagner, R.,** Association of vesicular stomatitis virus glycoprotein with virion membranes: characterization of the lipophilic tail fragment, *J. Virol.*, 16, 237, 1975.
32. **Cleveland, D. W., Fischer, S. G., Kirschner, M. W., and Laemmli, U. K.,** Peptide mapping by limited proteolysis in sodium dodecyl sulfate and analysis by gel electrophoresis, *J. Biol. Chem.*, 252, 1102, 1977
33. **Celis, J. E., Smith, J. D., and Brenner, S.,** Correlation between genetic and translational maps of gene 23 in bacteriophage T4, *Nature New Biol.*, 241, 130, 1973.
34. **Kang, C. Y. and Prevec, L.,** Proteins of vesicular stomatitis virus. III. Intracellular synthesis and extracellular appearance of virus specific proteins, *Virology*, 46, 678, 1971.
35. **Wagner, R., Kiley, M., Snyder, R., and Schnaitman, C.,** Cytoplasmic compartmentalization of the proteins and ribonucleic acid species of vesicular stomatitis virus, *J. Virol.*, 9, 672, 1972.
36. **Lafay, F.,** Envelope proteins of vesicular stomatitis virus: effect of temperature sensitive mutations in complementation groups III and V, *J. Virol.*, 14, 1220, 1974.
37. **Atkinson, P., Moyer, S., and Summers, D.,** Assembly of vesicular stomatitis virus glycoprotein and matrix protein into Hela cell plasma membranes, *J. Mol. Biol.*, 102, 613, 1976.
38. **Deutsch, V. and Berkaloff, A.,** Analyse d' un mutant thermolabile du virus de la stomatite vesiculaire, *Ann. Inst. Pasteur (Paris)*, 121, 101, 1971.
39. **Knipe, D., Lodish, H. F., and Baltimore, D.,** Analysis of the defects of temperature-sensitive mutants of vesicular stomatitis virus: intracellular degradation of specific viral proteins, *J. Virol.*, 21, 1140, 1977.
40. **Knipe, D., Baltimore, D., and Lodish, H. F.,** Maturation of viral proteins in cells infected with temperature-sensitive mutants of vesicular stomatitis virus, *J. Virol.*, 21, 1149, 1977.
41. **Hunt, L. A. and Summers, P. F.,** Glycosylation of vesicular stomatitis virus glycoprotein in virus infected Hela cells, *J. Virol.*, 20, 646, 1976.
42. **Robertson, J. S., Etchison, J. R., and Summers, D. F.,** Glycosylation sites of vesicular stomatitis virus glycoprotein, *J. Virol.*, 19, 871, 1976.
43. **Burge, B. W. and Huang, A. S.,** Comparison of membrane protein glycopeptides of sindbis and vesicular stomatitis virus, *J. Virol.*, 6, 176, 1970.
44. **Leavitt, R., Schlesinger, S., and Kornfeld, S.,** Impaired intracellular migration and altered solubility of nonglycosylated glycoproteins of vesicular stomatitis virus and sindbis virus, *J. Biol. Chem.*, 252, 9018, 1977.
45. **Gottlieb, C., Baenzinger, J., and Kornfeld, S.,** Deficient uridine diphosphate-N-acetylglucosamine; glycoprotein N acetylglucosaminyl-transferase activity in a clone of Chinese hamster ovary cells with altered surface glycoproteins, *J. Biol. Chem.*, 250, 3303, 1975.
46. **Schlesinger, S., Gottlieb, C., Feil, P., Gelb, N., and Kornfeld, S.,** Growth of enveloped RNA viruses in a line of Chinese hamster ovary cells with deficient N-acetylglucosaminyltransferase activity, *J. Virol.*, 17, 239, 197.
47. **Moyer, S. A. and Summers, D. F.,** Vesicular stomatitis virus envelope glycoprotein alterations induced by host cell transformation, *Cell*, 2, 63, 1974.
48. **Etchison, J. R. and Holland, J. J.,** Carbohydrate composition of the membrane glycoprotein of vesicular stomatitis virus grown in four mammalian cell lines, *Proc. Natl. Acad. Sci. U.S.A.*, 71, 4011, 1974.
49. **Moyer, S. A., Tsang, J. M., Atkinson P. H., and Summers, D. F.,** Oligosaccharide moities of the glycoprotein of vesicular stomatitis virus, *J. Virol.*, 18, 167, 1976.
50. **Kornfeld, R. and Kornfeld, S.,** Comparative aspects of glycoprotein structure, *Ann. Rev. Biochem.*, 45, 1976, 217.
51. **Waechter, C. J. and Lennarz, W. J.,** The role of polyprenol-linked sugars in glycoprotein synthesis, *Ann. Rev. Biochem.*, 45, 1976, 95.
52. **Segrest, J. P. and Jackson, R. L.,** Molecular weight determination of glycoproteins by polyacrylamide gel electrophoresis in sodium dodecyl sulfate, in *Methods in Enzymology*, 28B, Ginshug, V., Ed., Academic Press, New York, 1972, 54.
53. **Morrison, T. G., Stampfer, M., Baltimore, D., and Lodish, H. F.,** Translation of vesicular stomatitis virus messenger RNA by extracts from mammalian and plant cells, *J. Virol.*, 13, 62, 1974.
54. **Cohen, G., Atkinson, P., and Summers, D.,** Interactions of vesicular stomatitis virus structural proteins with Hela plasma membranes, *Nature New Biol.*, 231, 121, 1971.

55. **Morrison, T. and McQuain, C.**, Association of vesicular stomatitis virus membrane proteins and membranes in a cell free systm, *J. Virol.*, 21, 451, 1977.
56. **Kang, C. Y. and Prevec, L.**, Proteins of vesicular stomatitis virus. I. Polyacrylamide gel analysis of viral antigens, *J. Virol.*, 3, 404, 1969.
57. **Wagner, R. R., Schnaitman, T. C., Snyder, R. M., and Schnaitman, C. A.**, Protein composition of the structural components of veiscular stomatitis virus, *J. Virol.*, 3, 611, 1969.
58. **Huang, A. S., Baltimore, D., and Stampfer, M.**, Ribonucleic acid synthesis of vesicular stomatitis virus. III. Multiple complementary messenger RNA molecules, *Vitology*, 42, 946, 1970.
59. **Emerson, S. U. and Wagner, R. R.**, Dissociation and reconstit ution of the transcriptase and template activities of vesicular stomatitis B and T virions, *J. Virol.*, 10, 297, 1972.
60. **Emerson, S. U. and Yu, Y.**, Both NS and L proteins are required for *in vitro* RNA synthesis by vesicular stomatitis virus, *J. Virol.*, 15, 1348, 1975.
61. **Soria, M.**, Controlling Elements in the Replication of Vesicular Stomatitis Virus, Ph.D. thesis, Harvard University, 1974.
62. **Rettenmier, C., Dumont, R., and Baltimore, D.**, Screening procedure for complementation dependent mutants of vesicular stomatitis virus, *J. Virol.*, 15, 41, 1975.
63. **Lenard, J. and Compans, R.**, The structure of lipid containing viruses, *Biochim. Biophys. Acta*, 344, 51, 1974.
64. **Duc-Nguyen, H., Rose, H. M., and Morgan, C.**, An electron microscopic study of changes at the surface of influenza-infected cells as revealed by ferritin conjugated antibodies, *Virology*, 28, 404, 1966.
65. **Compans, R. W. and Choppin, P. W.**, The structure and assembly of influenza and parainfluenza viruses, in *Comparative Virology*, Maramorosch, K. and Kurstak, E., Eds., Academic Press, New York, 1971, 407.
66. **Hotchin, J. E., Cohen, S. M., Ruska, H., and Ruska, C.**, Electron microscopical aspects of hemadsorption in tissue cultures infected with influenza virus, *Virology*, 6, 689, 1978.
67. **Compans, R. W., Holmes, K. V., Dales, S., and Choppin, P. W.**, An electron microscopic study of moderate and virulent virus cell interactions of the parainfluenza virus SV5, *Virology*, 30, 411, 1964.
68. **Compans, R. W. and Dimmock, N. J.**, An electron microscopic study of single cycle infection of chick embryo fibroblasts by influenza virus, *Virology*, 39, 499, 1969.
69. **Wagner, R. R., Heine, J. W., Goldstein, G., and Schnaitman, C. A.**, Use of antiviral-antiferritin hybrid antibody for localization of viral antigen in plasma membrane, *J. Virol.*, 7, 274, 1971.
70. **Choppin, P. and Compans, R.**, Phenotypic mixing of envelop proteins of parainfluenza virus SV5 and vesicular stomatitis virus, *J. Virol.*, 5, 609, 1970.
71. **McSharry, J. J., Compans, R. W., and Choppin, P. W.**, Proteins of vesicular stomatitis virus and phenotypically mixed vesicular stomatitis virus-SV5 virions, *J. Virol.*, 8, 722, 1971.
72. **Zavada, J.**, VSV pseudotype particles with the coat of avian myeloblastosis virus, *Nature New Biol.*, 240, 122, 1972.
73. **Zavada, J.**, Pseudotypes of vesicular stomatitis virus with the coat of murine leukemia and of avian myeloblastosis viruses, *J. Gen. Virol.*, 15, 183, 1972.
74. **Huang, A. S., Besmer, P., Chu, L., and Baltimore, D.**, Growth of pseudotypes of vesicular stomatitis virus with N-tropic murine leukemia virus coats in cells resistant to N-tropic viruses, *J. Virol.*, 12, 659, 1973.
75. **Witte, O. N. and Baltimore, D.**, Mechanism of formation of pseudo-types between vesicular stomatitis virus and murine leukemia virus, *Cell*, 11, 505, 1977.
76. **Love, D. N. and Weiss, R. A.**, Pseudotypes of vesicular stomatitis virus determined by exogenous and endogenous avian RNA tumor viruses, *Virology*, 57, 271, 1974.
77. **Soria, M., Little, S. P., and Huang, A. S.**, Characterization of vesicular stomatitis virus nucleocapsids. I. Complementary 40S RNA molecules in nucleocapsids, *Virology*, 61, 270, 1974.
78. **Morrison, T. G., Stampfer, M., Lodish, H. F., and Baltimore, D.**, In vitro translation of vesicular stomatitis virus messenger RNAs and the existence of a 40S "plus" strand, in *Negative Strand Viruses*, Mahy, B. W. J. and Barry, R. D., Eds., Academic Press, New York, 293.
79. **Huang, A. S. and Manders, E. K.**, Ribonucleic acid synthesis of vesicular stomatitis virus. IV. Transcription by standard virus in the presence of defective interfering particles, *J. Virol.*, 9, 909, 1972.
80. **Sokol, F. and Clark, H F.**, Phosphoproteins structural components of rhabdoviruses, *Virology*, 52, 246, 1973.
81. **Inblum, R. L. and Wagner, R. R.**, Protein kinase and phosphoproteins of vesicular stomatitis virus, *J. Virol.*, 13, 113, 1974.
82. **Moyer, S. A., and Summers, D. F.**, Phosphorylation of vesicular stomatitis virus *in vivo* and *in vitro*, *J. Virol.*, 13, 455, 1974.
83. **Huang, A. S.**, Viral pathogenesis and molecular biology, *Bact. Rev.*, 41, 811, 1977.

84. **Brown, D. T. and Smith, J. F.,** Morphology of BHk-21 cells infected with sindbis virus temperature-sensitive mutants in complementation groups D and E, *J. Virol.*, 15, 1262, 1975.
85. **Brown, D. T and Riedel, B.,** Morphogenesis of vesicular stomatitis virus: electron microscope observations with freeze-fracture technique, *J. Virol.*, 21, 601, 1977.
86. **Hecht, T. and Summers, D. F.,** Interactions of vesicular stomatitis virus with murine cell surface antigens, *J. Virol.*, 19, 833, 1976.
87. **Strand, M. and August, T.,** Protein kinase and phosphate acceptor proteins in Rauscher murine leukemia virus, *Nature New Biol.*, 233, 137, 1971.
88. **Roy, P. and Bishop, D. H. L.,** Nucleoside triphosphate phosphotransferase. A new enzyme activity of oncogenic and non-oncogenic "budding" viruses, *Biochim. Biophys. Acta*, 235, 191, 1971.
89. **McSharry, J. J. and Wagner, R. R.,** Lipid composition of purified vesicular stomatitis viruses, *J. Virol.*, 7, 59, 1971.
90. **Klenk, H.-D. and Choppin, P. W.,** Glycolipid content of vesicular stomatitis virus grown in baby hamster kidney cells, *J. Virol.*, 7, 416, 1971.
91. **Compans, R. W. and Caliguiri, L. A.,** Replication of influenza virus: studies on synthesis of membrane components and on a purified viral RNA polymerase, in *Negative Strand Viruses*, Vol. 2, Mahy, B. W. J. and Barry, R. D., Eds., Academic Press, London, 1975.
92. **Singer, S. and Nicholson, G.,** The fluid mosaic model of the structure of cell membranes, *Science*, 175, 720, 1972.
93. **Mussgay, M. and Weibel, J.,** Electron microscopic studies on the development of vesicular stomatitis virus in KB cells, *J. Cell Biol.*, 16, 119, 1963.
94. **Birdwell, C. R. and Strauss, J. H.,** Maturation of vesicular stomatitis virus: electron microscopy of surface replicas of infected cells, *Virology*, 59, 587, 1974.
95. **Holmes, K. V.,** Scanning electron microscopic studies of virus infected cells. I. Cytopathic effects and maturation of vesicular stomatitis virus in L2 cells, *J. Virol.*, 15, 355, 1975.
96. **Orenstein, J., Johnson, L., Shelton, E., and Lazzarini, R. A.,** The shape of vesicular stomatitis virus, *Virology*, 71, 291, 1976.
97. **Fong, B. S., Hunt, R. C., and Brown, J. C.,** Asymmetric distribution of phoshatidyl-ethanolamine in the membrane of vesicular stomatitis virus, *J. Virol.*, 20, 658, 1976.
98. **Smith, J. F. and Brown, D. T.,** Envelopment of sindbis virus. Synthesis and organization of proteins in cells infected with wild type and maturation defective mutants, *J. Virol.*, 22, 622, 1977.
99. **Jones, K. M., Waite, M. R. F., and Bose, H. R.,** Cleavage of a viral envelopes precursor during the morphogenesis of sindbis virus, *J. Virol.*, 13, 809, 1974.
100. **Morrison, T. G.,** unpublished observations.

Chapter 7

RHABDOVIRUS GENETICS

A. Flamand

TABLE OF CONTENTS

I. INTRODUCTION

Genetic studies of rhabdoviruses have been reported for vesicular stomatitis virus (VSV), Indiana and New Jersey serotypes, as well as Cocal, Chandipura, rabies, and sigma viruses. The genome of each of these viruses is composed of a single-stranded RNA molecule (molecular weight of about 4.1×10^6 daltons) with a coding capacity for 5 or 6 polypeptides. The mechanism of replication of the RNA genome within a cell is not yet elucidated, but it is probable that the RNA-dependent RNA polymerase (transcriptase), which is found in the virions of all rhabdoviruses so far investigated, is responsible for both RNA replication and transcription.

VSV Indiana has been extensively studied due to its rapid growth cycle and ability to productively infect various cell types producing high titers of progeny virus. The virus is considered to be the prototype of the Rhabdoviridae family. Genetic and/or biochemical studies have been performed on several related but distinguishable isolates of VSV Indiana as discussed by Clewley et al.[1]

Various members of the Rhabdoviridae family are capable of infecting a wide variety of organisms, including plants, insects, fishes, birds, reptiles, and mammals. Several have been isolated from, and can multiply within, insect hosts as well as in mammals (see Chapter 1 Volume I). The maintenance of strain characteristics after multiplication in such different hosts is an intriguing question which will be discussed within this chapter, as will be more conventional genetic studies involving the isolation and classification of thermosensitive mutants.

II. GENETIC BEHAVIOR OF VSV INDIANA

A. Spontaneous Mutation Rate in Chick Embryo Fibroblasts

Spontaneous mutations are believed to arise primarily from errors during RNA replication. If this assumption is correct, then the spontaneous mutation rate is equal to the rate of template misreading. It produces a genome copy differing from its template by at least one nucleotide. An unfaithful copy may correspond to a lethal, unchanged, or modified phenotype. Since lethal or unchanged phenotypes cannot be easily detected by available procedures, the practical evaluation of the mutation rate is a minimum one based essentially on the appearance of modified phenotypes.

The rate of appearance of spontaneous thermosensitive viral phenotypes in chick embryo fibroblasts (CEF) has been measured for several different clones of the wild-type Orsay strain of prototype VSV Indiana grown in CEF cells. The history of this virus strain has been described by Clewley et al.[1] The procedure we have used for isolating thermosensitive (*ts*) viruses[2,3] involves the direct screening of plaques picked by a sterile needle and assaying the recovered virus at the permissive and nonpermissive temperatures of 30° and 39.8°C (or 39.3°C), respectively. We have found that wild-type VSV Indiana grows equally well at these temperatures and its plating efficiency at 39.8°C is about 0.3. Some 28 wild-type clones have been analyzed to determine the spontaneous mutation frequency. Each clone was recovered as a well-isolated, 2-day-old plaque using CEF cells incubated at 30°C. The virus retrieved from each plaque usually corresponded to some 10^6 PFU. As shown in Table 1, the results are statistically compatible with the hypothesis that each plaque contained essentially the same quantity of thermosensitive PFU, representing a mean percentage of 1.6%.

The homogeneous distribution of (*ts*) mutants and the absence of "jackpots" (i.e., clones giving many more *ts* mutants than others) corresponding to early or dominant mutations indicate that (1) the spontaneous mutation rate is high, (2) *ts* mutants are selected against even at 30°C, and (3) a stable frequency of mutants in a population may be established for this virus. Table 2 (upper panel) shows that, in fact, by two

TABLE 1

Number of Spontaneous *ts* Mutants in Different Clones of the Wild-Type Strain

Clone number	Number of plaques tested	Number of mutants
1	50	1
2	50	0
3	50	0
4	50	1
5	50	1
6	50	1
7	50	1
8	50	1
9	50	0
10	64	1
11	56	1
12	64	1
13	56	1
14	48	2
15	56	3
16	56	2
17	48	0
18	40	0
19	283	4
20	312	3
21	88	1
22	88	2
23	96	3
24	96	2
25	95	1
26	72	2
27	88	1
28	92	0
Total	2250	36

Reprinted from Flamand, A., Etude Génétique du Virus de la Stomatite Vésiculaire, these de Doctorat d'Etat, Orsay, 1974. With permission.

further passages at 30°C at MOI values of about 1, the percentage of VSV Indiana *ts* mutants does indeed stabilize, giving a mean value of 2.8%.

A question relevant to these studies concerns the number of successive cycles of virus infection that occurs in a plaque. An estimate of the *maximum* number of cycles that occurs can be provided by one-step growth curves. Figure 1 shows that before 3.5 hr, the probability is less than unity for an infected CEF cell producing at least one PFU. If this is also true under agar, then it suggests that at least 3.5 hr are required for an infected cell to infect its neighbors. Therefore, the *maximum* number of viral cycles in a 2-day-old plaque would be 48/3.5 = 14. Thus, for the experiment described in Table 2, the maximum number of successive infections leading to a stable mutant frequency corresponds to 14 + 2 (i.e., 16 cycles).

The observations on mutant frequency described above have been taken to imply that *ts* mutants are selected against even at permissive temperatures. Is this true? Re-

TABLE 2

Evolution of the Percentage of *ts* Mutants During Passages at 30 and 39.3°C

		First passage	Second Passage	Third passage	Fourth passage	Ninth passage
At 30°C		1.7 (2 *ts*, 115 wt)	2.8 (3 *ts*, 104 wt)	3.7 (6 *ts*, 146 wt)	2.9 (5 *ts*, 171 wt)	2.6 (5 *ts*, 191 wt)
	First experiment	5.0 (6 *ts*, 114 wt)	8.3 (10 *ts*, 110 wt)	5.3 (9 *ts*, 158 wt)	7.2 (19 *ts*, 245 wt)	2.1 (6 *ts*, 282 wt)
At 39.3°C	Second experiment	—	7.0 (27 *ts*, 357 wt)	—	—	0.9 (4 *ts*, 444 wt)

Note: The first passages were as follows: CEF petri dishes were infected with wild-type VSV Indiana. After two days at 30°C, a well-isolated plaque was picked up and resuspended in 1 mℓ of saline medium. CEF petri dishes were infected with the viral suspension and incubated for 24 hr at 30 or 39.3°C. Subsequent passages were made by infecting petri dishes of CEF with undiluted preceding passages. The plaques contained 1.4, 1.4, and 0.9% *ts* mutants, respectively.

Reprinted from Flamand, A., Etude Génétique du Virus de la Stomatite Vésiculaire, thèse de Doctorat d'Etat, Orsay, 1974. With permission.

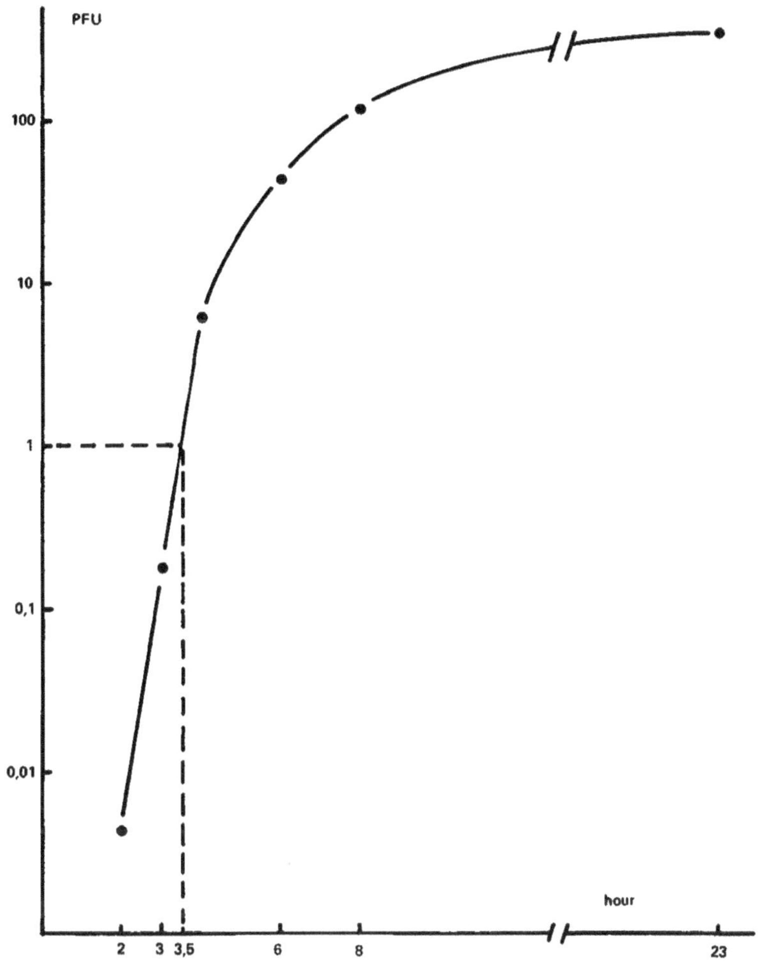

FIGURE 1. Mean viral production per CEF cell. (Reprinted from Flamand, A., Etude Génétique du Virus de la stomatite Vesiculaire, thèse de Doctôrat d'Etat, Orsay, 1974. With permission.)

construction experiments performed by Carvalho-Jeronimo[4] and Flamand,[5] showed that during successive multiplication cycles *ts* mutants were rapidly eliminated from mixed parental populations (Table 3). Although Youngner and Quagliana[6] reported that *ts* mutants selectively interfered with wild-type virus replication, in their experiments the MOIs used in the single and mixed virus infections were 1 for the *ts* mutant and 0.1 for the wild-type virus. Under those conditions, wild-type virus is produced primarily during the second infection cycle, which occurs either in uninfected cells (control) or in cells previously infected with mutant virus several hours beforehand (mixed infection). Since it is known that cytopathic effects are produced by mutants at both permissive and nonpermissive temperatures, this may explain why their wild-type virus replication was apparently inhibited. In a similar experiment to that of Youngner and Quagliana, we have found that wild-type virus is inhibited if cells are preinfected with a *ts* mutant 4 to 6 hr beforehand. However, if cells are simultaneously infected by a mutant and wild-type virus at multiplicities of one for both viruses, then both viruses are replicated in comparable amounts. As mentioned above, we have

TABLE 3

Evolution of *ts* Frequency During Passages at 30°C in an Artificial Mixture of Weight and *ts* Mutant

	Percent of mutant	
	In initial mixture	In ninth passage
*ts*O I 5, wt	40.8	4.8
	(62 *ts*, 90 wt)	(12 *ts*, 118 wt)
	40.8	6.2
	(62 *ts*, 90 wt)	(10 *ts*, 150 wt)
*ts*O I 8, wt	50.5	10.7
	(48 *ts*, 47 wt)	(6 *ts*, 50 wt)
*ts*O III 23, wt	20.0	8.5
	(24 *ts*, 96 wt)	(17 *ts*, 183 wt)

Reprinted from Flamand, A., Etude Génétique du Virus de la Stomatite Vésiculaire, these de Doctorat d'Etat, Orsay, 1974; and Carvalho-Jeronimo, M., Recherche d'un Effet de la Cellule Hôte sur l' Instabilité Génétique d'une Souche du Virus de la Stomatite Vésiculaire de Sérotype Indiana, thèse de 3° cycle, Orsay, 1974. With permission.

found that several successive multiplication cycles lead to a progressive elimination of a mutant from the population (Table 3).

Assuming that VSV viral multiplication is compatible with the model of Drake,[7] then the mutation rate may be calculated:

$$m = \frac{0.43\ f}{\log N - \log N_0}$$

where f is the frequency of mutants, N is the total size of the population, and N_o is the initial size of the population.

When studying a clone, $N_o = 1$ and

$$m = \frac{0.43\ f}{\log N}$$

i.e.,

$$m = \frac{0.43 \times 1.6 \times 10^{-2}}{6} = 1.1 \times 10^{-3}$$

This figure is a minimum value, since we have shown that selection acts against *ts* mutants. On the other hand, in these calculations, replication events which do not lead to the formation of infectious particles are not taken into account. Even if there are 10 or 100 times more genomes synthesized than PFU detected, the value of m does not differ considerably (0.98×10^{-3} and 0.86×10^{-3}, respectively). Even with these restrictions in mind, we can assume that *the probability of one ts mutation occurring per replication event is at least about 10^{-3}*, and perhaps considerably higher. Since some 87% of the VSV Indiana *ts* mutations are in Gene I (see later), this means that the average mutation rate for this gene is also at least about 10^{-3}. This value is very high when compared to the rates observed for DNA-based organisms, which are of the order of 10^{-6} to 10^{-8}.[7]

It is of interest to know if the genetic instability we have observed with VSV Indiana is a general property of all rhabdoviruses. For other VSV strains, we have obtained frequencies of 1.7% for a recently cloned VSV New Jersey serotype virus.[8a] Data obtained by Pringle and Wunner[8] and Pringle[9] indicate that, within a factor of 2, *ts* mutation frequencies in stocks of VSV Indiana, VSV New Jersey, and Cocal serotype viruses are not significantly different. In those studies, a lower nonpermissive temperature was used (39°C), which probably accounts for the slightly lower spontaneous mutation frequencies that were obtained. That this may be true is suggested by studies undertaken by this author using the VSV Indiana virus stock provided by Pringle, whereby at 39.8°C a 2% frequency of *ts* mutants was obtained.[9a]

Youngner et al.[21] have reported values for *ts* mutant frequencies of 4.4% in their VSV Indiana stock. Clark and Wiktor[10] also reported high frequencies of *ts* mutants in stocks of Lagos Bat and Mokola viruses when cultured in baby hamster kidney (BHK) cells. For example, they found 4 out of 32 Lagos Bat virus clones were thermosensitive. Bussereau and Flamand[11] reported frequencies of 0.3% for *ts* mutants in a recently cloned stock of rabies virus (two *ts* mutants out of 600 plaques analyzed). This observation does not necessarily mean that rabies virus is genetically more stable than the other viruses that have been studied, since a nonpermissive temperature of 38.5°C has been used for rabies virus due to difficulties in obtaining wild-type virus plaques at higher temperatures. As discussed in Section III, most VS Indiana *ts* mutants no longer exhibit a *ts* phenotype when the nonpermissive temperature is lowered even by 1 to 2°C, which indicates that as lower nonpermissive temperatures are used, one will obtain fewer *ts* mutants.

For sigma virus, mutation rates have been estimated to be between 10^{-3} and 10^{-4} by Ohanessian-Guillemain[12] and Vigier.[13] Using Piry virus, Brun has observed that several clones of this virus isolated on CEF cells contained approximately 5% *ts* mutants.[13a] Studies by Roy with Spring viremia of Carp virus (SVCV) indicated that there were 2% spontaneous *ts* mutants in clones of SVCV.[13b]

Thus, for all rhabdoviruses so far examined, the evidence obtained from various sources supports the hypothesis that genetic instability is a general feature of members of the family.

One explanation for the difference between DNA-based organisms and these RNA viruses could be that there is an efficient repair mechanism existing for DNA replication. However, it should be pointed out that a certain genetic instability leading to a rapid adaptation may be an advantage for a virus which has to adapt to new conditions (different hosts and ecological niches, etc.).

B. Influence of the Host on the Mutation Rate of VSV Indiana

The percentage of spontaneous *ts* mutants which are recovered from clones of VSV Indiana isolated in BHK-21, L, and HeLa cells, as well as in *Drosophila melanogaster*, has been studied by Carvalho-Jeronimo.[4] As in the case of CEF cells, it was found that virus clones isolated in the same cell line contained approximately the same proportion of *ts* mutants. The calculated mean percentages obtained were 3.8% for BHK-21 cells, 3.6% for L cells, and 0.7% for HeLa cells (Table 4). None of these values is significantly different to the percentages obtained in CEF cells. In each case, no change in the proportion of *ts* mutants was observed through nine successive passages in the appropriate cells. However, one virus clone isolated from *Drosophila* contained significantly more *ts* mutants than two other clones (17.3% as opposed to 1.1% and 2.0%). This observation may be explained by either inadvertant contamination of the original

TABLE 4

Percentage of *ts* Mutants in Clones of VSV Indiana Isolated in Different Cell Lines or in *Drosophila melanogaster*

Origin of clones	Clone number	Number of plaques tested	Number of *ts* mutants	% of *ts* mutants	Average %
BHK	1	224	8	3.6	3.8
	2	117	5	4.3	
	1	120	5	4.2	
L	2	72	2	2.8	3.6
	3	58	2	3.4	
Hela	1	120	1	0.8	0.7
	2	180	1	0.6	
Drosophila melanogaster	1	110	23	20.9	?
	2	264	3	1.1	
	3	48	1	2.0	

Note: All clones originated from VSV adapted to CEF.

Reprinted from Carvalho-Jeronimo, M., Recherche d'un Effet de la Cellule Hôte sur L' Instabilité Génétique d'une Souche du Virus de la Stomatite Vésiculaire de Sérotype Indiana, thèse de 3° cycle, Orsay, 1974. With permission.

clone, a selected mutational event early in the history of the clone, or an increased genetic instability of the virus when virus replication occurs in *Drosophila*. The latter hypothesis is unlikely since the percentage of *ts* mutants in the two other virus clones tested in *Drosophila* was not significantly different from that observed in CEF cells.

Two conclusions may be drawn from these results: (1) any influence of the host on the maturation rate of a virus is discreet, if it exists at all, and (2) when a CEF-adapted virus strain is transferred to another cell line, genetic instability does not significantly increase.

C. Influence of Temperature on the Spontaneous Mutation Rate of VSV Indiana

After repetitive cloning of the VSV Indiana Orsay strain at 30°C using CEF cells, we have observed that subsequent transfer to 39.3°C has consistently led to a *transient* increase in genetic instability (Table 2). The increased genetic instability reverts to initial levels by the ninth consecutive passage at 39.3°C. If spontaneous mutations are indeed the result of errors which occur during RNA replication, then the increased instability that we have observed is probably due to some change in the properties of the RNA polymerase at elevated temperatures. The subsequent return to genetic stability may be related to a selection of a modified polymerase which is more faithful at higher temperatures. It will be of interest to know if other strains of VSV Indiana or other rhabdoviruses show similar increased genetic instability when replicated at high temperatures.

D. Stability of Virus Strains During Successive Multiplications

Successive multiplications of VSV or other rhabdoviruses may be obtained either through virulent or persistent infections. Since the stability of a virus appears to differ under these two conditions, they will be discussed separately.

1. Virulent Infections

Infection of tissue culture cells by VSV Indiana is usually virulent, resulting in the destruction of the host cells and the production of massive amounts of infectious virus particles within a few hours. Virus propagation at low multiplicities of infection involves successive cycles of virus maturation and reinfection, so that the first virions produced by an infected cell have a selective advantage over the rest of the progeny. The behavior of the Orsay strain of VSV Indiana during successive passages at high and low temperatures has been studied by Flamand.[14]

a. *Successive Passages of VSV Indiana at Low Temperatures*

We have found that the properties of the VSV Indiana strain we have employed undergo no apparent evolution after nine successive passages at 30°C. The efficiencies of plating at each passage and the capacity to produce virus at 39.3°C remain unchanged; likewise, the percentage of *ts* mutants present in virus stocks also remains stable (at 2.8%). This observation is important since the stability of the wild-type virus at low temperatures is a prerequisite for genetic studies using *ts* mutants. For example, if passage at low temperatures resulted in a changed phenotype, then mutant virus studies could not be relied on. These results contrast the situation for two picornaviruses, EMC and poliovirus, reported by Lwoff and Lwoff.[15] They found that after nine successive passages at low temperatures, these viruses lost their ability to grow at elevated temperatures, probably due to the selection of cold adapted variants. In this context, it is of interest to note that no vectors are involved in the replication and transmission cycle of these viruses so that their replication processes presumably are not exposed to low temperatures. VSV Indiana, on the other hand, is classified as an arbovirus[16] since it can replicate in, and be transmitted by, insects.

b. *Successive Passages of VSV Indiana at High Temperatures*

After nine successive passages at high temperatures, we have found with the Orsay strain of VSV Indiana that virus production at 30°C or 39.3°C remains unchanged, although the plating efficiency at 39.3°C increases from 0.3 to 0.6. This property is maintained after subsequent cloning.

There are also other data which indicate that virus strains remain stable during successive manipulations. As shown by Clewley et al.,[1] different VSV Indiana isolates have been used by various investigators for their genetic studies.The Orsay strain was derived from the prototype VSV Indiana isolated in 1927 in Richmond, Ind., while Glasgow strain was derived from a virus which originated from the 1942 Colorado outbreak of the disease.[1] Both isolates have undergone multiple passages over several years in various cell lines and/or animals. Genome oligonucleotide sequence studies indicate that particular virus strains passaged in different laboratories over several years show little detectable evolution.[1] This indicates that natural populations of VSV Indiana remain genetically stable for considerable periods of time. The Orsay and Glasgow Indiana strains showed several differences at the genome level; however, since their respective *ts* mutants complement each other (see below), it is evident that the different isolates of the same VSV Indiana serotype virus have maintained genetic homology. It should be noted that differences in RNA sequences do not necessarily result in polypeptide changes, since the genetic code is degenerate and the nucleotide differences between strains might be located in noncoding regions of the RNA.

Notwithstanding this, these observations suggest that naturally occurring VSV Indiana strains have remained relatively stable — presumably due to selection in nature against variants.

2. *Persistent Infections*

Persistent infections with VSV Indiana can be established in cell cultures by various procedures as described in Chapter 3 Volume III. Insects persistently infected by rhabdoviruses have also been reported. Virus persistence in cell culture may either occur through low levels of virulent virus production keeping pace with cell division or through the division of cells containing viral material without the involvement of extracellular virus. In insects, no cell multiplication occurs except in germ lines, so that virus propagation probably requires both virus maturation and infection of additional cells.

It has been observed that in tissue culture cells, or insects, that are persistently infected with VSV Indiana there is an accumulation of *ts* mutant viruses and that the population acquires thermosensitive characteristics after several serial passages.[17-21] Youngner et al.[21] showed that eight *ts* mutants isolated from persistently infected cells belonged to complementation Group I. Although this is not unexpected since most *ts* mutants that have been isolated are Group I mutants (see below), it will be of interest to know if mutants representing the other complementation groups also occur. Nishiyama[18] in his studies also reported that, *unlike persistently infected cultures,* 20 serial *lytic* passages of the parental virus in L cells at 35°C gave a virus population that did not acquire a thermosensitive character — a finding analogous to the observations made in our laboratory.

A possible explanation for the evolution of dominant *ts* viruses could be that, in the absence of virus maturation, selection does not act against *ts* mutants, which thus tend to accumulate. Whether this is also true for insect infections, or in cultures maintained at low temperatures, is not known.

III. RHABDOVIRUS MUTANTS

Temperature-sensitive, plaque morphology, complementation-dependent, thermolabile, rifampin-sensitive, and host-restricted mutants of rhabdoviruses have been isolated (Table 5). Most rhabdovirus mutants have been isolated after mutagen treatment, the exceptions being the plaque morphology mutants isolated by Wagner et al.[23] and by Wertz and Levine,[24] as well as the rifampin-sensitive mutant described by Moreau[25] and 171 spontaneous *ts* mutants of the Orsay collection. The base analogue 5-fluorouracil (5-FU) has been found to be one of the most effective mutagens for producing *ts* viruses, although 5-azacytidine (5-AZAC), *N*-methyl-*N'*-nitro-*N*-nitrosoguanidine (NTG), ethylmethane sulfonate (EMS), nitrous acid (NA), proflavine (P), and ultraviolet light (UV) have also been used with apparent success, although in some studies UV irradiation did not appear to increase the frequency of *ts* mutants of VSV Indiana.[5]

Pringle[9,26] has reported that the yield of *ts* mutants of VSV Indiana is proportional to the mutagen dose, with mutant recoveries representing 46% of the progeny virus when 100 μg of 5-FU per milliliter are employed. Both Cocal and VSV New Jersey serotype viruses have been shown to be 3 or 35 times, respectively, more resistant to mutant production when 5-FU, NA, EMS, or P are used. Bussereau and Flamand[11] have shown that in using an NA dose which resulted in 99% inactivation of rabies virus, the frequency of *ts* mutant recovery among the survivors did not differ from the spontaneous mutation frequency (two *ts* mutants were obtained from 675 progeny plaques analyzed). In contrast, when a dose of 100 μg of 5-FU per milliliter was used, some 70 *ts* mutants were obtained from 2276 progeny clones analyzed. Clark and Koprowski[28] have isolated five *ts* mutants by mutagenizing rabies virus with NA, 5-AZAC, or 5-FU.

A. Thermosensitive (*ts*) Mutants

The majority of mutants that have been obtained for the different rhabdoviruses studied have been *ts* mutants. Some 436 VSV Indiana, 49 VSV New Jersey, 40 Cocal,

TABLE 5

Mutants and Variants of Rhabdoviruses

Virus	Serotype	Mutant designation	Phenotype	Mutagen[a]	Ref.
VSV	Indiana	—	Plaque morphology	None, Sp	23
	New Jersey	—	Plaque morphology	UV	55
	Indiana	L/s	Plaque morphology	None, Sp	24
	Indiana	*ts*	Temperature sensitive	None, Sp	2, 3, 5
	Indiana	*ts*	Temperature sensitive	5-FU, NA, EMS, P	27
	Indiana	*ts*	Temperature sensitive	-FU, 5-AZAC, EMS	9, 26
	Indiana	*ts*	Temperature sensitive	5-FU, NA, NTG	38
	New Jersey	*ts*	Temperature sensitive	UV, NA	56
	New Jersey	*ts*	Temperature sensitive	5-FU	30
	Cocal	*ts*	Temperature sensitive	5-FU	31
	Indiana	*tl*	Thermolabile	5-FU	40
	Indiana	*hr*	Host range (restricted in HeLa cells)	5-FU,NTG	34
	New Jersey	*hr*	Host range (restricted in CE cells)	5-FU	Pringle (unpublished)
	Indiana	*td* CE	Host range (conditionally restricted in CE cells)	5-FU	Pringle (unpublished)
	New Jersey	*td* CE	Host range (conitionally restricted in CE cells)	5-FU	35
	Cocal	*td* CE	Host range (conditionally restricted in CE cells)	5-FU	Pringle (unpublished)
	Indiana	Rif[+]	Rifampin sensitivity	None, Sp.	25
	Indiana	*cd*	Complementation dependent	NA, NTG	38
Chandipura		*ts*	Temperature sensitive	5-FU	31
		td CE	Host range (conditionally restricted in CE cells)	5-FU	Gadkari and Pringle (unpublished)
Rabies		*ts*	Temperature sensitive	5-FU	28
		ts	Temperature sensitive	5-FU	11
		—	Plaque size	None, Sp	57

TABLE 5 (continued)

Mutants and Variants of Rhabdoviruses

Virus	Serotype	Mutant designation	Phenotype	Mutaren[a]	Ref.
Sigma		*ts*	Temperature sensitive	5-FU	58
		Tr	Temperature resistance	None, Sp	59
		P^+/P^-	Host range	None, Sp	60
		—	"Plaque size"	None, Sp	61
		—	"Plaque type"	None, Sp	13
		rho	Replication defective	None, Sp	61
		ultrarho	Replication defective	None, Sp	61
		g^+/g^-	Germinal transmission	None, Sp	62
		pr^+/pr^-	Egg invasiveness	None, Sp	62
		V^+/V^-	Sperm invasiveness	None, Sp	63

[a] Sp, spontaneous; 5-FU, 5-fluorouracil; 5-AzaC, 5-azacytidine; NA, nitrous acid; NTG, *N*-methyl-*N*-nitro-*N*-nitrosoguanidine; EMS, ethylmethane sulfonate; P. proflavine; UV, ultraviolet light.

From Pringle, C. R., in *Comprehensive Virology*, Vol. 9, Fraenkel-Conrat, H. and Wagner, R. R., Eds., Plenum Press, New York, 1977, chap. 6. With permission.

30 Chandipura, 75 rabies, and 26 SVCV mutants have been isolated and studied in various laboratories. Genetic studies of sigma virus are presented in Chapter 5 Volume III and will not be discussed here. The majority of the mutants that have been analyzed for any one of these virus systems complement mutants of other groups isolated from the same virus. Presumably, therefore, they represent single gene mutations, most likely involving missense changes. Some multiple mutants (i.e., those that do not complement other mutants of the virus) have been described from mutagenized virus stocks and probably correspond to multiple missense mutations involving different gene products.

1. Nomenclature of ts Mutants

Cormack et al.[29] suggested the use of a laboratory prefix to designate the origin of the various VSV Indiana mutants that have been obtained: B for Bratislava, G for Glasgow, O for Orsay, M for Massachusetts, T for Toronto, and W for Winnipeg. To indicate which complementation group number is indicated in parentheses after the mutant isolate number, e.g., *ts* O 5 (Group I). Other conventions which have been used in the literature include ones in which the group number is placed before the mutant isolate number (e.g., *ts* O I 5), or Arabic numbers (1 to 6) are placed directly before the mutant isolate number (e.g., *ts*G-11, 12, . . . 110, 111, 112, etc).

Pringle and associates[30,31] have employed complementation group assignments of A to F for VSV New Jersey, α to γ for Cocal virus, and Ch I to Ch V for Chandipura virus. We have employed the VSV Indiana rule for rabies virus mutants since *ts* mutants have been obtained in at least two laboratories. Thus, rabies *ts* O 2 is *ts* mutant isolate number 2 obtained for rabies virus in Orsay.

2. Characteristics of ts Mutants

As a rule, all forms of intermediates exist between wild-type and strongly affected *ts* mutants. Usually leaky mutants are eliminated from *ts* mutant collections that are to be used for genetic studies. Mutants are generally considered to be suitable for study when their residual growth at nonpermissive temperature is 1%, 0.1%, or less than that obtained at permissive temperatures.

After cloning and multiplication at permissive temperatures, the frequency of revertants found in mutant virus stocks appears to be a function of the particular mutant. Our observations at Orsay may be summarized as follows: the frequency of revertants is usually undetectable ($<6 \times 10^{-5}$) for RNA^- mutants; approximately 5×10^{-4} for RNA^+ mutants of Groups III and V; and relatively high (2×10^{-3}) for Group II mutants.[32] High reversion frequencies for Group II, III, and IV mutants have been reported by Pringle,[9,22] who also found both low and moderate reversion frequencies for Groups I and V mutants, respectively. A direct comparison of these results is difficult, since the reversion frequency in a mutant stock depends on the number of successive cycles which occur after cloning and on the selective advantage of wild-type virus over the particular mutant that is being studied.

The behavior of *ts* mutants at various temperatures has been studied by Lafay.[33] The results obtained with four *ts* mutants are shown in Figure 2. Except for one mutant (*ts* O 44 [Group V]), viral production at 30°C is comparable to that of wild-type virus. Studies with other viruses indicate that at permissive temperature mutant maturation is often slower than that of the wild-type virus although they usually reach the same plateau titers. Mutant *ts* O 44 (Group V) is apparently both cryo- and thermosensitive, since maturation at 37°C is better than at 30°C or 39°C.

The differences observed between temperature-sensitive and the wild-type virus not only increase with increasing temperatures but also vary with the particular mutant

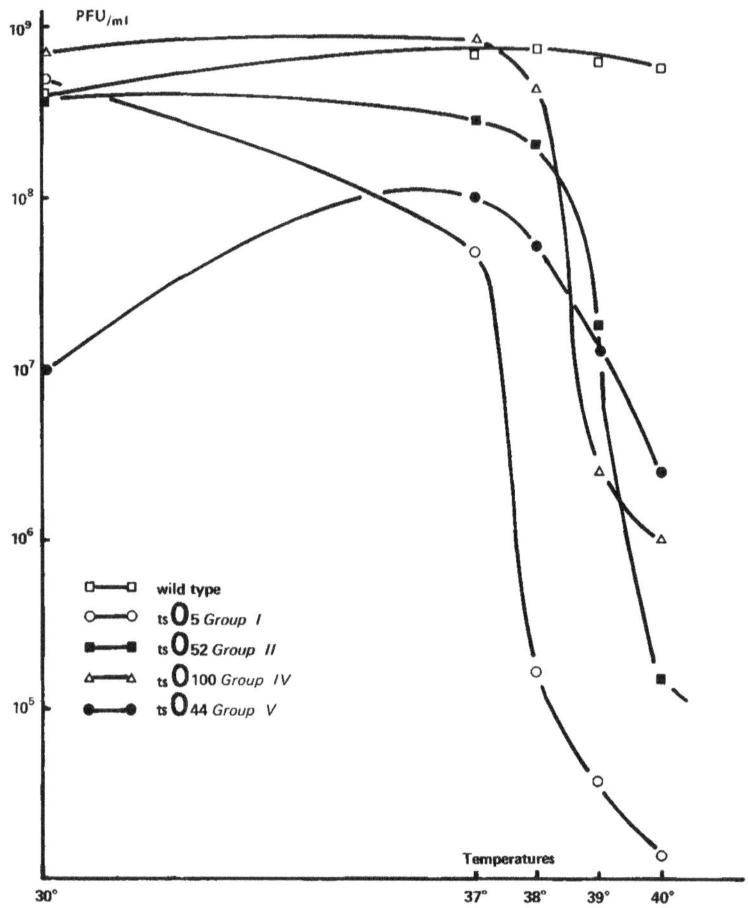

FIGURE 2. Multiplication of wild-type and *ts* mutants at t = 10 hr, at different temperatures. (From Lafay, F., Etude Physiologique des Mutants Thermosensibles du Virus de la Stomatite Vesiculaire, these de 3° cycle, Paris, 1972. With permission.)

studied. For example, at 38°C only one of the four mutants described in Figure 2, can be categorized as thermosensitive (*ts* O 5 [group I]). The other mutants at 38°C do not appear to be particularly thermosensitive. At 40°C all four mutants can be characterized as thermosensitive when compared to the wild-type virus. As mentioned previously, the observation of different thermosensitivities as a function of growth temperature could explain the variable percentages of *ts* mutants obtained by different investigators.

B. Other Mutants

The host-restricted (*hr*) mutants first isolated by Simpon and Obijeski[34] are of particular interest. After mutagenesis by 5FU or NTG, those investigators plated progeny virus clones on a mixture of CEF and HeLa cells. Several virus clones were obtained which were unable to grow on HeLa cells but which retained a capacity to grow on CEF cells. Most of these mutants also exhibited thermosensitive characteristics (i.e., were *hr/ts*). It is not known if the same mutation event was responsible for both phenotypes. All of the *hr* mutants described by Simpson and Obijesky[34] were RNA$^-$ in nonpermissive cells, indicating that transcription and/or replication events were defective. Subsequently, Pringle et al.[22] and Szilágyi et al.,[35,36] using a mixture of CEF and BHK-21 cells as indicators, isolated mutants of VSV Indiana, VSV New Jersey, Cocal, and Chandipura viruses which were restricted in their growth capability in CEF cells. Some of these mutants were temperature-dependent, host-range mutants, which formed plaques at both 31 and 39°C on BHK cells but only at 31°C on CEF cells. Biochemical investigations have established that the virion transcriptase is temperature-sensitive in vitro for this class of mutants. This finding is in agreement with the results obtained by Obijeski and Simpson.[37] The assignment to a complementation group would be an additional proof that the site of mutation is the transcriptase locus; however, complementation has not been reported for these mutants.

In view of the fact (see below) that some 87% of the known VSV Indiana *ts* mutants belong to Group I, Rettenmier and associates[38] devised a procedure for the selection of mutants representing other complementation groups. For their studies, mixed infections were produced using cell monolayers infected with mutant *ts* G 11 (Goup I) at an MOI of 5 and mutagenized wild-type VSV Indiana at an MOI of 0.01. Upon subsequent incubation at 38.5°C, plaques were obtained which were produced by either residual wild-type virus or complementation between a mutant and the *ts* G 11 infection. Screening these plaques led these investigators to isolate eight new *ts* mutants and two complementation-dependent mutants.

Since VSV Indiana *ts* O 45 (Group V) is mutated in its G protein[39] and is thermolabile in vitro, Závada[40] has devised a procedure for the isolation of thermolabile mutants. He screened for G protein mutants by growing virus at low temperatures in the presence of VSV antiserum, using the fact that G protein also carries neutralizing determinants. By this procedure three mutants were isolated which were thermolabile (i.e., were rapidly inhibited in vitro upon incubation at 38°C or 42°C) and thermosensitive (unable to grow at high temperatures). Complementation tests have indicated that these mutants are probably multiple mutants.[41]

Moreau[25] has reported the isolation of a rifampin-sensitive VSV mutant recovered from wild-type, rifampin-insensitive VSV Indiana. Revertants were also selected by successive plaque purification of mutant virus stocks grown in the presence of 100 μg of rifampin per milliliter. The transcriptase of the rifampin-sensitive virions apparently has a greater affinity for rifampin than that of wild-type or the revertant virus strains.

Small plaque mutants of VSV Indiana and VSV New Jersey serotype viruses have been isolated in several laboratories (see Table 5). In general, these variants are less virulent than the wild-type virus and show a longer lag period and longer time course of viral production than the wild-type virus.[23,42] Wertz and Levine[24] have shown that the viral RNA species synthesized in cells after infection with a VSV Indiana small plaque variant are of the same size classes as those obtained with wild-type viruses, although differences in the proportion of each size class of RNA were observed.

IV. COMPLEMENTATION

Genetic complementation has been demonstrated for *ts* mutants of VSV Indiana, VSV New Jersey, Cocal, and Chandipura viruses. Complementation of rabies virus has not yet been reported. The possibility that the 70 *ts* mutants isolated by Bussereau and Flamand[11] fall into the same complementation group cannot at present be eliminated.

For complementation tests, the cells undergoing mixed infection are incubated at a nonpermissive temperature and the harvest titrated at a permissive temperature. The complementation index (C) is equal to

$$C = \frac{(AB_{39.8})_{30}}{(A_{39.8})_{30} + (B_{39.8})_{30}}$$

where $(AB_{39.8})_{30}$ is the yield obtained at the nonpermissive temperature (39.8°C), assayed at the permissive temperature of 30°C, for the mixed mutant infection, and $(A_{39.8})_{30}$ and $(B_{39.8})_{30}$ are the yields obtained at the nonpermissive temperature (39.8°C) assayed at 30°C for the individual mutant viruses A and B, respectively.

In the absence of complementation, C is theoretically equal to 1, but values between 1 and 2 are generally considered as insignificant because of the statistical fluctuations of the test.

A. Intrastrain Complementation

1. VSV Indiana

Almost all of the 436 *ts* mutants obtained and analyzed by various laboratories fall into five or possibly six nonoverlapping complementation groups (Table 6). Complementation is often very efficient and high C values have been obtained. The yields obtained from mixed mutants virus infections often approach those reached with wild-type virus (Table 7). The actual complementation values obtained appear to be a function of the particular mutants employed. For example, we have found that mutants belonging to two RNA^- complementation groups usually give low C indexes, whereas when mutants from one RNA^- and one RNA^+, or two RNA^+, complementation groups are employed, the C indexes are high (Tables 7 and 8). Pringle[26] has shown that maximum complementation indexes are obtained when low multiplicities of infection are used (~5PFU per cell).

As a general rule, a mutant of one complementation group does not complement mutants of the same group, although it does complement mutants of another group. Exceptions to this rule have been observed by Flamand[3] for certain Group I *ts* mutants. Some mixed mutant virus infections of Group I give low but reproducible complementation indexes. This has been interpreted as evidence for intracistronic complementation. Similar results have been reported by Wong et al.[43] The differences between intracistronic complementation between different Group I mutants and intercistronic complementation between mutants of Groups I and IV are illustrated in Table 8.

Until recently, only five complementation groups were reported for VSV Indiana, corresponding to the identification of five gene products. However, mutant *ts* O 82 complements mutants of all other groups (Table 7). The high complementation indexes obtained for this mutant suggest that it has an RNA^+ phenotype. This has been confirmed by Martinet.[43a] Rettenmier et al.[38] have isolated two mutants which complement mutants representing Groups I through V; they have placed these mutants into a sixth complementation Group (VI). Whether the Group VI mutants of Rettenmier and associates and *ts* O 82 are comparable to each other remains to be established. Without further studies to eliminate the possibility that these mutants represent mutants which exhibit strong intracistronic complementation, the question of a sixth complementation group for VSV Indiana is undecided.

Pringle[22,26] and Cormack et al.[29] have demonstrated that complementation gives progeny not only having a *ts* phenotype, but also representing both parental genotypes. Cormack et al.[29] show that the proportion of each genotype obtained depended on the mutants being used. For mixed infections involving *ts* W 16B (Group IV) and Group I mutants, the proportions of the two parental genotypes in the progeny yield were equivalent to each other.

Some 87% of all the VSV Indiana *ts* mutants obtained (Table 6) — regardless of the source of origin, strain of virus, spontaneous or induced nature of the mutation

TABLE 6

Classification of Indiana *ts* Mutants into Complementation Groups

Laboratory	Origin	Cell	I	II	III	IV	V	VI	Unclassified	Total	Ref.
Orsay	Sp	CE	158	2	2	5	3	1	—	171	2, 3, 5
	5-FU, NG, NA	CE	18	—	2	1	1	—	—	22	65
Glasgow	5-FU, 5 AzaC, EMS	BHK	177	2	3	22	0	0	6	210	9, 64
Winnipeg	5-FU, NA, EMS, P	L	7	0	2	2	0	0	14	25	27
Massachusset	5-FU, NA, NTG	L	—	0	3	0	2	2	1	8	38
Total			360	4	12	30	6	3	21	436	

TABLE 7

Efficiency of Complementation Between VSV Indiana *ts* Mutants of Group I to VI

	Group I *ts*O 5	Group II *ts*O 52	Group III *ts*O 23	Group IV *ts*O 100	Group V *ts*O 45
Group I *ts*O 5	—				
Group II *ts*O 52	660 *5.0×10^7*	—			
Group III *ts*O 23	355 *1.2×10^7*	197 *1.5×10^7*	—		
Group IV *ts*O 100	14 *1.2×10^7*	36 *6.6×10^6*	26 *6.2×10^6*	—	
Group V *ts*O 45	3000 *3.0×10^6*	274 *2.2×10^7*	210 *3.0×10^6*	8 *2.9×10^6*	—
Group VI (?) *ts*O 82	15 *9.5×10^4*	87 *9.9×10^6*	394 *2.5×10^6*	115 *2.9×10^6*	214 *2.5×10^6*

Note: The values in the table are mean complementation indexes. The yields obtained in mixed infections expressed as PFU per Leighton tube are given in italics. In the same conditions (24 hr at 39.8°C) wild-type gave between 2 and 8×10^7 PFU per Leighton tube depending on the experiment.

— fall into the complementation Group I. Some 7% of the mutants are catagorized in Group IV, 3% in Group III, and 1% in each of the remaining groups. The existence of other complementation groups appears to be unlikely since the coding capacity of VSV Indiana cannot exceed one polypeptide more than the five already identified. If mutations occur randomly in the replication of the viral RNA genome, then it is reasonable to ask why the majority of the *ts* mutants are Group I mutants. As discussed in Chapter 8 in this volume, several investigators have shown that the defective gene product of group I mutants is the viral transcriptase. As discussed elsewhere in this book, the transcriptase is at least twice the size of the viral G protein, three times the size of the N protein, and four to six times larger than M or NS. Size differences are not, however, sufficient to explain the observed asymmetry of mutant distribution. Possibly amino acid substitution in N, NS, or M are often more lethal (or alternatively indifferent) than a substitution in the L polypeptide.

In general, the *ts* phenotype is not a host-dependent characteristic. Thus, most *ts* mutants isolated in CEF, BHK-21, or L cells retain their mutant phenotype when grown in the other cell lines. Cross-complementation between mutants isolated from different VSV Indiana strains and/or in different laboratories, using different cell types, has been achieved.[29,44]

Using a different approach, Deutsch and associates[45-47] attempted to complement *ts* mutants with UV-irradiated Indiana virions. Complementation occurred either by a reutilization of the proteins of the irradiated parent, termed nongenetic complementation by Deutsch, or by a functional survival of the gene. Group I and V mutants were rescued exclusively by nongenetic complementation, Group II mutants were rescued by both mechanisms, and mutants of Group IV primarily by functional survival of the gene. Group III mutants could not be rescued. If transcription is sequential, as discussed in Chapters 1 and 4 of this volume, then the intensity of complementation by survival of the gene should follow the gene order N>NS>M>G>L, which is more or less the case. However, it should be borne in mind that the possibility of nongenetic

TABLE 8

Complementation Between Mutants of Groups I and IV. Differences Between Inter- and Intracistronic Complementation

	Group I										Group IV	
	*ts*O 5	*ts*O 7	*ts*O 53	*ts*O 78	*ts*O 115	*ts*O 80	*ts*O 83	*ts*O 60	*ts*O 62	*ts*O 94	*ts*O 100	*ts*O 111
*ts*O 4	0.6	*14.2*	0.6	1.0	0.7	0.5	1.4	*4.0*	*11.0*	*3.0*	*3.1*	*3.1*
*ts*O 5	—	1.3	1	0.4	0.9	0.1	0.4	0.1	*2.3*	*8.8*	*14.0*	*8.1*
*ts*O 7	—	—	0.8	0.5	0.2	0.9	0.3	*7.1*	*7.6*	*3.9*	*5.5*	*5.5*
*ts*O 53	—	—	—	2.0	1.2	<2	1.9	1.0	*4.2*	*11.0*	*22.6*	*16.1*
*ts*O 78	—	—	—	—	*2.8*	1.8	<2	1.3	*4.9*	*2.0*	*44.1*	*20.8*
*ts*O 115	—	—	—	—	—	0.3	0.2	0.7	*5.9*	*7.1*	*4.4*	*10.5*
*ts*O 80	—	—	—	—	—	—	0.9	*2.7*	*4.2*	*7.7*	*111*	*61*
*ts*O 83	—	—	—	—	—	—	—	<2	*4.8*	*48.7*	*11.9*	*14.4*
*ts*O 60	—	—	—	—	—	—	—	—	*7.8*	*4.7*	*6.0*	*12.2*
*ts*O 62	—	—	—	—	—	—	—	—	—	2.0	1.9	0.8
*ts*O 94	—	—	—	—	—	—	—	—	—	—	0.2	0.3
*ts*O 100	—	—	—	—	—	—	—	—	—	—	—	0.4

Note: C indexes given in the table are the means from duplicate determination in a single experiment. Significant values are italicized.

Reprinted from Flamand, A., *J. Gen. Virol.*, 8, 187, 1970. With permission.

complementation, requiring liberation and reincorporation of parental proteins into progeny, is not likely to be the same for the five VSV structural proteins; e.g., it is unlikely with N, which remains tightly bound to the template. It is undoubtedly biologically significant that proteins of Group IV (N or NS) and III (M) cannot rescue mutants affected in the corresponding gene.

Deutsch[45] also showed that irradiated VSV Indiana T particles could rescue mutant *ts* O 45 (Group V) which is mutated in G, as a result of the reutilization of structural G proteins. Unirradiated T particles could not rescue the mutant, probably because they interfered with the replication of the mutant. Small T particles are, in general, incapable of complementing *ts* mutants. This is not surprising, since (1) most are devoid of competent transcriptase activity and (2) they inhibit viral replication (see Chapters 9 to 11 of this volume).

2. Other Rhabdoviruses

Pringle et al.[30] have isolated 49 *ts* mutants of VSV New Jersey and classified them in six nonoverlapping groups. Two Groups, A and B, contain 17 and 21 of the mutants, respectively. Pringle and Wunner[31] have isolated 40 *ts* mutants for Cocal virus. They have been classified in four groups, one of which contains 29 of the mutants. For Chandipura virus, some 30 *ts* mutants have been isolated and classified in five nonoverlapping groups.[31,31a] Of these mutants, 25 are in Group I.

B. Interstrain Complementation

Interstrain complementation is useful for determining to what extent viruses are related. It was reported by Pringle and Wunner[31] that, with the exception of VSV Indiana and Cocal viruses, no complementation occurred between mutants of different rhabdoviruses belonging to the vesiculovirus genus of the Rhabdoviridae family. Even for VSV Indiana and Cocal virus, the indexes of complementation obtained were low and were limited to Indiana RNA^+ mutants of Groups III and V and Cocal RNA^- mutants of group α and β. Since RNA^- Cocal mutants do not synthesize detectable amounts of protein per se (Pringle and Wunner),[31] the first step in the interstrain complementation process probably involved the utilization of VSV Indiana transcriptase enzymes (L and NS) by the Cocal template, resulting in Cocal G and M polypeptides in the cytoplasm of the doubly infected cells. By studying the progeny of such mixed virus infections, it should be possible to determine if the Cocal proteins do, in turn, complement the *ts* defects of VSV Indiana Group III and IV mutants. If, in each case, the progeny are exclusively composed of Cocal virions, then this would indicate that the complementation was asymmetric and that Cocal G and M proteins cannot replace the corresponding VSV Indiana proteins. Mixed VSV Indiana and Cocal virions in the progeny would indicate that complementation is symmetric and that the respective proteins of the two viruses can interrelate.

The fact that VSV Indiana *ts* mutants representing Groups I, II, and IV were not complemented by any of the Cocal *ts* mutants, even by RNA^+ mutants of group γ, suggests that the transcription and/or replication enzymes of Cocal virus cannot function on the VSV Indiana template. This is not consistent with the in vitro transcriptase complementation results of Bishop et al.[48] (see below).

The failure to observe complementation between VSV Indiana and VSV New Jersey (or between VSV Indiana and Chandipura; Cocal and VSV New Jersey; Cocal and Chandipura; VSV New Jersey and Chandipura) can be interpreted as evidence that none of the proteins of these four viruses can be successfully exchanged. Whether these results reflect the true capabilities of these viruses is not certain. For example, it is possible that the absence of complementation is due to heterotypic interference as reported for VSV Indiana and VSV New Jersey by Cooper[49] and Legault et al.[50] This heterotypic interference is not absolute, since one-step simultaneous infection can be obtained in certain circumstances as shown by Repik et al.[51] and by Deutsch.[46] Using highly sensitive hybridization techniques, Repik et al.[51] also showed that VSV New Jersey wild-type virus could partially complement the defect of VSV Indiana *ts* G 114 (Group I). Primary transcription of the genome of the mutant virus was detected in cells coinfected at nonpermissive temperatures with wild-type VSV New Jersey but not in the single mutant virus infection. No amplification of transcription occurred in the coinfection, suggesting that the replication of *ts* G 114 was not promoted by the presence of the wild-type VSV New Jersey virus.

Attempts to reconstitute transcriptase activity and infectivity from dissociated components in vitro have been successful with VSV Indiana and Cocal viruses, but have met with failure with heterologous combinations of VSV Indiana, VSV New Jersey, and Chandipura transcriptase components.[48] Thus, it may be concluded that the VSV New Jersey transcriptase is capable of performing transcription on the VSV Indiana template in vivo but not in vitro; also it can be concluded that the VSV New Jersey transcriptase is unable to participate in the replication process of the VSV Indiana genome in vivo. Such a situation may be considered as partial complementation between the gene products of the respective viruses. The second conclusion that can be made from these observations is that the Cocal transcriptase is functional, at least in vitro, with the genome of VSV Indiana, and vice versa.

For other rhabdovirus proteins, the stuation is not clear. For example, Deutsch[46] showed that the G protein of UV-irradiated New Jersey virions could rescue the VSV Indiana mutant *ts* O 45 (Group V), yielding infectious VSV Indiana particles which could be inactivated by anti-VSV New Jersey serum, suggesting that the VSV New Jersey G proteins were included in the membrane of VSV Indiana particles. If such complementation can occur under these particular situations, it is reasonable to ask why complementation does not occur between the VSV Indiana *ts* O 45 mutant and any of the VSV New Jersey *ts* mutants. One explanation may be that VSV New Jersey G protein cannot be furnished in large quantities by the New Jersey RNA^- groups A, B, E, and F since they are defective in transcription and/or replication (Lesnaw and Reichmann[52]), nor by Group C (or D), which is probably defective in G protein. Consequently, complementation is unlikely to occur between at least five of the six VSV New Jersey groups and the VSV G protein-defective *ts* O 45. The other RNA^+ VSV New Jersey complementation group is probably mutated in M protein. By analogy with VSV Indiana, if the G and M proteins are interdependent in their maturation processes, then a mutation in VSV New Jersey M may prevent the insertion of VSV Indiana G protein into the corresponding part of the infected cell membrane, thus preventing maturation of progeny viruses containing VSV New Jersey G and VSV Indiana M proteins.

Flamand and Bussereau[53] demonstrated that no complementation occurred between VSV Indiana *ts* mutants representing Groups I, III, and V and rabies virus. Phenotypically mixed virions containing G proteins of both viruses, however, were produced in mixed infections with VSV Indiana *ts* O 45 (group V) and rabies. The virions produced were extremely thermolabile, a problem which may complicate the complementation question.

In summary, the data currently available suggest that, in certain cases, transcriptase components and G proteins may be exchanged between different rhabdovirus serotypes, allowing some level of complementation between those viruses.

V. ABSENCE OF RECOMBINATION

Recombination is studied by performing mixed virus infections at permissive temperatures, followed by titration of the progency at nonpermissive temperatures. The recombination index (R) is equal to

$$R = \frac{(AB_{30})_{39.8}}{(A_{30})_{39.8} + (B_{30})_{39.8}}$$

where $(AB_{30})_{39.8}$ is the titer at the nonpermissive temperature (39.8°C) of the yield obtained at the permissive temperature (30°C) for the mixed mutant (A and B) infection, and $(A_{30})_{39.8}$ and $(B_{30})_{39.8}$ are the titers at the nonpermissive temperature of the individual mutant virus harvests grown at 30°C.

When titrated at nonpermissive temperatures, crosses involving complementing VSV Indiana *ts* mutants yielded plaques which were initially interpreted by Pringle[26] as recombinants. There were, however, certain indications suggesting that these plaques may, in fact, have arisen by complementation and not by recombination:

1. They were obtained in crosses involving only complementing *ts* mutants.
2. They were smaller and more irregular than wild-type plaques.
3. In most cases, their frequency decreased as the square of the dilution, suggesting a cooperation between independent particles.
4. They segregated during subsequent clonings.

It appears that these plaques originated from cells independently infected with both progeny *ts* mutants or from cells infected with polyploid or aggregated particles. True wild-type plaques were never more frequent in mixed mutant virus infections than in control single mutant infections. In crosses involving two *ts* mutants representing different complementation groups (and having low reversion frequencies), the frequency of wild-type progeny in the mixed infection was found to be lower than 2×10^{-5}. Wong and associates,[54] Pringle and Wunner,[8] and Flamand[5] have concluded that genetic recombination does not occur for VSV Indiana.

ACKNOWLEDGMENTS

I wish to thank Dr. P. Vigier, in whose laboratories our research has been done, and G. Leblon for reading the manuscript. The research on the genetic study of rhabdoviruses was supported in part by L.A. 40086 and NATO grant 994.

REFERENCES

1. **Clewley, J. P., Bishop, D. H. L., Kang, C. Y., Coffin, J., Schnitzlein, W. M., Reichmann, M. E., and Shope, R. E.,** Oligonucleotide fingerprints of RNA species obtained from Rhabdoviruses belonging to the vesicular stomatitis virus subgroup, *J. Virol.*, 23, 153, 1977.
2. **Flamand, A.** Etude des mutants thermosensibles du virus de la stomatite vésiculaire: mise au point d'un test de complementation, *CR Acad. Sci.*, 268, 2305, 1969.

3. **Flamand, A.**, Etude génétique du virus de la stomatite vesiculaire: classement de mutants thermosensibles spontanes en groupes de complementation, *J. Gen. Virol.*, 8, 187, 1980.
4. **Carvalho-Jeronimo M.**, Recherche d'un Effet de la Cellule Hôte sur l'Instabilité Génétique d'une Souche du Virus de la Stomatite Vésiculaire de Serotype Indiana, these de 3° cycle, Orsay, 1974.
5. **Flammand, A.**, Etude Génétique du Virus de la Stomatite Vésiculaire, these de Doctorat d'Etat, Orsay, 1974.
6. **Youngner, J. S. and Quagliana, D. O.**, Temperature-sensitive mutants of vesicular stomatitis virus are conditionally defective particles that interfere with and are rescued by wild-type virus, *J. Virol.*, 19, 102, 1076.
6a. **Flamand, A.**, unpublished observations.
7. **Drake, J. W.**, *The Molecular Basis of Mutation*, Holden-Day, San Francisco, 1969, chap. 5.
8. **Pringle, C. R. and Wunner, W. H.**, A comparative study of the structure and function of the VSV genome, in *Negative Strand Viruses and the Host Cell*, Vol. 2, Mahy, B. W. J., and Barry, R. D., Eds., Academic Press, New York, 1975, 707.
8a. **Flamand, A.**, unpublished results.
9. **Pringle, C. R.**, Genetic characteristics of conditional lethal mutants of vesicular stomatitis virus induced by 5-fluorouracil, 5-azacytidine and ethylmethane sulfonate, *J. Virol.*, 5, 559, 1970.
9a. **Flamand, A.**, unpublished data.
10. **Clark, H F. and Wiktor, T. J.**, Plasticity of phenotypic characters of rabies related viruses: spontaneous variation in plaque morphology, virulence and temperature-sensitivity characters of serially propagated Lagos bat and Mokola viruses, *J. Infect. Dis.*, 130, 608, 1974.
11. **Bussereau, F. and Flamand, A.**, Isolation and preliminary characterization of ts mutants of Rabies virus, in *Negative Strand Viruses and the Host Cell*, Mahy B. W. J. and Barry, R. D., Eds., Academic Press, New York, in press.
12. **Ohanessian-Guillemain, A.**, Etude de facteurs genetiques controlant les relations due virus sigma et de la Drosophile hôte, thèse de Doctorate d'Etat, Paris, 1963.
13. **Vigier, P.**, Contribution á l'étude de l'instabilité génétique du virus σ de la Drosophile, *Ann. Genet.*, 9, 1, 1965.
13a. **Brun, G.**, personal communication.
13b. **Roy, P.**, personal communication.
14. **Flamand, A.**, Genetical behaviour of vesicular stomatitis virus during successive passages at high and low temperatures, *Mutat. Res.*, 17, 177, 1973.
15. **Lwoff, A. and** Lwoff, M., Sur les facteurs du développement viral et leur rôle dans L'évolution de l'infection, *Ann. Inst. Pasteur*, 98, 174, 1960.
16. **Taylor, R. M.**, Catalog of Arthropod-Borne Viruses of the World, U.S. Government Printing Office, Washington, D.C., 1967.
17. **Ramseur,** J. M. and Friedman, R. M., Prolonged infection of interferon-treated cells by vesicular stomatitis virus: possible role of temperature sensitive mutants.and interferon, *J. Gen. Virol.*, 37, 523, 1977.
18. **Nishiyama, Y.**, Studies of L cells persistently infected with VSV: factors involved in the regulation of persistent infection, *J. Gen. Virol.*, 35, 265, 1977.
19. **Printz, P.**, Relationship of sigma virus to vesicular stomatitis virus, in *Advances in Virus Research*, Vol. 18, Academic Press, New York, 1973, 143.
20. **Mudd, J. A., Leavitt, R. W., Kingsbury, D. T., and Holland, J. J.**, Natural selection of mutants of vesicular stomatitis virus by cultured cells of *Drosophila melanogaster*, *J. Gen. Virol.*, 20, 341, 1973.
21. **Youngner, J. S., Dubovi, E. J., Quagliana, D. O., Kelly, M., and Preble, O. T.**, Role of temperature-sensitive mutants in persistent infections initiated with vesicular stomatitis virus, *J. Virol.*, 19, 90, 1976.
22. **Pringle, C. R.**, Genetics of Rhabdoviruses, in *Comprehensive Virology*, Vol. 9, Fraenkel-Conrat, H. and Wagner, R. R., Eds., Plenum Press, New York, 1977, chap. 6.
23. **Wagner, R. R.**, Levy, A. H., Snyder, **R.** M., **Raycliff,** G. A., and Hyatt, D. F., Biologic properties of two plaque variants of vesicular stomatitis virus (Indiana serotype), *J. Immunol.*, 91, 112, 1963.
24. **Wertz, G. W. and Levine, M.**, RNA synthesis of vesicular stomatitis virus and a small plaque mutant: effects of cycloheximide, *J. Virol.*, 12, 253, 1973.
25. **Moreau, M. C.**, Inhibition of a vesicular stomatitis virus mutant by rifampin, *J. Virol.*, 14, 517, 1974.
26. **Pringle, C. R.**, The induction and genetic characterization of conditional lethal mutants of vesicular stomatitis virus, in *The Biology of Large RNA Viruses*, Barry, R. D. and Mahy, B. W. J., Eds., Academic Press, New York, 1970, 567.
27. **Holloway, A. F., Wong, P. K. Y., and Cormack, D. V.**, Isolation and characterization of temperature-sensitive mutants of vesicular stomatitis virus, *Virology*, 42, 917, 1970.

28. **Clark, H F. and Koprowski, H.**, Isolation of temperature-sensitive conditional lethal mutants of "fixed" rabies virus, *J. Virol.*, 7, 295, 1971.
29. **Cormack, D. V., Holloway, A. F., and Pringle, C. R.**, Temperature-sensitive mutants of vesicular stomatitis virus: homology and nomenclature *J. Gen. Virol.*, 19, 295, 1973.
30. **Pringle, C. R., Duncan, I. B., and Stevenson, M.**, Isolation and characterization of temperature-sensitive mutants of vesicular stomatitis virus, New Jersey serotype, *J. Virol.*, 8, 836, 1971.
31. **Pringle, C. R. and Wunner, W. H.**, Genetic and physiological properties of temperature-sensitive mutants of Cocal virus, *J. Virol.*, 12, 677, 1973.
31a. **Pringle, C. R. and Gadkari**, unpublished results.
32. **Flamand, A. and Lafay, F.**, Etude des mutants thermosensibles du virus de la stomatite vésiculaire appartenant au groupe de complementation II, *Ann. Microbiol. (Inst. Pasteur)*, 124A, 261, 1973.
33. **Lafay, F.**, Etude Physiologique des Mutants Thermosensibles du Virus de la Stomatite Vésiculaire, thèse de 3° cycle, Paris, 1972.
34. **Simpson, R. W. and Obijeski, J. F.**, Conditional lethal mutants of vesicular stomatitis virus. I. Phenotypic characterisation of temperature sensitivity, *Virology*, 57, 357, 1974.
35. **Szilágyi, J. F. and Pringle, C. R.**, Virion transcriptase activity differences in host range mutants of vesicular stomatitis virus, *J. Virol.*, 16, 927, 1975.
36. **Szilágyi, J. F., Pringle, C. R., and Macpherson, T. M.**, Temperature dependant host range mutation in vesicular stomatitis virus affecting polypeptide L, *J. Virol.*, 22, 381, 1977.
37. **Obijeski, J. F. and Simpson, R. W.**, Conditional lethal mutants of vesicular stomatitis virus. II. Synthesis of virus-specific polypeptides in nonpermissive cells infected with "RNA" host-restricted mutants, *Virology*, 57, 369, 1974.
38. **Rettenmier, C. W., Dumont, R., and Baltimore, D.**, Screening procedure for complementation dependent mutants of vesicular stomatitis virus, *J. Virol.*, 15, 41, 1975.
39. **Lafay, F.**, Envelope proteins of vesicular stomatitis virus: effect of temperature-sensitive mutation in complementation groups III and V, *J. Virol.*, 14, 1220, 1974.
40. **Závada, J.**, Vesicular stomatitis virus pseudotype particles with the coat of avian myeloblastosis virus, *Nature (London) New Biol.*, 240, 122, 1972.
41. **Závada, J. and Závodksa.**, Complementation and phenotypic stabilisation of vesicular stomatitis virus temperature-sensitive mutants by avian myeloblastosis virus, *Intervirology*, 2, 25, 1974.
42. **Streckfuss, J. L. and Strechmeister, I. L.**, The intracellular development of vesicular stomatitis virus and two of its mutants, *Arch. Gesamte Virusforsch,*, 35, 208, 1971.
43. **Wong, P. K. Y., Holloway, A. F., and Cormack, D. V.**, Characterisation of three complementation groups of vesicular stomatitis virus, *Virology*, 50, 829, 1972.
43a. **Martinet, C.**, personal communication.
44. **Flamand, A. and Pringle, C. R.**, The homologies of spontaneous and induced temperature-sensitive mutants of vesicular stomatitis virus isolated in chick embryo and BHK-21 cells, *J. Gen. Virol.*, 11, 81, 1971.
45. **Deutsch, V.**, Nongenetic complementation of group V temperature-sensitive mutants of vesicular stomatitis virus by UV-irradiated virus, *J. Virol.*, 15, 798, 1975.
46. **Deutsch, V.**, Parental G protein reincorporation by a vesicular stomatitis virus temperature sensitive mutant of complementation group V at non-permissive temperature, *Virology*, 69, 607, 1976.
47. **Deutsch, V., Muel, B., and Brun, G.**, Action spectra for the rescue of temperature sensitive mutants of vesicular stomatitis virus by ultraviolet-irradiated virions at non-permissive temperature, *Virology*, 77, 294, 1977.
48. **Bishop, D. H. L., Emerson, S. U., and Flamand, A.**, Reconstitution of infectivity and transcriptase activity of homologous and heterologous viruses: vesicular stomatitis (Indiana serotype), Chandipura, vesicular stomatitis (New-Jersey serotype) and Cocal viruses, *J. Virol.*, 14, 139, 1974.
49. **Cooper, P. D.**, Homotypic non-exclusion by vesicular stomatitis virus in chick cell culture, *J. Gen. Microbiol.*, 19, 350, 1958.
50. **Legault, D., Takayesu, D., and Prevec, L.**, Heterotypic exclusion between vesicular stomatitis viruses of the New-Jersey and Indiana, *J. Gen. Virol.*, 35, 53, 1977.
51. **Repik, P., Flamand, A., and Bishop, D. H. L.**, Synthesis of RNA by mutants of vesicular stomatitis virus (Indiana serotype) and the ability of wild type VSV New-Jersey to complement the VSV Indiana ts G114 transcription defect, *J. Virol.*, 20, 157, 1976.
52. **Lesnaw, J. A. and Reichmann, M. E.**, RNA synthesis by temperature-sensitive mutants of vesicular stomatitis virus, New-Jersey serotype, *Virology*, 63, 492, 1975.
53. **Flamand, A. and Bussereau, F.**, phenotypic mixing between rabies virus and vesicular stomatitis virus, in *Negative Strand Viruses and the Host Cell*, Mahy, B. W. J. and Barry, R. D., Eds., Academic Press, New York, in press.

54. **Wong, P. K. Y., Holloway, A. F., and Cormack, D. V.**, Search for recombination between temperature-sensitive mutants of vesicular stomatitis virus, *J. Gen. Virol.*, 13, 477, 1971.
55. **Shechmeister, I. L., Streckfuss, J., and St. John, R.**, Comparative pathogenicity of vesicular stomatitis virus and its plaque type mutants, *Arch. Gesamte Virusforsch.*, 21, 127, 1967.
56. **Pittman, D.**, Temperature sensitive mutants of a rod-shaped RNA animal virus, *Genetics*, 52, 468, 1965.
57. **Matsumoto, S.**, Rabies virus, *Adv. Virus Res.*, 16, 257, 1970.
58. **Contamine, D.**, Etude de mutants thermosensibles du virus sigma, *Mol. Gen. Genet.*, 124, 233, 1973.
59. **Ohanessian-Guillemain, A.**, Etude génétique du virus héréditaire de la *Drosophile* (σ), mutatios et recombinaison génétique, *Ann. Genet.*, 1, 59, 1959.
60. **Guillemain, A.**, Découverte et localisation d'un géné empechânt la multiplication du virus de la sensibilite hereditaire au CO_2 chez D. M., *CR Acad. Sci.*, 236, 1085, 1953.
61. **Brun, G.**, Etude d'une Association du Virus et de Son Hôte la *Drosophile:* l'État Stabilisé, these Biol. Exp., Orsay, 1963.
62. **Duhamel, C.**, Etude de la sensibilité héréditaire à l'anhydride carbonique chez la *Drosophile:* description de quelques variants du virus, *CR Acad Sci.*, 239, 1157, 1954.
63. **Goldstein, L.**, Contribution a l'etude de la sensibilite hereditaire au gaz carbonique chez la *Drosophile:* mise en evidence d'une forme nouvelle du genoide, *Bull. Biol. Fr. Belg.*, 83, 177, 1949.
64. **Pringle, C. R. and Duncan, I. B.**, Preliminary physiological characterisation of temperature-sensitive mutants of vesicular stomatitis virus, *J. Virol.*, 8, 56, 1971.
65. **Lafay, F. and Flamand, A.**, unpublished results.

Chapter 8

GENE ASSIGNMENT AND COMPLEMENTATION GROUP

C. R. Pringle and J. F. Szilágyi

TABLE OF CONTENTS

I. CODING CAPACITY AND GENOME ORGANIZATION

The genome of vesicular stomatitis virus (VSV), the prototype of the rhabdovirus group, is a single-stranded, unsegmented RNA molecule with a molecular weight in the region of 4×10^6. Thus, the genome contains sufficient information for the synthesis of unique polypeptides with a total molecular weight of approximately 4×10^5. The known gene products of VSV are the five structural components of the virion (the L, G, N, NS, and M proteins). No nonstructural proteins have been identified unequivocally. The combined molecular weight of the polypeptides of the five structural proteins (approximately 3.9×10^5) accounts for almost all of the coding capacity of the genome of VSV, assuming that the gene sequences are orientated in a nonoverlapping linear sequence. Overlapping gene sequences and reading of the same nucleotide sequence in different phases are known to occur in both bacterial (φX 174)[1] and mammalian (SV 40)[2] DNA-containing viruses and effectively expand the coding capacity of the genome. Both must place constraints on mutability, however. This type of organizational complexity has not been recognized so far in RNA-containing viruses, apart from the synthesis of the A-prime protein of Qβ by occasional read-through of the UGA termination signal of the coat protein gene.[3] The uniformly high mutability of rhabdoviruses (see Chapter 7 of this volume) suggests that it is unlikely to occur.

Temperature-sensitive (*ts*) mutants of both the Indiana and the New Jersey serotypes of VSV have been classified into six complementation groups, indicating that the genome of VSV consists of at least six functional units and that there are six corresponding gene products.[4] Thus, at least one virus-specified polypeptide remains to be identified. It is the purpose of this chapter to review the evidence for gene assignments in VSV and to highlight some of the remaining problems. In principle, it should be possible to equate the five known gene products of VSV (the structural polypeptides) with their genetic determinants (five of the six complementation groups) since there is no evidence of high molecular weight precursor polypeptides and post-translational proteolytic cleavage during the synthesis of VSV proteins. It is likely, therefore, that transcription of the VSV genome yields monocistronic messenger RNAs, which are translated directly into polypeptides with only secondary post-translational modification such as glycosylation, phosphorylation, sulfation, etc. It is a reasonable assumption that each complementation group represents a separate transcriptional unit of the genome. However, the precise mode of transcription of the VSV genome by the virus-specified transcriptase is unknown, and uncertainty surrounds the role of the leader RNA sequence in transcription and the significance of the complementarity of the terminal sequences of the viral genome.[5] Therefore, the effect of mutation in nonpolypeptide specifying regions of the genome cannot be predicted at present, but it is conceivable that mutation in a nonpolypeptide specifying region, such as the leader sequence, could have pleotrophic phenotypic effects through an effect on messenger RNA processing.

This discussion of gene assignments relates only to the cytoplasmic VSV-like rhabdoviruses, since the available data are virtually confined to the Indiana, Cocal, and New Jersey serotypes of VSV and to Chandipura virus. The genetic organization of other members of the rhabdovirus group, such as those rhabdoviruses of plants which appear to multiply in the nucleus of the cell, may be entirely different.[6]

II. PHENOTYPIC PROPERTIES OF THE COMPLEMENTATION GROUPS OF VSV AND CHANDIPURA VIRUS

Mutants have been classified as RNA-negative or RNA-positive by measuring cumulative incorporation of radio-labeled uridine in the presence of actinomycin-D at restrictive temperature. This incorporation measures amplification of RNA synthesis (transcription and replication) and distinguishes mutants with defects in assembly and maturation functions (RNA-positive mutants) from those with defects in RNA synthesis (RNA-negative mutants). The RNA-negative phenotype embraces mutants which are defective in transcription and/or replication. In general, synthesis of viral proteins intracellularly correlated with the RNA phenotype as illustrated in Table 1. With certain exceptions, it was found that mutants in the same complementation group exhibited the same gross phenotype, in support of the hypothesis that each complementation group corresponded to one transcriptional unit of the genome. The preponderance of RNA-negative mutants can be attributed to the fact that three of the five virion polypeptides are components of the transcribing nucleoprotein (TNP) complex[7] and account for about 80% of the coding capacity of the genome.

III. REQUIREMENTS FOR ASSIGNMENT

The genome of VSV is an uninterrupted nucleotide sequence of negative-stranded polarity, which is transcribed from the 3′-end into a noncoding leader sequence of approximately 48 nucleotides, followed by sequences coding for the N, NS, M, G, and L polypeptides, in that order.[5] Nontranscribed (and/or nontranslated) sequences may separate the polypeptide specifying regions, but it is assumed that the five (or six) complementation groups of VSV correspond to mutations in the transcribed polypeptide-specifying regions. Hence, in the assignment of VSV genes, the gene product can be either a monocistronic messenger RNA or a polypeptide. Much can be deduced from the study of the gross physiological properties of *ts* mutants, and this approach has been reviewed in detail elsewhere.[4,8] Unequivocal assignment, however, requires demonstration of a direct effect on the gene product (mRNA or polypeptide). In addition, it must be shown that reversion of the *ts* phenotype is accompanied by restoration of a normal gene product and that the observed effect is a common property of mutants in the same complementation group. It may also be important, particularly in the case of thermolabile mutants, to avoid bias in selection of revertants, by isolation of spontaneous revertants. In no single case so far have all of these criteria been met.

Much of the difficulty in assignment is a consequence of the fact that the complementation groups have been defined using *ts* mutants, which if isolated under optimal conditions should represent missense mutation at single sites in the viral genome. A consequent single amino acid change in a polypeptide (or a nucleotide in a polynucleotide) may not be detectable without direct fingerprinting, and at least partial sequencing, of the gene product. Consequently, the assignment of complementation groups has been less definitive than anticipated by the apparent simplicity of the system and the limited number of gene products. Nevertheless, four of the five genes determining the structural proteins of the Indiana serotype of VSV have been assigned to complementation groups. There is a sufficient number of discrepant observations, however, to justify continued caution.

IV. ASSIGNMENT OF THE COMPLEMENTATION GROUPS OF VSV INDIANA

Six complementation groups of VSV Indiana have been described (see Chapter 7 in this volume). Groups I to V are well characterized; the sixth group has only recently been described. Mutant *ts* O82(VI), classified by Flamand as representing a sixth complementation group, is an RNA-positive mutant and is distinct from the RNA-negative

TABLE 1[a]

Complementation Grouping and Phenotype

Virus	Serotype	Complementation group	Number in group	RNA synthesis		Protein Synthesis		Possible assignment[b]
				Phenotype	Number examined	Phenotype	Number examined	
VSV	Indiana (315 mutants)	I	260	−	23	−	3	L
		II	4	−	3	−	2	None
				+	1	+	1	
		III	12	+	10	+	6	M
		IV	30	−	13	−	4	N
		V	6	+	5	+	5	G
				−	1	−	1	
		VI	3	+	1	N.D.[c]	0	None
				−	2	−	2	
VSV	Cocal (40 mutants)	α	29	−	2	−	2	None
		β	2	−	2	−	2	None
		γ	3	+	2	+	2	G
		δ	6	−	1	−	1	None
				+	1	+	1	
VSV	New Jersey (48 mutants)	A	17	−	5	−	1	None
		B	21	−	7	−	1	None
		C	4	+	4	+	1	None
		D	1	+	1	+	1	N
		E	3	−	2	−	1	None
				+	1	N.D.	0	
		F	2	−	2	−	1	None
Chandipura	— (50 mutants)	ChI	44	−	14	−	6	None
		ChII	2	+	2	+	2	None
		ChIII	1	−	1	+	1	None
		ChIV	1	−	1	+	1	None

ChV	I	+	I	+	I	M
ChVI	I	−	I	+	I	None

[a] Compiled from Flamand (Chapter 7) and References 4, 8, 13, 22, 43, and 47.
[b] Discussed in Sections IV, V, and VI.
[c] No data

mutants classified by Rettenmier et al.[9] as Group VI. The latter mutants resemble Group IV, and their status is dubious for other reasons.

A. Complementation Group I

The least equivocal assignment is the identification of Group I with the gene which codes the L polypeptide of the virion. The initial observation that the polymerase activity of three of six mutants chosen at random from complementation Group I was defective in vitro at the restrictive temperature for these mutants suggested that the mutational lesion involved the virion-associated RNA transcriptase. The same in vitro phenotype was observed irrespective of whether detergent-treated virus or purified virion cores (transcribing nucleoprotein) were used. Restoration of polymerase activity with reversion of the *ts* phenotype confirmed the specificity of the phenotype.[10] The virions of these three mutants were more thermolabile than normal. Subsequently, Hunt et al.[11] showed by dissociation and reassociation of pellet and supernatant fractions from wild-type and *ts* mutant virus that the defective component was present in the supernatant and not the template fraction. These experiments confirmed that the L polypeptide was essential for transcriptase activity. It was concluded, therefore, that complementation Group I comprised mutants with lesions in the L polypeptide and that thermolability of the polymerase was responsible for the *ts* phenotype. This conclusion was reinforced by the experiments of Repik et al.,[12] who showed that primary transcription did not occur in cells infected with mutant *ts* G114(I) maintained at restrictive temperature. Mutant *ts* G114(I) is the most thermolabile of the Group I mutants, however, and some in vivo primary transcription was detected in the case of less thermolabile Group I mutants.

Thus, some of the criteria for assignment of the L protein-coding gene to Group I have been satisfied. Nevertheless, three of six mutants studied in detail were thermostable and exhibited normal levels of in vitro polymerase activity. It has been assumed, but not demonstrated, that an aberrant nonfunctional product was synthesized under these conditions. The wild-type revertants of the polymerase defective mutants were isolated from single plaques appearing on monolayers at 39°C.[10] In view of the thermolability of these mutants, this was a less satisfactory procedure than isolation of revertants without selection, and co-reversion of enzyme lability and virion heat lability cannot be discounted entirely.

The major discrepant finding has been the observation of Knipe et al.[13] that the N polypeptide of *ts* G12(I), one of the three thermostable Group I mutants referred to previously, was more rapidly degraded at 39°C when compared with wild-type or other Group I mutants. However, wild-type revertants were not isolated or characterized to exclude the possibility that this phenotype was determined by an independent non-*ts* mutation.

Other mutants (the temperature-dependent host-range phenotype — *td* CE) have been shown to have a mutational lesion in the L polypeptide by similar methodology.[14,15] These mutants have a conditionally host-restricted phenotype and will be considered in the section dealing with mutational effects on the activity of the virion polymerase and the problem of gene assignment (see Section VI) because despite the precise identification of the L protein gene, the assignment of other components of the viral core is more difficult.

B. Complementation Group II

No satisfactory assignment has been made in the case of Group II mutants. Group II has become equated with the NS polypeptide-specifying gene by a process of elimination, since the other four virion polypeptides have been associated with Groups I,

III, IV, and V.[8] There is no direct evidence, however, in favor of this assignment, and the enigmatic sixth group could equally be equated with the NS polypeptide-specifying gene. Indeed, Combard et al.[16] reported that the nucleocapsid from *ts* O52(II) virions or mutant-infected cells at 31°C was less stable than wild-type nucleocapsid when exposed at 39.5°C. The RNA component of the nucleocapsid became sensitive to pancreatic ribonuclease at 39.5°C, and it was concluded that this was a consequence of a defective N polypeptide. In their interpretation, therefore, complementation Group II represented the N polypeptide-specifying gene.

C. Complementation Group III

Mutants in Group III synthesized RNA in normal or supranormal amounts at restrictive temperature, and all five structural polypeptides were detected in infected cells.[8] Nucleocapsid structures accumulated in infected cells at restrictive temperature, but no virions were released. None of the three Group III mutants isolated from the Indiana-C strain was heat labile. Experiments by Lafay[17] suggested that the M protein was the site of the mutational lesion. M polypeptide radio-labeled at 39°C did not appear in virions which were released from infected cells after downshift to permissive temperature, suggesting that the M polypeptide synthesized at 39°C was defective and responsible for the failure of maturation.

Knipe et al.[13] have shown that polypeptide M was rapidly degraded at 39°C in cells infected by mutants *ts* M301(III) or *ts* G33(III) (possibly the same mutant) but not in cells infected by wild-type virus or *ts* mutants belonging to the other complementation groups. No other polypeptides were similarly affected during the 60-min chase period. Furthermore, it was shown that the instability was an intrinsic property of the M protein because it could not be rescued by wild-type virus. It was concluded, therefore, that the M polypeptide was the site of the mutational lesion and that complementation Group III represented the gene coding for the M protein. No studies of revertants of Group III mutants were reported, however.

Buller[18] observed mobility differences by electrophoresis in gradient gels between the intracellular M polypeptide of wild-type virus and four mutants of Group III: *ts* G31, *ts* G32, *ts* G33, and *ts* O23. Reversion of the *ts* phenotype of mutant *ts* O23(III) was accompanied by a change in electrophoretic mobility of the M polypeptide, but the wild-type mobility was not restored. Apparently, the normal function of the M polypeptide had been restored by mutation at another site within the M protein gene.

The evidence is consistent for assignment of the M protein gene to complementation Group III, since all members of the group which have been examined showed the same defect in the M polypeptide. The sole discordant observation is the report by Knipe et al.[13] that the M protein of mutant *ts* M502(V) was degraded at a rate nearly equal to that of *ts* M301(III). However, mutant *ts* M502(V) is atypical in other respects (see Section IV.E).

D. Complementation Group IV

Mutants in Group IV are RNA-negative mutants, but they displayed normal transcriptase activity in vitro at 39°C,[10] and primary transcription but not secondary transcription could be detected in infected BHK-21 cells at 39°C.[19] Dissociation and reassociation experiments, using a Group IV mutant (*ts* W16B(IV)) isolated by Holloway et al.,[20] showed that neither L nor NS was able to restore in vitro polymerase activity and suggested that the N protein might be the site of the mutational lesion.[21]

More recently, Knipe et al.[13,22] observed that N protein labeled at 39°C in L cells infected by mutant *ts* G41(IV) was more rapidly degraded than N protein in cells infected with wild-type virus during a 60-min chase period, whereas at 31°C the mutant and wild-type N proteins were equally stable. However, two other mutants, *ts* G12(I) and *ts* M601(VI), also exhibited a similar instability of the N polypeptide at 39°C. (However, *ts* M601(VI) failed to complement *ts* G41(IV),[9] and the status of this mutant is dubious.) Unfortunately, no other mutants in Group IV or revertants of *ts* G41(IV) were studied. Little and Huang[23] independently reported a reduction in N protein in *ts* G41(IV)-infected Chinese hamster ovary (CHO) cells at 39°C.

Freeman et al.[24] have described a method of mapping *ts* mutants which definitively established *ts* G41(IV) as an N protein mutant. The method has been used successfully with reovirus and SV 40 and, in theory, is applicable to any *ts* mutation.

Briefly, it was established that the four VSV mRNAs could be separated in polyacrylamide gel as homoduplexes formed with virion RNA.[25] If heteroduplexes are formed between RNA from a *ts* mutant and its wild-type (or, in this case, revertant), the site of the mutational lesion can be identified by the formation of an altered heteroduplex.

Heteroduplexes formed between *ts* G41(IV) virion RNA and mRNA from cells infected with a revertant of *ts* G41(IV) migrated identically to homoduplexes. However, after digestion with ribonuclease T1 at low salt concentration, a new duplex (X) was resolved. Since the N homoduplex was more resistant to ribonuclease digestion than the N heteroduplex, it was concluded that duplex X was generated by cleavage of a single base mismatch within the N duplex. If these findings can be reproduced with other Group IV mutants, the assignment of the N protein gene to complementation Group IV will be definitively established.

Freeman et al.[24] have pointed out, however, that the success of fine-mapping by mismatching in heteroduplexes is limited by the mechanism of reversion and by the ability of ribonuclease to recognize different degrees of mismatching. Reversion can occur by a base change at the site of the original mutation, at a different site in the same coding triplet, or at a more remote site. The degree of mismatch will vary according to the base pair involved; a C-U pairing giving a weak mismatch, an A-G pairing a strong mismatch, with A-C and G-U pairings intermediate. Therefore, only an indeterminate proportion of mutant-revertant heteroduplexes may be mappable. Nevertheless, this approach provides the most definitive method of establishing gene assignments so far devised.

E. Complementation Group V

Mutant *ts* M502(V), unlike other Group V mutants, appeared to be totally defective for RNA transcription and replication at 39°C. This mutant grew poorly, however, and synthesized subnormal amounts of all species of viral RNA at 31°C,[13,22] and the atypical RNA-negative phenotype may be a consequence of multiple mutation. The other mutants were RNA-positive mutants, and synthesis of the five structural proteins was unimpaired at 39°C.[26,27] Two of the four Group V mutants isolated by Flammand[28] were thermolabile in vitro, and Printz and Wagner[26] attributed the *ts* phenotype to the thermolability of the virion. Deutsch and Berkaloff[30] reported that the envelope of the virion of thermolabile mutant *ts* O45(V) was unstable at 39°C and suggested that an envelope component was defective. Závada[31] demonstrated that mutant *ts* O45(V) could be complemented by avian myeloblastosis virus, which indicated also that the defect was in the viral envelope. However, the other two Orsay Group V mutants were thermostable and indistinguishable from wild-type.

More recently, Lafay[17] has obtained results from pulse-chase experiments with both thermostable and thermolabile mutants which suggest that a step in the synthesis of the G polypeptide is the site of the lesion. In the case of the four Orsay mutants, G protein synthesized at 39°C became incorporated into virions after downshift to 31°C. Knipe et al.[13,22] showed that the G polypeptide of mutants *ts* M501(V) and *ts* O45(V) did not undergo the change in mobility (the G1 → G2 transition) associated with the addition of sialic acid;[32] i.e., the G protein remained underglycosylated at 39°C, and it did not appear at the cell surface. Again, it was observed that G protein synthesized at 39°C could move to the cell surface once the temperature was lowered. It was concluded, therefore, that these two Group V mutants were defective in maturation of the G protein and/or its movement to the cell surface at 39°C because the failure to achieve complete glycosylation at 39°C may be a result of the inability of the protein to migrate to the proper site in the cell for glycosylation rather than an inability to act as receptor for carbohydrate residues.

All the evidence supports the assignment of the G protein gene to complementation Group V, but study of revertants remains to be done.

F. Complementation Group VI

The Group VI mutants isolated by Rettenmier et al.[9] and characterized by Knipe et al.[13,22] are RNA-negative mutants, one of which (*ts* M601[VI]) resembles *ts* G41(IV). Mutant *ts* M601(VI) possessed an N protein which was degraded rapidly, like *ts* G41(IV), and, in fact, *ts* M601(VI) and *ts* G41(IV) did not complement one another. However, mutant *ts* M602(VI) had an N polypeptide of normal stability, and although no viral RNA synthesis was detected at 39°C, the virion polymerase was not defective in vitro at 39°C.[32a] The Group VI mutant *ts* O82(VI) isolated by Flamand is an RNA-positive mutant. The heterogeneity of the phenotype of the mutants presently classified as Group VI makes any assignment impossible.

In conclusion, the evidence favors the assignment of the L, G, N, and M protein genes of VSV Indiana to complementation Groups, I, V, IV, and III, although several discrepancies remain to be reconciled. A problem which has not been considered so far is the phenotype exhibited by multimutant clones. Both Pringle and Wunner[8,33] and Rettenmier et al.[9] have described clones defined as multiple mutants by their inability to complement two or more complementation groups. The double mutant *ts* M5/6-03(V,VI) showed no defect in RNA synthesis despite the fact that the Massachusetts Group VI mutants have RNA-negative phenotypes. Similarly, Pringle isolated multiple-mutant clones which failed to complement Group I,[8] yet were able to synthesize RNA at 39°C.[46] These observations are difficult to accommodate in a model of mutations located in sequentially, but independently, transcribed genes.

V. GENE ASSIGNMENT IN OTHER RHABDOVIRUSES

One assignment can be made in VSV New Jersey, VSV Cocal, and Chandipura virus.

A. VSV Cocal, Complementation Group Γ

Buller[18] showed that *ts* γ1 behaved rather like mutants of Group V of VSV Indiana. Mutant *ts* γ1 is a thermolabile, RNA-positive mutant, and the thermolability is a property of the envelope rather than the core. The transition from G1 to G2 did not take place in *ts* γ1-infected cells at 39°C. However, G polypeptide synthesized at 39°C became fully glycosylated and incorporated into virions when the temperature of incubation was lowered to 31°C. Therefore, it is probable that complementation group Γ represents the gene coding for the G protein of VSV Cocal.

B. VSV New Jersey, Complementation Group D

Wunner and Pringle[34] observed that the electrophoretic mobility of the N polypeptide of *ts* D1 migrated slightly faster than wild-type. Reversion of the *ts* phenotype was accompanied by restoration of normal N mobility. An alteration of the mobility of the G polypeptide was also observed in this mutant, which was unaffected by reversion of the *ts* phenotype. Therefore, it was concluded that complementation Group D corresponded to the N protein gene of VSV New Jersey.

Mutant *ts* D1 is an RNA-positive mutant,[29] which contrasts with the assignment of the N protein gene to an RNA-negative group (Group IV) in VSV Indiana. The discussion of the in vitro polymerase activity of mutants of VSV New Jersey (Section VI) also suggests that the genome organization of these two viruses may be differet.

C. Chandipura Virus, Complementation Group ChV

Mutant *ts* Ch851, the sole representative of complementation group ChV, resembles the Group III mutants of VSV Indiana. In pulse-chase experiments it has been observed that the M polypeptide is unstable,[47] in contrast to wild-type and mutants representing the other five complementation groups of Chandipura virus. Mutant *ts* Ch851 is an RNA-positive mutant like the Group III mutants of VSV Indiana. Therefore, the M protein gene can be assigned to complementation group ChV of Chandipura virus.

VI. EFFECTS OF MUTATION ON THE VIRION POLYMERASE AND THE PROBLEM OF GENE ASSIGNMENT

A. General Considerations

All the polypeptides required for transcription are contained within the transcribing nucleoprotein (TNP) component,[7] as demonstrated by its infectivity. The TNP is composed of the virion RNA and the three polypeptides L, N, and NS; therefore, it is almost certain that these three polypeptides alone are required for transcription.[35] Figure 1 illustrates the specificity of in vitro transcription directed by the TNP complex.

The precise roles of each of these polypeptides in transcription are as yet unknown. Dissociation and reconstitution experiments have shown that polypeptides L and NS are essential for transcription and almost certainly are part of the transcriptase.[36] The role of polypeptide N in transcription is not clear, however, since RNA transcriptase activity cannot be restored by reconstitution when all three polypeptides are removed. Theoretically,therefore, mutations affecting transcription should be confined to three of the six complementation groups.

In the simplest case, mutation affects the polypeptide which is involved in transcription to make it incapable of synthesizing RNA at 39°C. It is also possible, however, that RNA synthesis does take place at 39°C and that the mutated transcriptase produces nonfunctional RNA. Furthermore, it is likely that polypeptides L, NS, and N take part in more than one stage of RNA synthesis and that mutation at certain sites in such multifunctional polypeptides can affect transcription while mutation at other sites affects other functions, such as replication. It is also possible that a polypeptide which is part of the transcriptase may function normally but yet be incapable of forming a complex with the virion RNA at 39°C and so create a block to secondary, as opposed to primary, transcription. A polypeptide which is part of the transcriptase may also play some part in the assembly of the virion through interaction with the coat proteins. Thus, if the site of interaction is affected, the mutated polypeptide could cause a defect in assembly. On the other hand, it is also possible that a polypeptide which is not part of the transcriptase complex can affect transcription at 39°C by obstruction of the RNA template.

There are two ways of establishing whether mutation has affected the transcriptase. The first is to assay the enzyme at 31° and 39°C using purified virus or virion cores (TNP) and to compare the amounts of RNA synthesized at the two temperatures; the

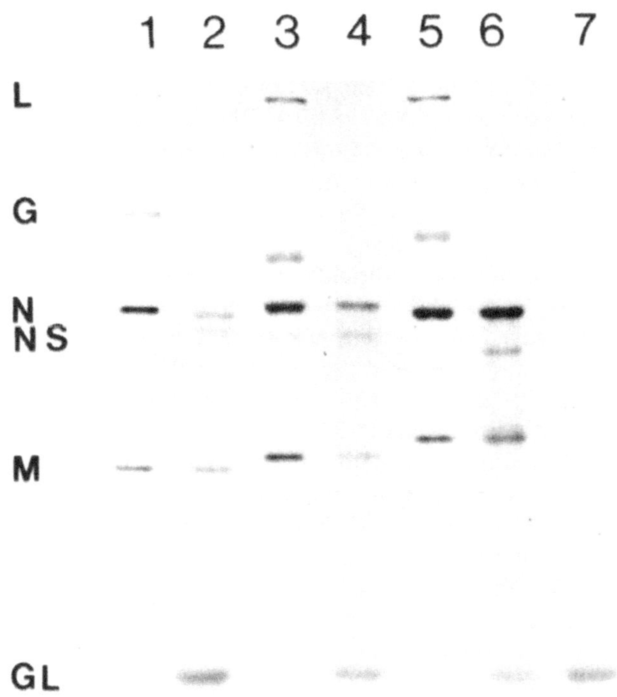

FIGURE 1. Polypeptides synthesized in response to in vitro-produced RNAs by TNP of various rhabdoviruses. Track 1, polypeptides of virions of Chandipura virus; track 2, in vitro-synthesized Chandipura polypeptides in a transcription-translation mixture containing Chandipura TNP; track 3, virion polypeptides of VSV New Jersey; track 4, in vitro-synthesized polypeptides in a transcription-translation mixture containing VSV New Jersey TNP; track 5, virion polypeptides of VSV Indiana; track 6, in vitro-synthesized polypeptides in a transcription-translation mixture containing VSV Indiana TNP; track 7, polypeptides synthesized *in vitro* without the addition of any transcription mixture. (Reproduced from Preston, C. M. and Szilágyi, J. F., *J. Virol.*, 21, 1002, 1977. With permission.)

other is to measure the amount of RNA synthesized in infected cells where protein synthesis and cellular RNA synthesis are inhibited. Both methods determine whether or not the mutant can synthesize RNA at the restrictive temperature.

The advantage of the in vitro assay is its relative simplicity which enables many mutants to be tested. Temperature-shift experiments can distinguish failure to synthesize RNA at 39°C due to conformational change of the polypeptide from irreversible thermal inactivation. The advantage of in vivo assay is that it both confirms the result of the in vitro experiments under physiological conditions and also allows the study of other stages of RNA synthesis, such as replication.

However, interpretation of the results is complicated by the fact that most of the *ts* mutants of VSV Indiana and VSV New Jersey fall into two complementation groups (Groups I and IV in the case of VSV Indiana; Groups A and B in the case of VSV New Jersey), while the other groups are represented by only limited numbers of mutants. This may explain the apparent or real discrepancy in the number of complementation groups which contain RNA-negative mutants, since VSV Indiana contains three such groups (I, III, and IV) and VSV New Jersey four (A, B, E, and F) (see Table 1).

Other mutational effects on the virion polymerase of VSV besides temperature sensitivity are known. Host-restricted mutants with defective polymerase activity have been described,[14,15,37,38] and the best characterized of these — the *td* CE phenotype — will be discussed in more detail. A unique rifampin-sensitive mutant of VSV has been studied by Moreau[39] and was reported to exhibit unusual DNA-like polymerase activity.[40] However, this interesting observation awaits independent confirmation and will not be considered here.

B. Polymerase Function and the *ts* Mutants of VSV Indiana

To determine whether mutation affected virion-associated RNA transcriptase, Szilágyi and Pringle[10] assayed the transcriptase activity of randomly selected mutants at 31° and 39°C in vitro and compared the kinetics of RNA synthesis to those obtained with wild-type virus (Figure 2). The RNA-positive mutants of complementation Groups III and V synthesized RNA at 39°C similarly to wild-type virus, showing that the polypeptides affected by the mutation in these groups were unlikely to be involved in transcription. The two mutants of complementation Group II, mutants *ts* G21 and *ts* G22, both synthesized RNA at 39°C similar to wild-type virus, which indicated that the transcriptase was not affected by the mutation. In the case of the RNA-negative *ts* G21, however, another stage of RNA synthesis, possibly replication, was affected by the mutation. Mutants of the RNA-negative complementation Group IV synthesized RNA at 39°C, and so it was unlikely that the transcriptase of these mutants was affected by this mutation.

On the basis of in vitro RNA synthesis at 39°C, six randomly selected mutants of complementation Group I fell into two categories. Three mutants — *ts* G13(I), *ts* G16(I), and *ts* G114(I) — synthesized only very little RNA at 39°C, and synthesis ceased altogether soon after it started. The mode of inhibition was studied by heat-treating incubation mixtures which contained all the ingredients except $MgCl_2$ and assaying the residual transcriptase activity at 31°C. This technique showed that these three mutants lost their transcriptase activity after incubation periods as short as 15 min at 39°C. Thus, the heat lability of the transcriptase of these mutants was due to the irreversible thermal inactivation of the transcriptase. The heat lability of these mutants could also be shown by loss of infectivity after exposure of virions at 50°C for 15 min. Hunt et al.[11] showed, by dissociation and reconstitution experiments, that the heat labile polypeptide of *ts* G114(I) was polypeptide L.

The other three mutants of complementation Group I — *ts* G11, *ts* G12, and *ts* G15 — synthesized RNA in vitro at 39°C, although the amount of RNA synthesized was somewhat lower than the amount synthesized by wild-type virus.

In these mutants it is possible that either the RNA product was nonfunctional or that polypeptide L was involved in replication as well as transcription, and that mutation affected replication rather than transcription.

C. Polymerase Function and the *ts* Mutants of VSV New Jersey

The *ts* mutants of VSV New Jersey are intrinsically interesting because they comprise six genetically well-defined complementation groups, four of which contain RNA-negative mutants.

Lesnaw and Reichmann[41] studied primary transcription in infected cells using one mutant from each complementation group. They detected primary transcription in two of the RNA-negative mutants, *ts* A1 and *ts* E1, as well as in the two RNA-positive mutants, *ts* C1 and *ts* D1. However, primary transcription was not observed in the RNA-negative mutants *ts* B1 and *ts* F1.

They concluded that mutation in complementation groups B and F affected polypep-

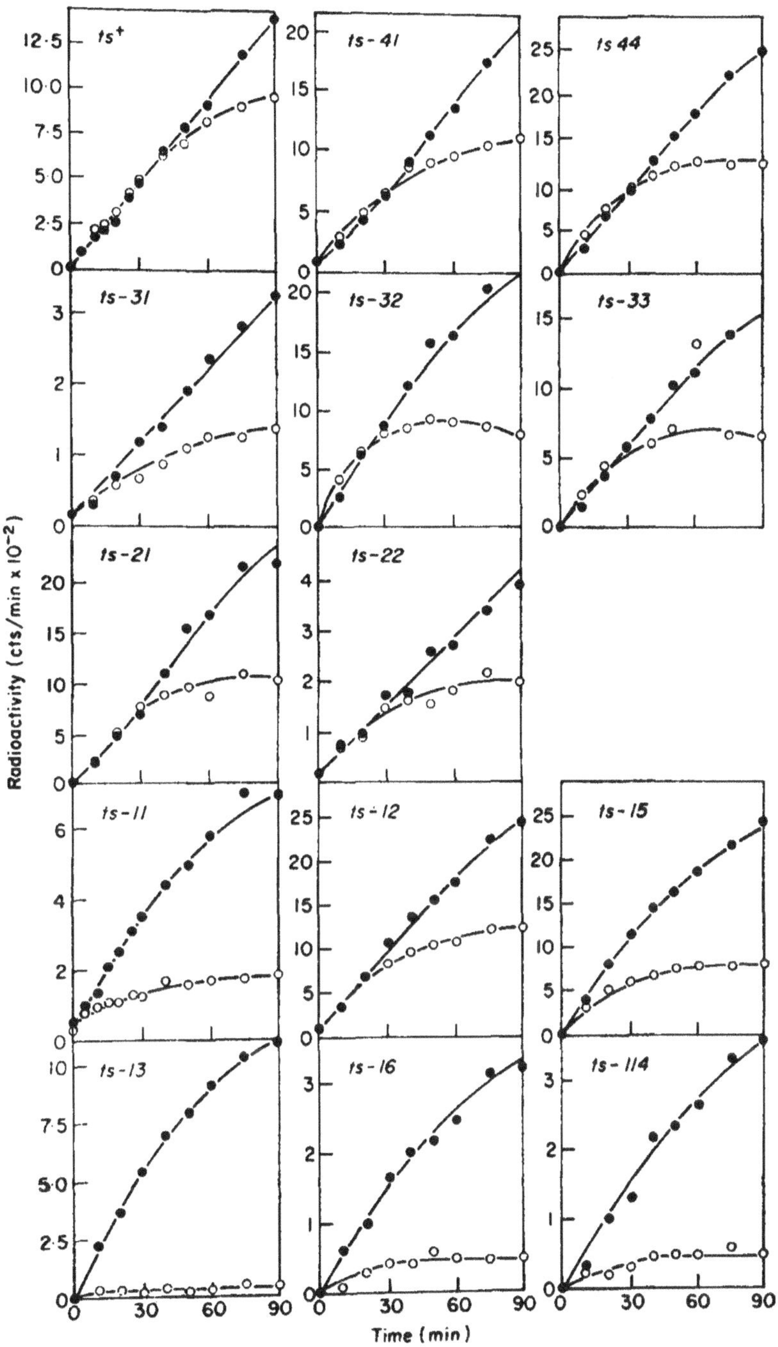

FIGURE 2. In vitro RNA synthesis at 31°C (—•—•—) and at 39°C (—O—O—) by *ts* mutants of VSV Indiana. (Reproduced from Szilágyi, J. F. and Pringle, C. R., *J. Mol. Biol.*, 71, 281, 1972. With permission.)

tides involved in transcription: group B was probably similar to Group I of VSV Indiana, but group F had no counterpart. The mutants of the other RNA-negative complementation groups, A and E, were defective in replication rather than transcription and were presumed to be comparable to Groups IV and II, respectively. Since RNA synthesis in the RNA-positive complementation groups appeared to be normal, they presumed that they were comparable to Groups III and V of VSV Indiana.

A survey of the *ts* mutants of VSV New Jersey has been undertaken[45] to determine the effect of mutation on in vitro transcription. The RNA-positive mutants of groups C and D synthesized complementary RNA in vitro similar to wild-type virus. These results confirm that these mutants are not defective in any stages of RNA synthesis (see Table 1) despite the fact that Group D has been equated with the N protein gene (see Section V.B).

In vitro RNA synthesis at 39°C by the RNA-negative *ts* A1 was even more active than by wild-type virus. It is additionally unlikely that transcription was affected since primary transcription was observed also in vivo.[41] The mutation presumably affected replication, and in this respect, mutant *ts* A1 has similarities to the Group IV mutants of VSV Indiana. On this basis, the genetic defect should reside in polypeptide N.

All three mutants comprising complementation group E were investigated. Mutant *ts* E1 synthesized very little RNA at 39°C, and the inability of this mutant to synthesize RNA at 39°C was due to irreversible thermal inactivation of transcriptase activity. This conflicts with the observation of Lesnaw and Reichmann that primary transcription occurred in vivo at 39°C. However, the other two mutants (*ts* E2 and *ts* E3) did synthesize RNA in vitro at 39°C in amounts comparable to those synthesized by wild-type virus. The transcriptase activity of both these mutants was heat stable.

Therefore, there are differences in the mode of RNA synthesis at 39°C by the Group E mutants. Firstly, in the case of *ts* E1, transcription was inhibited in vitro. Secondly, mutant *ts* E3 transcribed complementary RNA normally but failed to replicate in vivo. Thirdly, both transcription and replication were observed with *ts* E2. Consequently, in Group E the mutated polypeptide appears to play a multifunctional role, because it appears to take part both in transcription and replication and presumably also, in the case of *ts* E2, in virus maturation. Group E is unique in that there are three functions affected by the mutation; consequently, it cannot be equated with any one complementation group of VSV Indiana. Group II may be comparable to Group E, since one mutant (*ts* G22) resembled *ts* E3 and the other (*ts* G21) resembled *ts* E2. Since there are only two mutants in Group II, it is possible that mutants with phenotypes similar to *ts* E1 remain to be isolated. Dissociation and reconstitution experiments might identify the mutated polypeptide.

In Group F, one of the two mutants (*ts* F1) synthesized very little RNA at 39°C, while the other (*ts* F2) synthesized RNA in amounts comparable to wild-type virus. The inability of *ts* F1 to synthesize RNA at 39°C was due to irreversible thermal inactivation. Thus, in this group also, the polypeptide affected by mutation appeared to be involved both in transcription and in replication and to have a multifunctional role. Since mutants of Groups E and F complement each other, it is unlikely that the same polypeptide is involved. Complementation Group F does not appear to have a counterpart in VSV Indiana.

Mutant *ts* B1 from the RNA-negative Group B synthesized RNA at 39°C only to a very limited extent. In this respect, it was similar to *ts* E1 and *ts* F1, but temperature-shift experiments showed that unlike these mutants the transcriptase activity of *ts* B1 was immediately and fully restored when the temperature of incubation was lowered from 39° to 31°C.

Thus, the inability of this mutant to synthesize RNA at 39°C was due presumably to a conformational change of the mutated polypeptide and not to its thermal inactivation. In this respect, *ts* B1 resembled the temperature-dependent host range mutant *td* CE3 (see Section VI, E). Since the mutation in mutant *td* CE3 is known to affect polypeptide L, it is possible that this polypeptide is affected in mutant *ts* B1 also.

Serotype	Complementation Groups	Phenotype		
		Transcription Negative Replication Negative	Transcription Positive Replication Negative	Transcription Positive Replication Positive
VSV Indiana	IV		*ts* G41	
	I	*ts* G13 (heat labile)	*ts* G11	
	II		*ts* G22	*ts* G21
VSV New Jersey	E	*ts* E1 (heat labile)	*ts* E3	*ts* E2
	F	*ts* F1 (heat labile)	*ts* F2	
	B	*ts* B1 (stable)		
	A		*ts* A1	

FIGURE 3. A diagrammatic representation of the phenotypes of mutants of VSV in complementation groups containing mutants with defects in RNA synthesis.

Furthermore, since Group B is the largest VSV New Jersey complementation group with 21 mutants, it is possible that it corresponds to Group I of VSV Indiana where mutation also affected polypeptide L.

D. Summary and Conclusions

The data obtained from study of *ts* mutants of both serotypes of VSV are summarized in Figure 3. The general conclusions which can be drawn are the following:

1. Mutation affected transcription in three complementation groups of VSV New Jersey (Groups B, E, and F) and only in one (Group I) in VSV Indiana. Since three polypeptides — namely L, N, and NS — are required for transcription, presumably these are the polypeptides affected by the mutation in Groups B, E, and F of VSV New Jersey. Precise assignment must await dissociation and reconstitution experiments. Provisionally we suggest that Group B represents the L protein gene, while Groups E and F correspond to the genes for the two other core polypeptides N and NS. Groups E and F cannot be equated directly with any of the Indiana groups.
2. The temperature sensitivity of the transcriptase activity appeared to be of two kinds: an irreversible thermal inactivation as in the case of some of the mutants of Group I of VSV Indiana and Groups E and F of VSV New Jersey or a reversible change in conformation (Group B of VSV New Jersey).

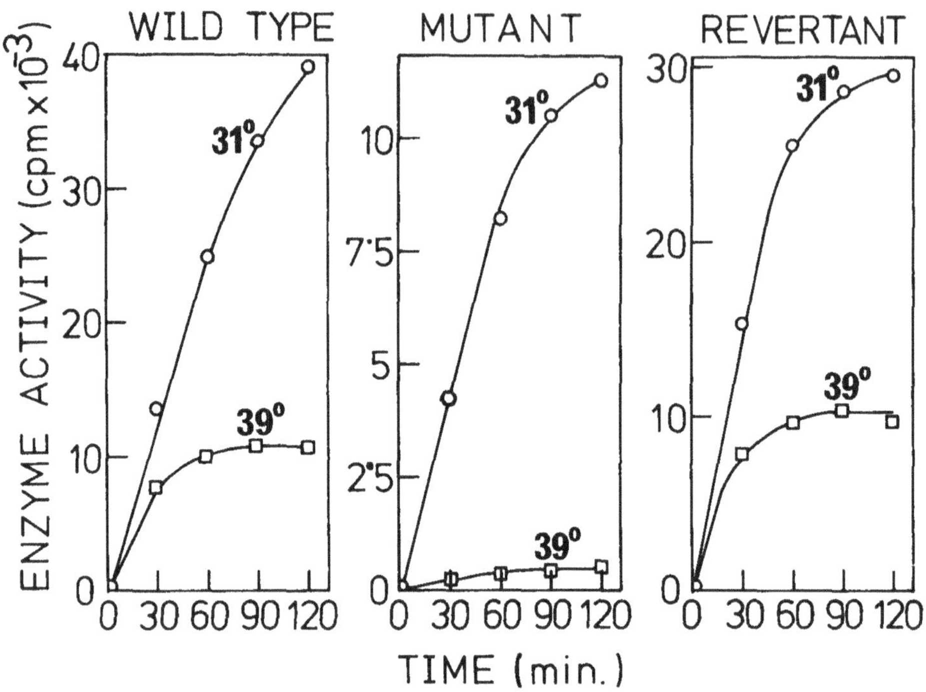

FIGURE 4. In vitro RNA synthesis by mutant *td*CE3, a revertant, and wild-type VSV New Jersey at 31°C (—O—O—) and at 39°C (—□—□—). A, RNA synthesis by wild-type; B, RNA synthesis by mutant *td*CE3; C, RNA synthesis by revertant line *td*CE/R1 isolated from *td*CE3. (Reproduced from Szilágyi, J. F. and Pringle, C. R., *J. Virol.*, 16, 927, 1975. With permission.)

3. The site of the mutational lesion in Groups E and F of VSV New Jersey appeared to be polypeptides with multifunctional roles involved both in transcription and replication, and in the case of Group E, possibly even in assembly. These mutants provide clear evidence for the interdependence of transcription and replication.
4. Group A of VSV New Jersey resembled Group IV of VSV Indiana. The evidence outlined in Section IV, D suggested that polypeptide N was affected in Group IV. However, since we have argued that polypeptides L, N, and NS are accounted for by Groups B, E, and F, it is possible that Group A represents the gene coding for the presumptive sixth polypeptide.

E. The *td*CE Phenotype

The temperature-dependent host range (*td*CE) mutants of VSV[14,44] are characterized by their ability to grow in BHK cells both at 31° and 39°C and only at 31°C in secondary chick embryo cells. Two of these *td*CE mutants, *td*CE1 and *td*CE2, synthesized RNA normally at the restrictive temperature, while the other two, *td*CE3 and *td*CE4, synthesized only very small amounts. A spontaneous revertant of *td*CE3 was isolated which grew normally in chick embryo cells at 39°C. This revertant was able to synthesize RNA in vitro at 39°C similarly to wild-type virus; therefore, it was concluded that the virion transcriptase was defective in mutant *td*CE3 (Figure 4).

If the complete in vitro incubation mixture was incubated at 39°C for 30 min or longer and the temperature was then lowered to 31°C, transcription was resumed immediately (Figure 5). This indicated that the inability of this mutant to synthesize RNA at the restrictive temperature was due presumably to a change in the conformation of the mutated polypeptide. In this respect, the mode of inhibition of transcription is similar to that found in the conventional *ts* mutant of VSV New Jersey *ts*B1.

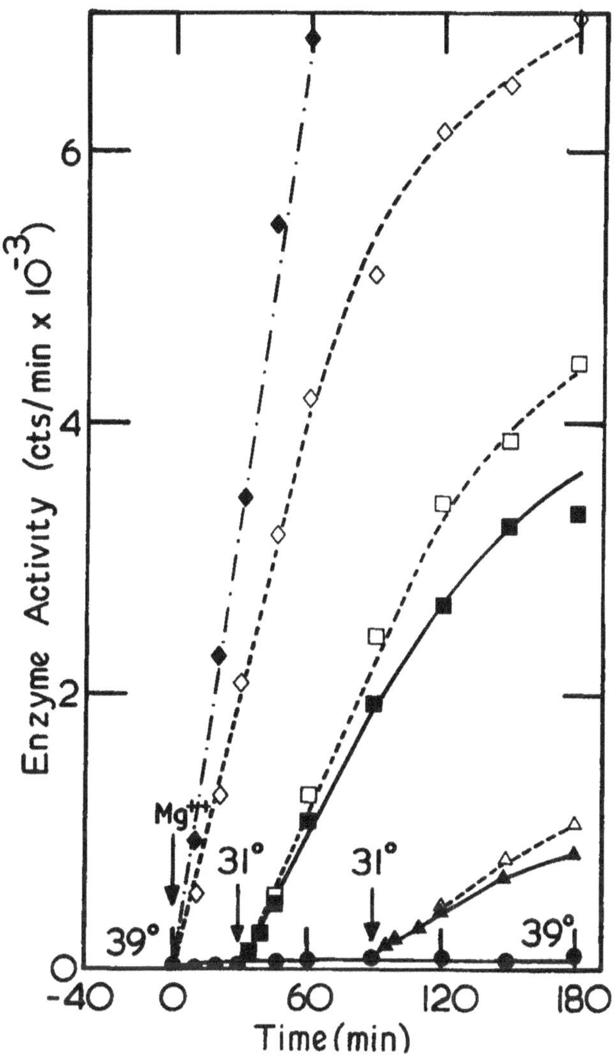

FIGURE 5. Effect of temperature shift from 39°C on the transcriptase activity of the mutant *td* CE3. Three reaction mixtures were preincubated at 39°C for 40 min, then $MgCl_2$ was added. The first one was incubated at 39°C for a further 30 min (■); the second was incubated for a further 90 min (▲) and then downshifted to 31°C; the third was incubated at 39°C for the entire 180 min of the experiment (•). As controls, identical reaction mixtures were preincubated for 1 min (♦), 40 min (◇), 70 min (□), or 130 min (△) at 39°C. $MgCl_2$ was added and residual transcriptase activity was assayed at 31°C. (Reproduced from Szilágyi, J. F. and Pringle, C. R., *J. Virol.*, 16, 927, 1975. With permission.)

The temperature sensitivity of the transcriptase of *td* CE3 was the same regardless of whether the mutant was grown in BHK or CE cells. Furthermore, the transcribing nucleoprotein complex (TNP) prepared from this virus also showed temperature sensitivity in vitro. It was concluded, therefore, that the polypeptide affected by the mutation was either L, N, or NS. Dissociation and reconstitution experiments (Figure 6) established that L, and not N or NS, was the site of the mutational lesion.[15]

To interpret these observations, we postulated the possible interaction of a host fac-

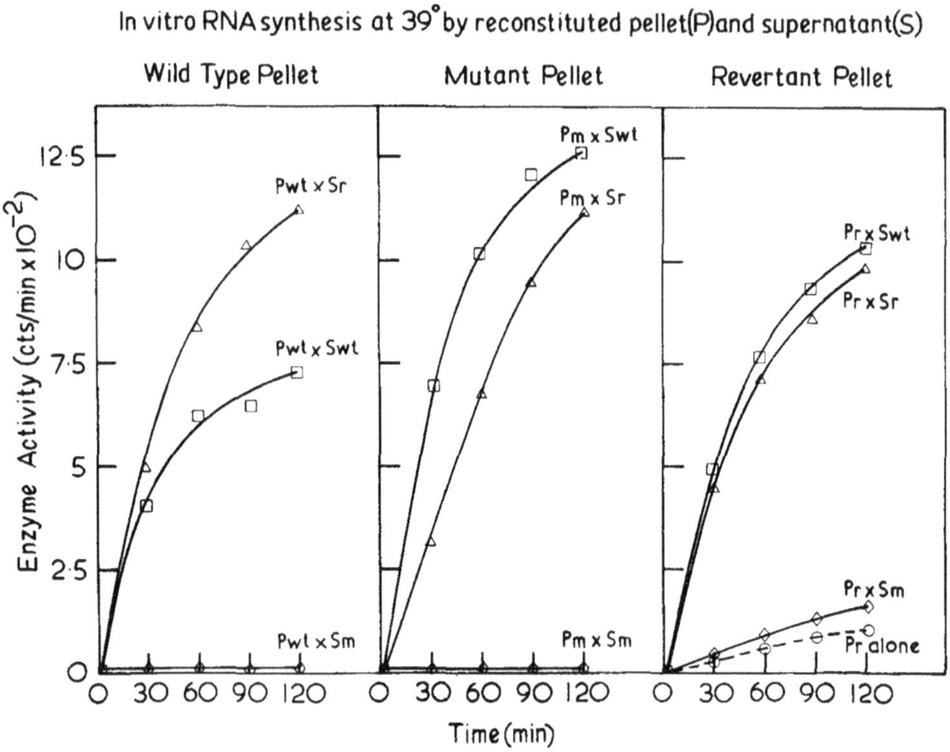

FIGURE 6. In vitro RNA synthesis at 39°C after reconstitution of 2.0 *M* LiCl pellets with 0.9 M LiCl supernatants. The 2.0 *M* LiCl pellets of wild-type, mutant *td* CE3, and revertant *td* CE/R1 were reconstituted with the 0.9 M LiCl supernatants of wild type (□), *td* CE3 (◊), and *td* CE/R1 (Δ), and the RNA transcriptase activity was assayed at 39°C. There was no detectable activity associated with any of the 2.0 *M* LiCl pellets or the 0.9 *M* LiCl supernatants, with the exception of the revertant pellet (0———0). Only the supernatants contain the L polypeptide. (Reproduced from Szilágyi, J. F., Pringle, C. R., and Macpherson, T. M., *J. Virol.*, 22, 381, 1977. With permission.)

tor with polypeptide L, which in some way enabled the transcriptase to overcome its heat sensitivity in BHK cells. We presumed that such a factor was either absent in CE cells or did not function.

None of the *td* CE mutants complemented conventional *ts* mutants of Groups A, B, E, and F, or each other; therefore, at present they do not provide information which facilitates assignment of the L protein gene to any complementation group.

F. Postscript

The overlaps in gene assignment and the ambiguity concerning the number of complementation groups may be a consequence of the small numbers of mutants in the minority groups and a confusion of intragenic and intergenic complementation. For instance, the complementation between Group D of VSV New Jersey (an N gene assignment, according to Section V.B) and Groups E and F (N or NS assignments, according to Section VI.D) might in one case be intragenic complementation. Indeed, Group D complemented poorly with Groups B, E, and F and Group B poorly with Group F.[29] Since Group D is represented by a single mutant, this question is difficult to resolve.

In a study of complementation between *ts* mutants of Chandipura virus,[42,43] Gadkari and Pringle have observed reproducible complementation between several individual pairs of mutants within the majority RNA-negative group ChI which contains 44 of

TABLE 2

Chandipura *ts* Mutants: Mean Complementation Indices and Ranges

Complementation Group (Number of Mutants)

	Ch I		Ch II	Ch III	Ch IV	Ch V	CH VI
	A	B					
	(18)	(10)	(2)	(1)	(1)	(1)	(1)
ts ChI	1.1	5.6	155	27	142	345	48
	(0.5—2.2)	(3.3—15)	(142—167)				
All subgroup	1.9	14.8	134	104	171	336	99
IA mutants	(0.3—17)	(0.1—375)	(10—1000)	(23—400)	(10—1000)	(33—1000)	(6—483)
All subgroup		5.9	143	61	54	312	15
IB mutants		(0.2—27)	(5—500)	(8—150)	(15—210)	(50—1500)	(6—41)

Note: The values in the table are mean complementation indexes for all combination of mutants, with the ranges in parentheses.

the 50 mutants (Table 1). Indeed, group ChI can be subdivided into two subgroups using *ts* Ch1 as the prototype mutant[42] for the group (Table 2). However, if only a few mutants from subgroup B had been available, it would have been necessary to create at least a seventh group. If complementation Group ChI corresponds to Group I of VSV Indiana, the extensive intragenic complementation suggests that the functional L protein is a multimeric structure.

REFERENCES

1. **Sanger R., Air, G. M., Barrell, B. G., Brown, N. L., Coulson, A. R., Fiddes, J. C., Hutchison, C. A., III, Slocombe, P. M., and Smith, H.,** Nucleotide sequence of bacteriophage ϕX 174 DNA, *Nature (London)*, 265, 687, 1977.
2. **Contreras, R., Rogiers, R., Van de Vourde, A., and Fiers, W.,** Overlapping of the VP_2-VP_3 gene and the VP_1 gene in the SV40 genome, *Cell*, 12, 529, 1977.
3. **Weiner, A. M. and Newbold, J. E.,** Phage QB-natural read-through at the UGA termination signal of the coat protein cistron, *Nature New Biol.*, 234, 26, 1971.
4. **Pringle, C. R.,** Genetics of Rhabdoviruses, in *Comprehensive Virology*, Vol. 9, Fraenkel-Conrat, H. and Wagner, R. R., Eds., Plenum Press, New York, 1977, 239.
5. **Banarjee, A. K., Abraham, G., and Colonno, R. J.,** Vesicular stomatitis virus: mode of transcription, *J. Gen. Virol.*, 34, 1, 1977.
6. **Knudsen, D. L.,** Rhabdoviruses, *J. Gen. Virol.*, (Symp. Suppl.), 20, 105, 1973.
7. **Szilágyi, J. F. and Uryvayev, L.,** Isolation of an infectious ribonucleoprotein from vesicular stomatitis virus containing an active RNA transcriptase, *J. Virol.*, 11, 279, 1973.
8. **Pringle, C. R.,** Conditional lethal mutants of vesicular stomatitis virus, *Curr. Top. Microbiol. Immunobiol.*, 69, 85, 1975.
9. **Rettenmier, C. W., Dumont, R., and Baltimore, D.,** Screening procedure for complementation dependent mutants of vesicular stomatitis virus, *J. Virol.*, 15, 41, 1975.
10. **Szilágyi, J. F. and Pringle, C. R.,** Effect of temperature-sensitive mutations on the virion-associated RNA transcriptase of vesicular stomatitis virus *J. Mol. Biol.*, 71, 281, 1972.
11. **Hunt, D. M., Emerson, S. U., and Wagner, R. R.,** RNA^- temperature-sensitive mutants of vesicular stomatitis virus: L protein thermosensitivity accounts for transcriptase restriction of group I mutants, *J. Virol.*, 18, 596, 1976.

12. **Repik, P., Flamand, A., and Bishop, D. H. L.**, Synthesis of RNA by mutants of vesicular stomatitis virus (Indiana serotype) and the ability of wild type New Jersey to complement the VSV Indiana *ts* G114 transcription defect, *J. Virol.*, 20, 157, 1976.
13. **Knipe, D., Lodish, H. F., and Baltimore, D.**, Analysis of the defects of temperature-sensitive mutants of vesicular stomatitis virus: intracellular degradation of specific viral proteins, *J. Virol.*, 21, 1140, 1977.
14. **Szilágyi, J. F. and Pringle, C. R.**, Virion transcriptase activity differences in host range mutants of vesicular stomatitis virus, *J. Virol.*, 16, 927, 1975.
15. **Szilágyi, J. F., Pringle, C. R., and Macpherson, T. M.**, Temperature-dependent host range mutation in vesicular stomatitis virus affecting polypeptide L, *J. Virol.*, 22, 381, 1977.
16. **Combard, A., Printz-Ané, C., Martinet, C., and Printz, P.**, Temperature-sensitive defect of vesicular stomatitis virus in complementation group II, *J. Virol.*, 21, 913, 1977.
17. **Lafay, F.**, Envelope proteins of vesicular stomatitis virus: effect of temperature-sensitive mutation in complementation groups III and V, *J. Virol.*, 14, 1220, 1974.
18. **Buller, R. M. L.**, Biological and Biochemical Characterization of Vesicular Stomatitis Virus and Temperature-Sensitive Maturation Mutants, Ph.D. thesis, University of Glasgow, 1975.
19. **Flamand, A. and Bishop, D. H. L.**, Primary in vivo transcription of vesicular stomatitis virus and temperature-sensitive mutants of five vesicular stomatitis virus complementation groups, *J. Virol.*, 12, 1238, 1973.
20. **Holloway, A. F., Wong, P. K. Y., and Cormack, D. V.**, Isolation and characterization of temperature-sensitive mutants of vesicular stomatitis virus, *Virology*, 42, 917, 1970.
21. **Ngan, J. S. C., Holloway, A. F., and Cormack, D. V.**, Temperature-sensitive mutants of vesicular stomatitis virus: comparison of the *in vitro* RNA polymerase defects of group I and group IV mutants, *J. Virol.*, 14, 765, 1974.
22. **Knipe, D. M., Baltimore, D., and Lodish, H. F.**, Maturation of viral proteins in cells infected with temperature-sensitive mutants of vesicular stomatitis virus, *J. Virol.*, 21, 1149, 1977.
23. **Little, S. P. and Huang, A. S.**, Synthesis and distribution of vesicular stomatitis virus — specific polypeptides in the absence of progeny production, *Virology*, 81, 37, 1977.
24. **Freeman, G. J., Rao, D. O., and Huang, A. S.**, Genome organization of vesicular stomatitis virus: mapping of *ts* G41 and the defective interfering T particle, in *Negative Strand Viruses and the Host Cell*, Mahy, B. W. J. and Barry, R. D., Eds., Academic Press, London, 261—270, 1978.
25. **Freeman, G. J., Rose, J. K., Clinton, G. M., and Huang, A. S.**, RNA synthesis of vesicular stomatitis virus. VII. Complete separation of the mRNA's of vesicular stomatitis virus by duplex formation, *J. Virol.*, 21, 1094, 1977.
26. **Printz, P. and Wagner, R. R.**, Temperature-sensitive mutants of vesicular stomatitis virus: synthesis of virus-specific proteins, *J. Virol.*, 7, 651, 1971.
27. **Wunner, W. H. and Pringle, C. R.**, Protein synthesis in BHK-21 cells infected with vesicular stomatitis virus. I. *Ts* mutants of the Indiana serotype, *Virology*, 48, 104, 1972.
28. **Flamand, A.**, Etude génétique du virus de la stomatite vésiculaire: classement de mutants thermosensibles spontanés en groupes de complementation, *J. Gen. Virol.*, 8, 187, 1970.
29. **Pringle, C. R., Duncan, I. B., and Stevenson, M.**, Isolation and characterization of temperature-sensitive mutants of vesicular stomatitis virus, New Jersey serotype, *J. Virol.*, 8, 836, 1971.
30. **Deutsch, V. and Berkaloff, A.**, Analyse d'un mutant thermolabile du virus de la stomatite vésiculaire (VSV), *Ann. Inst. Pasteur Paris*, 121, 101, 1971.
31. **Zaváda, J.**, Pseudotypes of vesicular stomatitis virus with the coat of murine leukemia and of avian myeloblastosis viruses, *J. Gen. Virol*, 15, 183, 1972.
32. **Knipe, D. M., Lodish, H. F., and Baltimore, D.**, Localization of two cellular forms of the vesicular stomatitis viral glycoprotein, *J. Virol.*, 21, 1121, 1977.
32a. **Szilágyi, J. F. and Pringle, C. R.**, unpublished data.
33. **Pringle, C. R. and Wunner, W. H.**, A comparative study of the structure and function of the VSV genome, in *Negative Strand Viruses*, Vol. 2, Mahy, B. W. J. and Barry, R. D., Eds., Academic Press, London, 1975, 707.
34. **Wunner, W. H. and Pringle, C. R.**, A temperature-sensitive mutant of vesicular stomatitis virus with two abnormal virion proteins, *J. Gen. Virol.*, 23, 97, 1974.
35. **Preston, C. M. and Szilagyi, J. F.**, Cell-free translation of RNA synthesized *in vitro* by a transcribing nucleoprotein complex prepared from purified vesicular stomatitis virus, *J. Virol.*, 21, 1002, 1977.
36. **Emerson, S. U. and Yu, Y. H.**, Both NS and L proteins are required for *in vitro* RNA synthesis by vesicular stomatitis virus, *J. Virol.*, 15, 1348, 1975.
37. **Simpson, R. W. and Obijeski, J. F.**, Conditional lethal mutants of vesicular stomatitis virus. I. Phenotypic characterization of single and double mutants exhibiting host restriction and temperature sensitivity, *Virology*, 57, 357, 1974.

38. **Obijeski, J. F. and Simpson, R. W.**, Conditional lethal mutants of vesicular stomatitis virus. II. Synthesis of virus-specific polypeptides in nonpermissive cells infected with "RNA⁻" host-restricted mutants, *Virology*, 57, 369, 1974.
39. **Moreau, M-C.**, Inhibition of a vesicular stomatitis virus mutant by rifampin, *J. Virol.*, 14, 517, 1974.
40. **Moreau, M-C., Sanzey, B., and Chany, Ch.**, In vitro transcription of a vesicular stomatitis virus mutant, *Inserm Colloque*, 47, 103, 1975.
41. **Lesnaw, J. A. and Reichmann, M. E.**, RNA synthesis by temperature-sensitive mutants of vesicular stomatitis virus, New Jersey serotype, *Virology*, 63, 492, 1975.
42. **Pringle, C. R. and Wunner, W. H.**, Genetic and physiological properties of temperature-sensitive mutants of Cocal virus, *J. Gen. Virol.*, 12, 677, 1973.
43. **Gadkari, D. A. and Pringle, C. R.**, in preparation.
44. **Pringle, C. R.**, The *td*CE and *hr*CE phenotypes: host range mutants of vesicular stomatitis virus in which polymerase function is affected, *Cell*, 15, 597, 1978.
45. **Szilágyi, J. F. and Pringle, C. R.**, Effect of temperature-sensitive mutation on activity of the RNA transcriptase of vesicular stomatitis virus New Jersey, *J. Virol.*, 30, 692—700, 1979.
46. **Pringle, C. R., Duncan, I. B. and Szilágyi, J. F.**, unpublished observations.
47. **Gadkari, D. A. and Pringle, C. R.**, in preparation.

Chapter 9

HOMOLOGOUS INTERFERENCE BY DEFECTIVE VIRUS PARTICLES

G. P. Faulkner and Robert A. Lazzarini

TABLE OF CONTENTS

I. INTRODUCTION

Standard virus repeatedly passaged at high multiplicity produces defective particles which interere with viral replication. These defective or "incomplete" particles were first discovered by von Magnus[1,2] using influenza virus-infected embryonated eggs. In the intervening years since that discovery, examples of defective interfering (DI) particles and homologous interference have been observed in almost every class of animal virus. The diversity of the molecular biological strategies[3] represented among the different classes of viruses suggests that the mechanisms by which the different DI particles interfere are also diverse. An analysis of the comparative biology of DI particles and a provocative discussion of their possible role in limiting natural viral infections can be found in three comprehensive reviews.[4-6] The present review has a much more limited scope and focuses on the structure of DI particles and the parameters affecting the interference phenomenon as it occurs in rhabdovirus infections. The closely related topics of the genesis of the DI particles and the genetic content of the DI genome as well as the transcriptive abilities of DI particles are extensively discussed in Chapters 10, 11, and 12 of this volume. Although dealing with interference in all rhabdovirus infections, this chapter out of necessity will draw almost exclusively from the vesicular stomatitis virus (VSV) literature. To our knowledge, DI particles and the interference phenomenon have been investigated in only two other rhabdovirus systems: rabies[7-14] and infectious hematopoietic necrosis virus.[15]

II. THE NATURE OF INTERFERING PARTICLES

Viral interference due to coinfection with a homologous or closely related virus can be subdivided into at least three types depending upon the nature of the interfering virus particle. These are interference by unconditional defective (DI) particles, conditional defective particles (temperature-sensitive mutants),[16] and nondefective heterotypic viruses.[17] As yet, there is no clear indication whether the same or different mechanisms are involved in these three types of interference. Although this chapter will deal only with interference by DI particles, many aspects of the discussion may apply equally to interference elicited by the other two types of particles.

The unconditional defective particles that have been characterized are all smaller than the standard virus and do not contain all of the genetic information present in the standard virus.[18-27] They have been termed deletion mutants, but in a strict sense, they may not be simple deletion mutants since many of the DI particles examined contain base sequences not present in the parent virus genome (see below). Full-length unconditional defective particles (point mutations) have not been observed during high multiplicity passage although the spontaneous mutation rate of VSV is high and these particles are likely to be produced at substantial rates. The absence of these particles may be more apparent than real: full-length interfering particles are difficult to identify in a population that may include DI particles as well as standard virus. On the other hand, their absence may be real and signify that this type of particle lacks some important characteristic that is necessary for its rapid accumulation and enrichment during repeated passage.

The physical characterization of DI particles that accumulate during high multiplicity passage is complicated because frequently complex mixtures of DI particles are generated in this method.[12,21,26,28,29] Nonetheless, at least 12 to 15 different preparations of VSV DI particles have been obtained which are sufficiently enriched in one or two DI particles that their characterization has been possible. As yet, it has not been possible to clone rhabdovirus DI particles, although a promising method has been developed for cloning reovirus DI particles.[30]

Early electron micrographic studies[31,32] of the morphology of VSV DI particles showed that they had many, if not all, of the structural components of the standard virus. They were truncated bullet forms that were enclosed by a membranous envelope which was covered with a fringe of protein projections. When the phosphotungstic acid penetrated sufficiently, a helically wound nucleocapsid was revealed similar to that seen in electron micrographs of the standard particles.[33,34] Early serologic studies by Huang et al.[32] showed that the surface antigens of the two particles were closely related and antisera directed against the DI particles were effective in neutralizing VSV. Even the relative amounts of the individual viral proteins were similar for the DI and standard particles when estimated by very sensitive double label sodium dodecyl sulfate (SDS) polyacrylamide gel electrophoresis.[35] The example shown in Figure 1 is the coelectrophoresis of ^{14}C-labeled DI-011 particle proteins with ^{3}H-labeled vesicular stomatitis (VS) virion proteins.[36]

Analysis of the particle RNA revealed the first major difference between the DI and infectious virus particles. The DI particle contained RNA that is smaller than that found in the infectious particle.[18] Subsequent comparisons showed that the lengths of the infectious VSV and the DI particles were roughly proportional to the size of the RNA they contained.[20]

Using viral mRNA, Leamnson and Reichmann[24] as well as Stamminger and Lazzarini[25] probed the structure of DI RNA by hybridization analyses. Several important conclusions came out of these studies. First, it was shown that DI particles contain viral, not host, RNA and, therefore, are not psuedovirions. Second, the RNA of the DI particles corresponded to discrete regions of the VSV genome and were not simply a random assortment of RNAs of the same size. Third, most DI particles contained RNA that corresponded to the 5′ end of the VSV genome and did not contain a complete viral gene. Consequently, it appears that interference does not depend upon the introduction of some viral gene by the DI particle. Although most of the DIs contain information from the 5′ end of the VSV genome, a large DI particle obtained from the heat-resistant strain of VSV contained information from the 3′ end. At first this type of DI was thought to be very unusual, but now there are two additional 3′ end DI particles. The first was derived by Adler and Banerjee;[37] the second was derived by Reichmann and Schnitzlein[38] using the thermal labile VSV mutant O/45.

Several lines of evidence suggest that the genomes of DI particles are not merely fragments of the parent genome. Attempts to form biologically active DI genomes by sonicating and fragmenting the VSV genome isolated from the standard virus were unsuccessful.[12] Furthermore, if DI particles were formed by random fragmentation of the standard virus genome, one might expect that the terminal nucleotides of at least some of the DI RNAs would be different from those found on the parental virus genome. Yet, analysis of both the 3′ and 5′ terminal trinucleotide sequences of three different DI particles revealed that they are all identical to those found in the standard VSV genome.[39-43]

Most DI particles are not simple deletion mutants since their genomes contain new sequences not present in the parent virus. The presence of these sequences was first observed in the most extreme case — DI-011, the prototype of a DI particle whose RNA contains covalently linked complementary sequences.[26] At least 80% of the sequences in the RNA obtained from this particle (approximately 2000 bases) are self-complementary. The RNA from infectious virus does not contain these complementary sequences, and even after self-annealing, only 0.3% of the RNA is resistant to digestion by ribonuclease and, therefore, is possibly in a duplex structure. Other DI particles that contain extensive stretches of covalently linked complementary sequences have been observed subsequently.[44,45]

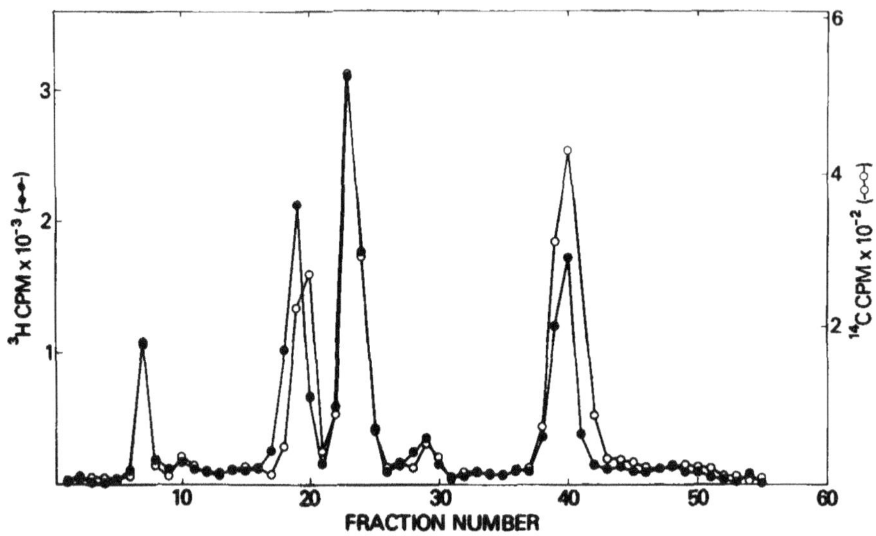

FIGURE 1. Coelectrophoresis of ^{14}C-DI$_{011}$ and ^{3}H-VSV proteins in 10% polyacrylamide gels containing sodium dodecylsulfate.

The complementary sequences are not always as apparent as they are in those extreme examples. Three species of DI particles (DI-111, DI-411, DI-611) isolated in parallel with DI-011 contained RNAs that showed respectively 15, 6, and 3% self-complementarity.[26,46] Although these DIs were obtained from the same strain of VS (IND), other isolates of the Indiana and New Jersey serotypes of VSV also give rise to DIs containing self-complementary RNA.[45,47] Kolakofsky working with Sendai virus[48] and Perrault and Leavitt[45,49] working with VSV have presented electron micrographic evidence that the complementary sequences are located at the ends of the DI RNA. When visualized in the electron microscope, the RNA was seen to be circular with a short, double-stranded RNA panhandle. By virtue of their resistance to ribonuclease in high ionic strength buffers, these panhandles can be freed of the single-stranded RNA attached to them and purified. The panhandles that have been examined range from 60 to 180 nucleotides in length.[49,50]

RNA from the standard VSV virus does not snap back or circularize. Furthermore, in the case of Sendai RNA all attempts to demonstrate complementarity between the ends of the virion RNA have failed.[48,50] Both Kolakofsky and colleagues[50,51] and Perrault and colleagues[52] have presented evidence that the sequences of the 5′ end of the virion RNA are conserved in the DI RNA and are present at its 5′ end. Consequently, since the DI has complementary ends and the virion RNA does not, the 3′ end of the DI RNA must be different from that of the virion RNA. Corroborative evidence for the presence of sequences near the 3′ end of the DI RNA that are not found in VSV RNA has been obtained from the analysis of oligonucleotide fragments generated with ribonuclease T_1.[53,54] However, it has not been possible to obtain this type of corroborative evidence with all DI RNAs. Kang and colleagues[55] have been unable to clearly identify any oligonucleotide in digests of DI RNAs that is not present in digests of VSV RNA.

A model for the genesis of DI particles that contain complementary sequences has been proposed by Kolakofsky and colleagues[50,51] and by Huang and colleagues.[53,54] This model is particularly appealing because it explains a number of seemingly unrelated observations. The model (Figure 2) proposes that the formation of the progenitor

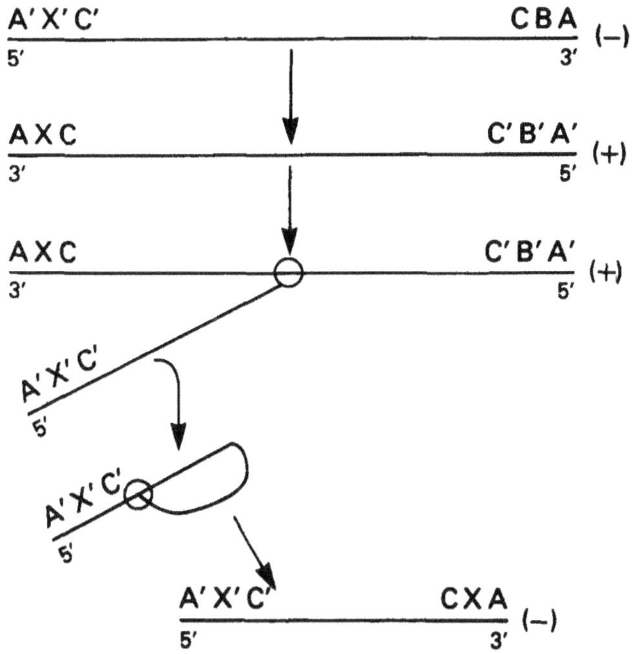

FIGURE 2. Model for the genesis of DI RNPs during the replication of VSV (after Kolokafsky[50,51] and Huang.[53,54] The letters A, B, C, and X represent nucleotide sequences whose complements are A′, B′, C′, and X′. The open circle represents the viral replicase. The parent minus strand RNP, which has short complementary sequences at its 3′ and 5′ termini (top line), serves as a template for the full-length plus strand RNP (second line). The latter, in turn, serves as a template for progeny minus strands. Occasionally during this event (third line), the replicase detaches from the template with its daughter (minus) strand still attached. If the replicase resumes synthesis on its own daughter strand, the product minus strand will have termini that are complementary over a longer length than the parental RNP.

DI ribonucleoprotein particle (RNP) occurs during an aberrant replicative event in which the replicase, which had been copying one template, detaches from that template with its newly synthesized daughter strand in tow and resumes synthesis on a second template. In order to account for the observation that RNA from most DI particles does not contain sequences from the 3′ half of the parent viral genome, this model stipulates that the aberration occurs during the copying of the plus strand RNA into progeny minus strand RNAs. The failure to copy the 5′ end portion of the plus strand template accounts for the missing sequences on the 3′ end of the progeny minus strands. The free replicase still attached to the daughter strand can resume synthesis if it engages an appropriate template. The template in highest concentration in the vicinity of the free replicase is likely to be the daughter strand attached to it. Resumption of synthesis using the 5′ end of the daughter strand as a template yields an RNA in which the 3′ and 5′ ends are complementary. The length of this complementary stretch obviously depends on the point at which the replicase reattaches and begins copying, Thus, DI RNAs with either large or small stretches of complementary sequences are accounted for by this model. The model also stipulates that the complementary regions must be at the termini of the RNAs.

It should be stressed that not all DIs have RNAs that differ from the VSV RNA at the 3′ terminus. The long DI (DI-LT) particle derived from the heat-resistant (HR) strain of VSV has a genome that is homologous with the 3′ half of the VSV genome. Not only do the 12 to 17S mRNAs anneal to DI-LT RNA,[24,25,27] but so does the 68-nucleotide long leader RNA — the transcript of the precise 3′ terminal sequence of VS.[56] Given this homolog at the 3′ end between VSV and the DI-LT, it is not surprising that the DI-LT genome is a competent template for mRNA synthesis, both in vivo[57-59] and in vitro.[56] The Kolakofsky model for generation of DIs might be expanded to include DIs derived from the 3′ end of the VSV genome if one allowed that the aberrant replicative event takes place while the replicase copies a minus strand RNA. If errors such as those shown in Figure 2 occur while copying the minus strand template, the resulting DI genome should have maintained the 3′ end intact and should have a 5′ terminal sequence that is complementary to the 3′ terminal sequence. Recent work on the genomic structure of DI-LT suggests that this is not the case — the 5′ and the 3′ end of the DI seem to be identical with those of the VSV genome.[41,56] Thus, DI-LT may represent a true simple internal deletion mutant. This type of DI apparently arises less frequently than those with complementary termini, but the events leading to their formation may be very simiar to those of Kolakofsky's model. A replicase after copying the 3′ end of the template may detach with its daughter strand and resume synthesis nearer the 5′ end of the same or a different full-length template. Perhaps the difference in the apparent frequency of appearance of the two types of DIs is related to the differences in effective concentration of the second template in the vicinity of the detached replicase. The newly synthesized daughter strand being attached to the replicase is likely to be always present in the highest concentration.

III. PARAMETERS AFFECTING ENRICHMENT OF DEFECTIVE INTERFERING PARTICLES

The reduced yield of infectious virus and the appearance of large amounts of DI particles in culture fluid from infections in which the inoculum was a high multiplicity passage virus are the result of at least three phenomena. First, the synthesis of the progenitor DI RNP must take place in the cytoplasm of the infected cell. This single particle is amplified in subsequent rounds of replication. Finally, when the DI concentration is sufficiently high in the inoculum, there is a sharp reduction in the yield of progeny infectious particles (interference). The combined effect of these latter two processes can be estimated as enrichment of the DI in the progeny.

Three prominent parameters affecting the sequence of events taking place during high multiplicity passage are the nature of the host cell, time of harvest, and the multiplicity of infection.

A. Nature of the Host Cell

The effects of the host cell on the course of events during repeated passage of virus was first documented by Choppin et al.[60] using influenza virus. They demonstrated that the von Magnus phenomenon occurred in MDBK cells to a much lower degree than in other cells. This difference was due in part to the poor replication of DI particles in the infected cells rather that the restricted genesis of DI particles. Subsequently, a number of different cell lines have been shown to behave as "low interference" hosts for VSV.[4,12,28] In some of these cell lines, added purified DI particles not only do not efficiently interfere with the replication of the helper virus, but they themselves are poorly replicated in the mixedly infected cell. It is possible that poor viral penetration of the cell may be involved in some cases, since the efficiency of plating of VSV on MDCK cells, a low interference cell line, is 20 to 50 times lower than it is on BHK-21 cells, a high interference line.[61] Once infected, the MDCK cell is a competent host for VSV.

When seeds of several types of DIs are present in one virus stock, the host cell may play an important role in determining the outcome of subsequent passages of the stock. Holland et al.[12] have shown that the composition of progeny DI mixtures from cells infected with seed stocks containing multiple DIs depend upon the cell line employed. The dependence is so marked that when infected with the same inoculum, some cell types yield DI classes that were virtually absent from the progeny obtained when a second cell type was employed. The biochemical basis for these host effects is unknown. However, a number of recent studies point to the involvement of host factors in the replication of VSV.[62-65]

B. Time of Harvest

The single-step growth curve of VSV in most cell lines spans approximately 6 to 8 hr. Beyond that time, little additional virus is released from the infected cell. This time course seems to apply to the release of standard virus whether or not DI particles are present in the infection. The release of DI particles occurs more slowly and takes 16 to 18 hr.[66] This slow release of DI particles does not appear to result from restricted replication of the RNA since DI nucleocapsids are abundantly present in mixedly infected cells. Rather, it suggests that the assembly of the DI particle may be the limiting process.[66] Because of the difference in the time course of standard and DI particle release from the same cell, harvest made before 16 to 18 hr post infection will select against DI particles.

C. Multiplicity of Infection

DI particles are frequently observed in virus stocks that have been passaged at high multiplicity and are rarely observed in those that have been passaged at very low multiplicities. This dependence of enrichment upon multiplicity is the result of several distinct processes. First, the growth of virus inoculated at very low multiplicities is a two-step process. In the first cycle, only a very small fraction of the cells are infected, and even if the inoculum contains equal numbers of DIs and infectious particles, virtually no cell is mixedly infected. Consequently, the yield of the first step must be greatly enriched in the infectious virus. In the second step, the newly released virus infects the remaining cells including those which have absorbed DI particles earlier. The composition of the final yield of progeny particles will vary depending upon the composition and multiplicity of the original inoculum, but at sufficiently low multiplicities the progeny are impoverished rather than enriched in DI particles when compared to the original inoculum.

In the example shown in Figure 3,[68] monolayers of BHK-21 cells were infected at different multiplicities with an inoculum that had a DI/VSV particle ratio of 1. The composition of the progeny of these infections is presented as a function of the input multiplicity. At multiplicities below one, the progeny particles are impoverished for DI particles; above that multiplicity, the progeny are enriched in DIs. While there are several interrelated processes in mixedly infected cells that contribute to the general shape of the curve shown in Figure 3, two processes are particularly prominent. The first is obvious: as the multiplicity of infection increases, the number of cells mixedly infected in both the first and second step increases. Consequently, the number of progeny DI particles increases, and because of interference in the second step, the number of progeny infectious particles decreases.

One might expect that once the multiplicity is high enough to insure infection of all cells with both DI and helper virus in the first cycle, further increases would be without

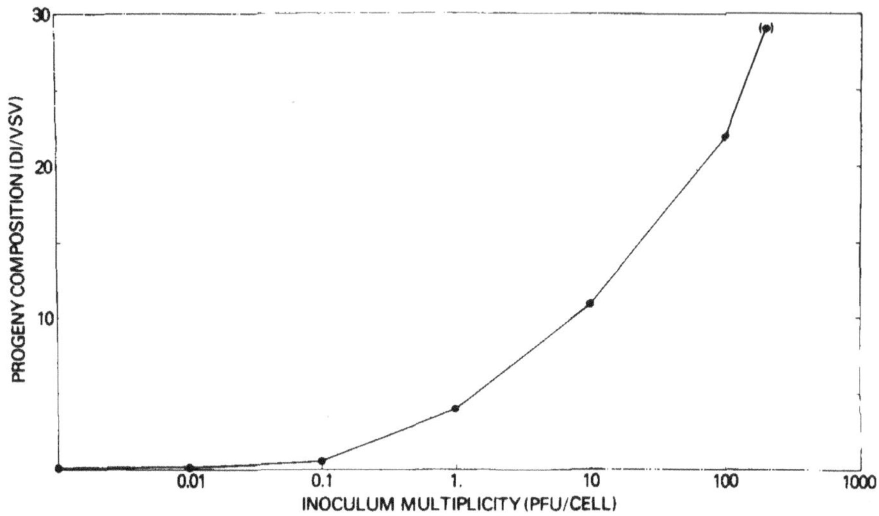

FIGURE 3. The effect of multiplicity of infection on the enrichment of DI particles. BHK-21 monolayers were infected with the indicated amounts of a 1:1 mixture of DI-LT and VSV particles. Since the measured plating efficiency of VSV is approximately 10%, these mixtures contained ten DI particles per PFU. The composition of the progeny was determined as before.[66, 69]

effect since the multiplicities of both the DI and standard virus increase in parallel. Clearly, this is not the case. Increasing the multiplicity of the mixed inoculum leads to a marked enrichment in DIs in the progeny as shown in Figure 3. This departure from the anticipated can be attributed to a second characteristic — the inability to reverse interference by increasing the number of infectious particles. This characteristic is illustrated in the two panels of Figure 4.[68] Monolayers of BHK-21 cells were infected with either 10 or 100 DI particles per cell and the indicated amounts of infectious virus. The lower line in each panel shows the composition of the infecting inoculum; the upper line, that of the progeny. The divergence of the curves is a measure of how poorly increased numbers of standard particles reverse interference. Consequently, there is a marked increase in enrichment with increasing input of infectious virus as indicated by the numbers in the parentheses. It is not clear at present whether the ineffectiveness of increasing the level of infectious virus results from some intracellular event or whether the particles are unable to penetrate into the interior of the cell in sufficient numbers to effect the interference phenomenon.

IV. MECHANISMS OF INTERFERENCE

Investigations of the biologic properties of DI particles have limited the number of possible mechanisms of autointerference but have not, as yet, pointed to a particular mechanism. Early experiments by Huang and Wagner[70] indicated that DI particles interfere with some intracellular event rather than with virus penetration or absorption, since UV-inactivated DI particles were shown to be incapable of autointerference. The same conclusion was drawn from other studies in which it was shown that DI particles which were added 2 hr after the infectious virus were still able to interfere with the production of progeny infectious virus.

The specific intracellular event that is sensitive to autointerference is unknown at present. Although a number of studies have been directed at this question, the intracellular events are so interrelated and interdependent that it has not been possible to

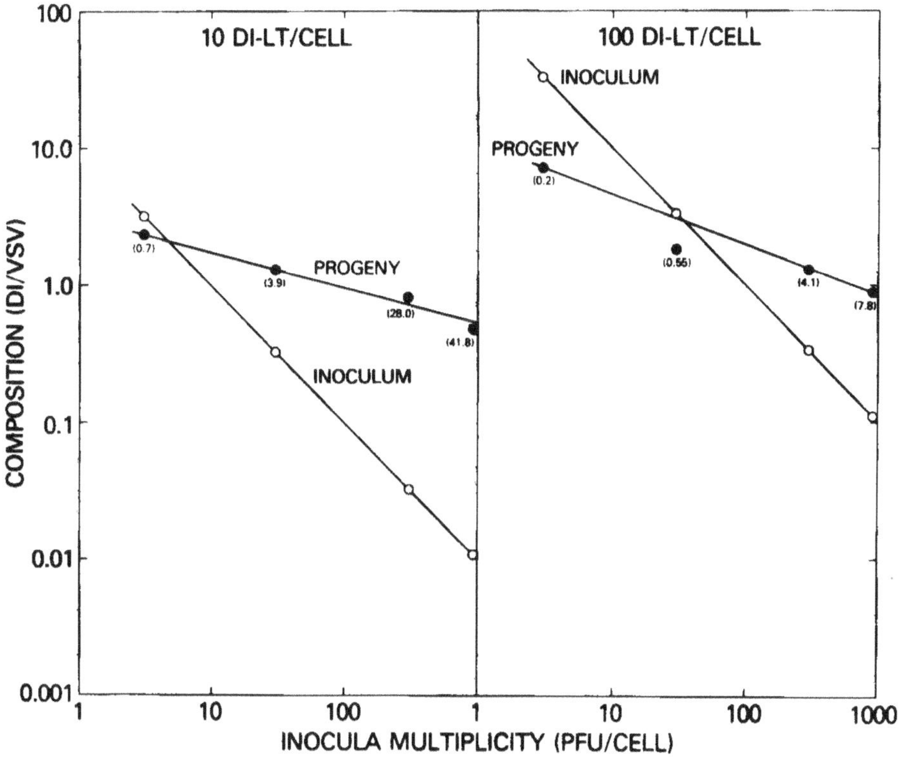

FIGURE 4. The effect of multiplicity of VSV on the composition of the progeny from mixedly infected cells. Monolayers of BHK-21 cells were infected with the indicated amounts of clonally purified VSV and either 10 (left panel) or 100 (right panel) DI-LT particles per cell. The composition of the progeny was determined as before.[66, 69]

isolate the pivotal process. Nonetheless, the data suggest that viral transcription and translation are not directly affected by the presence of DI particles. Most DI particles that contain information from the 5′ end of the VSV genome are capable of only a very limited transcription and produce a short (46-nucleotide) "leader"-like RNA in vitro.[23,71-74] Since these particles do not contain a single complete gene,[25,27] it is not surprising that cells infected with DI particles alone do not produce a recognizable viral potein. Of particular importance to the present topic is the observation that DI particles do not interfere with the primary transcription of the helper virus.[71,75] Since the biochemical events of primary and secondary (amplified) transcription are thought to be identical, these results indicate that DI particles do not directly interfere with viral transcription. DI particles do, however, interfere wih the production of the templates for secondary transcription. Stamminger and Lazzarini[46] and Khan and Lazzarini[66] have compared the relative amounts of DI and VSV nucleocapsids found in cells infected with VSV alone or mixedly infected with DI particles. Although the amount of nucleocapsid did not change with the degree of interference, the composition did. At high levels of autointerference, there was less standard-sized nucleocapsid and correspondingly more DI nucleocapsids. Since the standard nucleocapsid is the template for functional viral mRNA, DI particles do affect the level of viral transcription in an indirect way.

One might expect that a transcribing DI particle such as DI-LT would not affect the rate of transcription of the most abundant mRNAs (12 to 17S) since both standard and DI nucleocapsid are competent templates for these. Nonetheless, transcribing and nontranscribing DI particles affected transcription equally in monolayers of BHK-21 cells suggesting that in these cells the transcriptase, not the template nucleocapsid, is limiting.[46] More recently, Moyer[67] has compared the rate of tanscription in autointerfered infections of suspension BHK-21 cells and has found that the rate is higher when DI-LT was the interfering particle than when a nontranscribing particle was used. This apparent discrepancy with the previous study may only signal that in suspension cells the template RNP transcriptase is the limiting factor.

Unfortunately, there are no reported direct comparisons between the rates of viral protein synthesis in VSV infected and mixedly infected cells. Nonetheless, the translation of viral mRNA appears not to be affected by the presence of DI particles. Even at moderate levels of autointerference (i.e., three- to tenfold reduction in progeny infectious particles), the amount of viral proteins released into the culture fluid in the form of progeny DI and standard particles and soluble viral proteins is not substantially different from that found in control infections. Since the level of intracellular viral mRNA is appreciably reduced under these conditions, the efficiency of translation of the remaining mRNA into released viral proteins is increased, not decreased, by autointerference.

Available evidence suggests that viral assembly and maturation may be inefficient processes in cells mixedly infected with DI and helper viruses. Only a small fraction of the intracellular nucleocapsid is packaged into progeny virions, even in control, noninterfered infections.[66,67,76] However, in the case of VSV the process is very specific, and only minus strand RNA matures into progeny particles. In mixedly infected cells, both the efficiency and specificity of the process are sharply diminished. DI nucleocapsids accumulate intracellularly to a greater extent than their full-length counterparts.[66,67,77] It is possible that this accumulation results from the observed slower release of progeny DI particles compared to infectious particles.[66] Even the specificity for packaging only the minus strand frequently breaks down in mixedly infected cells. RNA prepared from DI particles often contains variable amounts of plus strand RNA above that contributed by the self-complementary termini and the covalently linked plus and minus strand RNAs.[78] It is unclear whether these complications of viral assembly are the result of some specific affect of DI particles on an otherwise competent process or whether the process is intrinsically less efficient in assembling and maturing the small DI nucleocapsids.

The synthesis of viral genome is frequently proposed as the site at which DI particles exert their interference.[79,80] The sensitivity of standard nucleocapsid synthesis to the presence of DI particles has been observed in a number of laboratories. As cited above, even the composition of the intracellular nucleocapsid varies regularly with increasing levels of infecting DI and autointerference.[46] These observations are consistent with the idea that some component involved in the synthesis of viral nucleocapsid is limiting and can be competed for by the nucleocapsids of the infecting DI particles.

The synthesis of nucleocapsids appears to be a linear process. To date, no biochemical or electron micrographic evidence has been obtained to suggest a circular intermediate in this synthetic event. Complex mechanisms for linear synthesis of progeny genomes are possible in which synthesis is initiated at internal points along the genome (for example, like those that operate during RNA tumor virus replication). However, several lines of evidence suggest that VSV replication and transcription are initiated at the precise 3′ termini of the respective templates. In the case of in vitro transcription, the only major tri- (or di-) phosphate-terminating RNA that has been identified is the leader RNA which is the complement of the 3′ end of the VSV genome.[80-82] Virtually all mRNA transcripts are capped with a 7-methyl guanylate residue at the 5′ terminus and may arise from the processing of a larger precursor mRNA.[83-85] Similarly, the 5′ ends of the replicated DI or VSV genomes also bear a tri-phosphate terminus,[37,40,42,86] indicating that the termini were not created by processing concatameric molecules, but represent the initiation points for the synthesis of the RNA strand. If replication of both DI and VSV are linear events in which a full-length (+) RNP is synthesized which, in turn, serves as a template for progeny (−) RNPs, then the replicase initiation sites on the DI and VSV genomes should be at their respective 3′ termini and should be very similar, if not identical, to each other. RNA base sequence analysis by Keene et al.[87] has confirmed the presence of nearly identical sequences at the 3′ terminus of two strains of VSV and DI particles derived from them. These sequences are

YpXpApYpGpGpUpUpUpGpUpUpUpGpUpCpUpUpCpGpU–OH VSV

ApYpYpYpGpGpUpUpUpUpGpUpGpGpUpCpUpUpCpGpU–OH DI

The regions of exact homology are indicated by the dash rectangles. The sequences of the DI and VSV RNA beyond those shown above diverge abruptly.

Thus, the analysis of the 3′ termini is compatible with the notion that autointerference results from a competition between the VSV and the DI genomes for a limited amount of replicase. It should be emphasized that an entirely similar argument can be developed using the N protein rather than the replicase as the factor competed for. Assembly of specific (ordered) RNA protein complexes often involves nucleation at a specific site and propagation from that site. Thus, the reassembly of capsid proteins on naked tobacco mosaic virus (TMV) RNA is initiated by rate-limiting reaction in which the first capsid binds to a specific site on the RNA. Subsequent capsid proteins associate in such a way that the nucleation site grows in both directions until the ends of the RNA are reached.[88] If a similar mechanism operates during VSV RNP synthesis, the conserved sequence shown above may form part of the specific site necessary for the nucleocapsid nucleation. If the supply of nascent N protein is limiting during viral replication, the replication of the DI nucleocapsid would diminish the synthesis of progeny VSV. Thus, although the biochemical mechanisms by which DI particles interfere with the replication of standard virus remain obscure, recent work has limited the number of possible models and has focused attention on the replication and assembly processes as possible sites of autointerference.

REFERENCES

1. **von Magnus, P.,** Propagation of the RP-8 strain of influenza A virus in chick embryos. III. Properties of the incomplete virus produced in serial passages of undiluted virus, *Acta Pathol. Microbiol. Scand.*, 29, 157, 1951.
2. **von Magnus, P.,** Incomplete forms of influenza virus, *Adv. Virus Res.*, 2, 59, 1954.
3. **Baltimore, D.,** Expression of animal virus genomes, *Bacteriol. Rev.*, 35, 235, 1971.
4. **Huang, A. S. and Baltimore, D.,** Defective viral particles and viral disease processes, *Nature (London)*, 226, 325, 1970.
5. **Huang, A. S.,** Defective interfering viruses, *Ann. Rev. Microbiol.*, 27, 101, 1973.
6. **Huang, A. S. and Baltimore, D.,** Defective interfering animal viruses, in *Comprehensive Virology*, Vol. 10, Fraenkel-Conrat, H. and Wagner, R. R., Eds., Plenum Press, New York, 1977, 73.
7. **Yoshino, K., Taniguchi, S., and Arai, K.,** Autointerference of rabies virus in chick embryo fibroblasts, *Proc. Soc. Exp. Biol. Med.*, 123, 387, 1966.

8. **Sokol, R., Kuwert, E., Wiktor, J. J., Hummeler, K., and Koprowski, H.**, Purification of rabies virus grown in tissue culture, *J. Virol.*, 2, 836, 1968.
9. **Crick, J. and Brown, F.**, An interfering component of rabies virus which contains RNA, *J. Gen. Virol.*, 22, 147, 1974.
10. **Kawai, A., Matsumoto, S., and Tanabe, K.**, Characterization of rabies virus recovered from persistently infected BHK cells, *Virology*, 67, 520, 1975.
11. **Holland, J. J. and Villareal, L. P.**, Purification of defective interfering T particles of vesicular stomatitis and rabies viruses generated *in vivo* in brains of newborn mice, *Virology*, 67, 438, 1975.
12. **Holland, J. J., Villareal, L. P., and Breindl, M.**, Factors involved in the generation and replication of rhabdoviruses defective T particles, *J. Virol.*, 17, 805, 1976.
13. **Wiktor, T. J., Dietzschold, B., Leamnson, R. N., and Koprowski, H.**, Induction and biological properties of effective interfering particles of rabies virus, *J. Virol.*, 21, 626, 1977.
14. **Kawai, A. and Matsumoto, S.**, Interfering and noninterfering defective particles generated by a rabies small plaque variant virus, *Virology*, 76, 60, 1977.
15. **McAllister, P. E. and Pilcher, K. S.**, Autointerference in infectious hematopoietic necrosis virus of almonid fish, *Proc. Soc. Exp. Biol. Med.*, 145, 840, 1974.
16. **Younger, J. S. and Quagliana, D. O.**, Temperature-sensitive mutants of vesicular stomatitis virus are conditionally defective particles that interfere with and are rescued by wild-type virus, *J. Virol.*, 19, 102, 1976.
17. **Legault, D., Takayesu, D., and Prevec, L.**, Heterotypic exclusion between vesicular stomatitis viruses of the New Jersey and Indiana serotypes, *J. Gen. Virol.*, 35, 53, 1977.
18. **Huang, A. S. and Wagner, R. R.**, Comparative sedimentation coefficients of RNA extracted from plaque-forming and defective particles of vesicular stomatitis virus, *J. Mol. Biol.*, 22, 381, 1966.
19. **Petric, M. and Prevec, L.**, Vesicular stomatitis virus — a new interfering particle, intracellular structures, and virus-specific RNA, *Virology*, 41, 615, 1970.
20. **Reichmann, M. E., Pringle, C. R., and Follet, E. A. C.**, Defective particles in BHK cells infected with temperature-sensitive mutants of vesicular stomatitis virus, *J. Virol.*, 8, 154, 1971.
21. **Bishop, D. H. L. and Roy, P.**, Kinetics of RNA synthesis by vesicular stomatitis virus particles, *J. Mol. Biol.*, 57, 513, 1971.
22. **Bishop, D. H. L. and Roy, P.**, Properties of the product synthesized by vesicular stomatitis virus particles, *J. Mol. Biol.*, 58, 799, 1971.
23. **Roy, P. and Bishop, D. H. L.**, Genome homology of vesicular stomatitis virus and defective T particles and evidence for the sequential transcription of the virion ribonucleic acid, *J. Virol.*, 9, 946, 1972.
24. **Leamnson, R. N. and Reichmann, M. E.**, The RNA of defective vesicular stomatitis virus particles in relation to viral cistons, *J. Mol. Biol.*, 85, 551, 1974.
25. **Stamminger, G. and Lazzarini, R. A.**, Analysis of the RNA of defective VSV particles, *Cell*, 3, 85, 1974.
26. **Lazzarini, R. A., Weber, G. H., Johnson, L. D., and Stamminger, G. M.**, Covalently linked message and anti-message (genomic) RNA from a defective vesicular stomatitis virus particle, *J. Mol. Biol.*, 97, 289, 1975.
27. **Schnitzlein, W. M. and Reichmann, M. E.**, The size and the cistronic origin of defective vesicular stomatitis virus particle RNA's in relation to homotypic and heterotypic interference, *J. Mol. Biol.*, 101, 307, 1976.
28. **Perrault, J. and Holland, J.**, Variability of vesicular stomatitis virus autointerference with different host cells and virus serotypes, *Virology*, 50, 148, 1972.
29. **Kang, C. Y., Glimp, T., Clewley, J. P., and Bishop, D. H. L.**, Studies on the generation of vesicular stomatitis virus (Indiana serotype) defective interfering particles, *Virology*, 84, 142, 1978.
30. **Spandidos, D. A. and Graham, A. F.**, Complementation of defective reovirus by ts mutants, *J. Virol.*, 15, 954, 1975.
31. **Hackett, A. J.**, A possible morphological basis for the autointerference phenomenon in vesicular stomatitis virus, *Virology*, 24, 51, 1964.
32. **Huang, A. S., Greenawalt, J. W., and Wagner, R. R.**, Defective T particles of vesicular stomatitis virus. I. Preparation, morphology, and some biologic properties, *Virology*, 30, 161, 1966.
33. **Simpson, R. W. and Houser, R. E.**, Structural components of vesicular stomatitis virus, *Virology*, 29, 654, 1966.
34. **Nakai, T. and Howatson, A. F.**, The fine structure of vesicular stomatitis virus, *Virology*, 35, 268, 1968.
35. **Wagner, R. R., Schnaitman, T. A., and Snyder, R. M.**, Structural proteins of vesicular stomatitis viruses, *J. Virol.*, 3, 395, 1969.
36. **Johnson, L. D. and Lazzarini, R. A.**, unpublished data, 1974.

37. **Adler, R. and Banerjee, A. K.**, Analysis of the RNA species isolated from defective particles of vesicular stomatitis virus, *J. Gen. Virol.*, 33, 51, 1976.
38. **Reichmann, M. E. and Schnitzlein, W. M.**, unpublished data, 1976.
39. **Keene, J. D., Rosenberg, M., and Lazzarini, R. A.**, Characterization of the 3′ terminus of RNA isolated from vesicular stomatitis virus and from its defective interfering particles, *Proc. Natl. Acad. Sci. U.S.A.*, 74, 1353, 1977.
40. **Johnson, L. D. and Lazzarini, R. A.**, The 5′ terminal nucleotide of RNA from vesicular stomatitis virus defective interfering particles, *Virology*, 77, 863, 1977.
41. **Epstein, D., Dirksen, M. L. D., Lazzarini, R. A., and Rosenberg, M.**, unpublished data, 1976.
42. **Hepti, E. and Bishop, D. H. L.**, The 5′ nucleotide sequence of vesicular stomatitis viral RNA, *J. Virol.*, 15, 90, 1975.
43. **Banerjee, A. and Rhodes, D. P.**, 3′ terminal sequence of vesicular stomatitis virus genome RNA, *Biochem. Biophys. Res. Commun.*, 68, 1387, 1976.
44. **Perrault, J.**, Cross-linked double-stranded RNA from a defective vesicular stomatitis virus particle, *Virology*, 70, 360, 1976.
45. **Perrault, J. and Leavitt, R. W.**, Characterization of snap-back RNA's in vesicular stomatitis defective interfering particles, *J. Gen. Virol.*, 38, 21, 1977.
46. **Stamminger, G. M. and Lazzarini, R. A.**, RNA synthesis in standard and autointerfered vesicular stomatitis virus infections, *Virology*, 77, 202, 1977.
47. **Khan, S. R., Adachi, R., and Lazzarini, R. A.**, unpublished data, 1976.
48. **Kolakofsky, D.**, Isolation and characterization of Sendai virus DI-RNA's, *Cell*, 8, 547, 1976.
49. **Perrault, J. and Leavitt, R. W.**, Inverted complementary terminal sequences in single-stranded RNA's and snap-back RNA's from vesicular stomatitis defective interfering particle, *J. Gen. Virol.*, 38, 35, 1977.
50. **Leppert, M., Kort, L., and Kolakofsky, D.**, Further characterization of Sendai virus DI-RNA's: a model for their generation, *Cell*, 12, 539, 1977.
51. **Kolakofsky, L.**, in *Negative Strand Virus and the Host Cell*, Mahy, B. W. J. and Barry, R. D., Eds., Academic Press, Cambridge, 1978, 539.
52. **Perrault, J., Semler, B. L., Leavitt, R. W., and Holland, J. J.**, Inverted complementary terminal sequences in defective interfering particle RNA's of vesicular stomatitis virus and their possible role in autointerference, in *Negative Strand Virus and the Host Cell*, Mahy, B. W. J. and Barry, R. D., Eds., Academic Press, Cambridge, 1978, 527.
53. **Huang, A. S.**, Viral pathogenesis and molecular biology, *Bacteriol. Rev.*, 41, 811, 1977.
54. **Freeman, G. J., Rac, D. D., and Huang, A. S.**, Genome organization of vesicular stomatitis virus: mapping ts G41 and the defective interfering T particle, in *Negative Strand Virus and the Host Cell*, Mahy, B. W. J. and Barry, R. D., Eds., Academic Press, Cambridge, 1978, 261.
55. **Kang, C. Y., Glimp, T., Clewley, J. P., and Bishop, D. H. L.**, Studies on the generation of vesicular stomatitis virus (Indiana serotype) defective interfering particle, *Virology*, 84, 142, 1978.
56. **Colonno, R. J., Lazzarini, R. A., Keene, J. D., and Banerjee, A. K.**, *In vitro* synthesis of messenger RNA by a defective interfering particle of vesicular stomatitis virus, *Proc. Natl. Acad. Sci. U.S.A.*, 74, 1884, 1977.
57. **Chow, J. M., Schnitzlein, W. M., and Reichmann, M. E.**, Expression of genetic information contained in the RNA of a defective interfering particle of vesicular stomatitis virus, *Virology*, 77, 579, 1977.
58. **Johnson, L. D. and Lazzarini, R. A.**, Replication of viral RNA by a defective interfering vesicular stomatitis virus particle in the absence of helper virus, *Proc. Natl. Acad. Sci. U.S.A.*, 74, 4387, 1977.
59. **Johnson, L. D. and Lazzarini, R. A.**, Synthesis of both + and − strand viral RNA in cells injected with only VSV DI particles, in *Negative Strand Viruses and the Host Cell*, Mahy, B. W. J. and Barry, R. D., Eds., Academic Press, Cambridge, in press.
60. **Choppin, P. W.**, Replication of influenza virus in a continuous cell line: high yield of infectious virus from cells inoculated at high multiplicity, *Virology*, 39, 130, 1969.
61. **Faulkner, G. P. and Lazzarini, R. A.**, unpublished data, 1977.
62. **Nowakowski, M., Bloom, B. R., Ehrenfield, E., and Summers, D. F.**, Restricted replication of vesicular stomatitis virus (VSV) in human lymphoblastoid cells, *J. Virol.*, 12, 1272, 1973.
63. **Obijeski, J. F. and Simpson, R. W.**, Conditional lethal mutants of vesicular stomatitis virus. II. Synthesis of virus-specific polypeptides in non-permissive cells infected with "RNA" host-restricted mutants, *Virology*, 57, 369, 19u4.
64. **Szilágyi, J. F., Pringle, C. R., and Macpherson, T. M.**, Temperature-dependent host range mutation in vesicular stomatitis virus affecting polypeptide L, *J. Virol.*, 22, 381, 1977.
65. **Szilágyi, J. F. and Pringle, C. R.**, A possible host factor requirement for the transcription of vesicular stomatitis virus RNA, in *Negative Strand Viruses and the Host Cell*, Mahy, B. W. J. and Barry, R. D., Eds., Academic Press, Cambridge, 1978, 285.

66. **Khan, S. R. and Lazzarini, R. A.**, The relationship between autointerference and the replication of a defective interfering particle, *Virology*, 77, 189, 1977.
67. **Moyer, S.**, unpublished data, 1977.
68. **Faulkner, G. P. and Lazzarini, R. A.**, unpublished data, 1978.
69. **Adachi, T. and Lazzarini, R. A.**, Elementary aspects of autointerference and the replication of defective interfering virus particles, *Virology*, in press, 1978.
70. **Huang, A. S. and Wagner, R. R.**, Defective T particles of vesicular stomatitis virus. II. Biologic role in homologous interference, *Virology*, 173, 1966.
71. **Perrault, J. and Holland, J. J.**, Absence of transcriptase activity or transcription inhibiting ability in defective interfering particles of vesicular stomatitis virus, *Virology*, 50, 159, 1972.
72. **Reichmann, M. E., Villareal, L. P., Kohne, D., Lesnaw, J., and Holland, J. J.**, RNA polymerase activity and poly(A) synthesizing activity in defective T particles of vesicular stomatitis virus, *Virology*, 58, 240, 1974.
73. **Emerson, S. O. and Wagner, R. R.**, Dissociation and reconstitution of the transcriptase and template activities of vesicular stomatitis B and T virions, *J. Virol.*, 10, 297, 1972.
74. **Emerson, S. U., Dierks, P. M., and Parsons, J. T.**, *In vitro* synthesis of a unique RNA species by a T particle of vesicular stomatitis virus, *J. Virol.*, 23, 708, 1977.
75. **Huang, A. S. and Mander, E. K.**, Ribonucleic acid synthesis of vesicular stomatitis virus. IV. Transcription by standard virus in presence of defective interfering particles, *J. Virol.*, 9, 909, 1972.
76. **Kang, C. Y. and Prevec, L.**, Proteins of vesicular stomatitis virus. III. Intracellular synthesis and extracellular appearance of virus-specific proteins, *Virology*, 46, 678, 1971.
77. **Palma, E. L. and Huang, A. S.**, Cyclic production of vesicular stomatitis virus caused by defective interfering particles, *J. Infect. Dis.*, 129, 402, 1974.
78. **Roy, P., Repik, P., Hefti, E., and Bishop, D. H. L.**, Complementary RNA species isolated from vesicular stomatitis (HR strain) defective virions, *J. Virol.*, 11, 915, 1973.
79. **Stampfer, M., Baltimore, D., and Huang, A. S.**, Ribonucleic acid synthesis of vesicular stomatitis virus I. Species of ribonucleic acid found in Chinese hamster ovary cells infected with plaque forming and defective particles, *J. Virol.*, 4, 154, 1969.
80. **Palma, E. L., Perlman, S. M., and Huang, A. S.**, Ribonucleic acid synthesis of vesicular stomatitis virus. VI. Correlation of defective particle RNA synthesis with standard RNA replication, *J. Mol. Biol.*, 85, 127, 1974.
81. **Colonno, R. J. and Banerjee, A. K.**, A unique RNA species involved in initiation of vesicular stomatitis virus RNA transcription in vitro, *Cell*, 8, 197, 1976.
82. **Colonno, R. J. and Banerjee, A. K.**, Mapping and initiation studies on the leader RNA of vesicular stomatitis virus, *Virology*, 77, 260, 1977.
83. **Abraham, G., Rhodes, D. P., and Banerjee, A. K.**, The 5′ terminal structure of the methylated mRNA synthesized *in vitro* by vesicular stomatitis virus, *Cell*, 5, 51, 1975.
84. **Moyer, S. A., Abraham, G., Adler, R., and Banerjee, A. K.**, Methylated and blocked 5′ termini in vesicular stomatitis virus *in vivo* mRNAs, *Cell*, 5, 59, 1975.
85. **Rose, J. K.**, Heterogenous 5′ terminal structures occur on vesicular stomatitis virus mRNA's, *J. Biol. Chem.*, 250, 8098, 1975.
86. **Hefti, E. and Bishop, D. H. L.**, The 5′ sequence of VSV viral RNA and its transcription product RNA, *Biochem. Biophys. Res. Commun.*, 66, 785, 1975.
87. **Keene, J. D., Schubert, M., Rosenberg, M., and Lazzarini, R. A.**, A comparative study of nucleotide sequences at the 3′ termini of ribonucleic acids from vesicular stomatitis virus and its defective interfering particles, in *ICN-UCLA Symposia and Persistent Virus Infections*, Stevens, J. and Todaro, G., Eds., Academic Press, New York, 1978.
88. **Ohno, T., Takahishi, M., and Okada, Y.**, Assembly of tobacco mosaic virus *in vitro:* elongation of partially reconstructed RNA, *Proc. Natl. Acad. Sci. U.S.A.*, 74, 552, 1977.

Chapter 10

RNA SYNTHESIS BY DEFECTIVE INTERFERING VESICULAR STOMATITIS VIRUS PARTICLES

Leslye D. Johnson and Robert A. Lazzarini

TABLE OF CONTENTS

I. INTRODUCTION

Defective interfering (DI) particles of vesicular stomatitis virus (VSV) are generated and enriched during consecutive undiluted passages of the virus.[1] Their name derives from their inability to produce progeny in the absence of standard VSV (B particles) and from their ability to reduce the yield of B particles in mixed infections with VSV. DI particles contain the same viral proteins in approximately the same relative amounts as are found in the standard infectious virus.[2-4] They differ from VSV in that their genomic RNA is smaller, ranging from one tenth to one half the size of VSV genomic RNA.[5]

Although the original characterization studies of VSV DI particles described above were performed on only a few species of such virions, they still remain correct. However, recent close examinations of the VSV genome and its molecular biology permit further classification of the DI virions. Transcription of the VSV genome was found to start at the 3′ end[6,7] and to produce a 4S leader RNA[8] and mRNAs of 12 to 17.5S and 31S.[9] These mRNA species have now been separated from one another and used to prime protein-synthesis reactions.[10-14] These studies led to the assignment of proteins with specific mRNAs. In addition, the genes have been ordered on the RNA strand by studying the effect of UV irradiation on synthesis of mRNAs or proteins.[6,7] These data are summarized in Chapters 3, 4, and 5 of this book.

With this information, more meaningful studies of DI particles were possible. Using as probes mRNAs synthesized by VSV, many DI particles have been examined to determine what part of the VSV genome is present in their RNAs. From these results, VSV DI particles can be divided into two general classes. The most prevalent DI particles are those which have lost the 3′ half and retained only a portion of the 5′ half of the VSV genome, i.e., the region coding for L protein mRNA.[5,15,16] Different DI virions have different amounts of this part of the genome — one fifth to almost all of it. Some RNAs of this DI particle class were shown to have genomic (−) and complementary (+) segments that were covalently attached.[17,18] Two sizes of complementary sequences have been described — one being about 5% of the nucleotides (panhandles,)[19] and the other 90% (hairpins,)[17,18]

The genome of the other class of DI particles is the same as the 3′ half of VSV RNA,[15,16] and the 5′ half is not present. There are few representatives in this class. To date the best characterized has been the long DI particle (DI-LT) generated by the heat-resistant strain of VSV_{IND}.[20] The name "LT" refers to the size of the DI virion and thus "LT" particles could actually have any portion of the VSV genome, even that of the 5′ half.[21] The RNA of the 3′ end DI-LT particle has sequences complementary to all four of the mRNAs synthesized from the 3′ half of VSV.[16]

From this information on the genomes of the various DI particles and on the order of transcription of VSV several predictions can be made for the behavior of the two classes of DI virions. Since transcription begins at the 3′ terminus of VSV RNA, only those DI RNAs having the 3′ sequence crucial for this process will be transcribed. Of the two DI types, the RNAs for the 5′ half DI particles are unlikely candidates for the transcription process since sequence studies show that they have retained only the first eight nucleotides of the 3′ terminus of VSV.[22] If by chance transcription occurs, even a full-length copy of the RNA would be too small to code for a functional L protein. On the other hand, the 3′ end DI RNA has a greater chance of having all the crucial 3′ terminal nucleotides, of being transcribed, and of producing competent mRNA species since it has retained sequences complementary to four mRNAs.

This review will examine the data pertaining to RNA synthesis by these two classes of DI particles.

II. RNA SYNTHESIS BY 5′ END DI PARTICLES

In general, the results with 5′ end DI particles fit the predictions. Several laboratories have found that in vitro transcription of disrupted DI virions occurred at 1 to 10% the rate of VSV.[21,23,24] This low level of activity was due to an incompetent DI ribonucleoprotein (RNP) template and not due to inactivity of the NS and L proteins since they were active in combination with the standard VSV RNP.[25] Recently, close examination revealed that this low level of synthesis was due to production of a unique 2 to 4S RNA species by the DI virion rather than to the synthesis of standard mRNAs by a B-virion contaminant.[21,23]

Of the studies thus far reported, that of Emerson et al.[21] on a 5′ end DI particle derived from the San Juan isolate of VSV_{IND} is the most complete. Their purified DI particle synthesized RNA at less than 10% the rate of B virions and produced a transcript less than 4S in size. Analyses by gels and by nucleotide composition indicated a chain length of 43 to 45 bases. This product did not contain large tracts of poly(A): it was completely sensitive to RNase, did not bind to oligo(dT) and had no adenylate tracts longer than 5 to 6. The product RNA annealed to DI RNA only when the latter had been subjected to limited alkali digest. This may relate to the ability of 5′ end DI particle RNAs to form panhandles due to self-complementary ends.[19] Although the 5′ terminus of the product RNA is complementary to the 3′ end of standard VSV for three nucleotides neither B virion RNA nor alkali fragmented B virion RNA formed hybrids with labeled DI product RNA. This suggested that the 3′ end of DI particle RNA may not be the same as that of the standard viral particle RNA beyond the first 3 nucleotides. Recently, Keene et al.[22] showed that DI and B virion RNAs are homologous for only eight nucleotides at the 3′ terminus. A new model for DI particle generation has been proposed which accommodates these data.[26,27]

The significance of this small DI product RNA is obscure at present. Conceivably, it may be the counterpart of the VSV leader RNA or the product of an abortive transcription or replication event.

Earlier, Reichmann et al.[23] studied several DI particles and extensively examined one from the Mudd-Summers stock of VSV. Later this DI virion was shown to be a 5′ end particle.[5,15,16] The rate of transcription of this DI virion was only a fraction that of the B virion. The product was unique and sedimented faster than a 5S marker. Unlike the small RNA product of the Emerson et al. study,[21] this RNA product was insensitive to ribonuclease suggesting the presence of tracts of poly(A). The undigested labeled product did not anneal to DI particle RNA or to B virion RNA. However, neither of these genomic RNAs was treated with alkali prior to annealing.

Although the above results seem definitive and consistent with the current information on 5′ end DI particles, widely disparate data exists in the early literature. Bishop and Roy,[24] using a virion population that was 90% 5′ end DI particles and 10% standard virions, found no transcriptase activity or products attributable to the DI virions. However, if such DI particle RNPs are transcribed at 1 to 10% the rate of B particle RNPs, the products of the DI portion of the mixture could easily be missed.

The results of Mori and Howatson[28] represent the other extreme. They tested two purified DI particles (MT and ST) derived from their strain of VSV_{IND} for in vitro transcription. Both of these particles synthesized more RNA than the standard virion. The product made by the MT particle was 5 to 15S in size (the same as made by B virions); the ST particle made a 5S RNA. The authors classified neither of these DI particles, but using the same parental stock others have generated a DI population dominated by a 5′ end DI particle.[5,16] However, the possibility remains that this DI species was neither of those used by Mori and Howatson since different cell lines used for virus growth favor the expression of different DI particles in the same seed stock.[13] Additionally, different clonal isolates of the same parental stock have been found to yield different DI virions.[29]

The in vivo RNA synthetic abilities of some 5′ end DI particles have been compared with those of the standard virion in cultured cells. Uridine incorporation by cells infected only with DI particles was the same as that by uninfected cells.[30,31] When RNA from DI particle-infected cells was analyzed on sucrose gradients, no viral specific RNA was observed. Thus, neither the small RNA product nor any DI RNA product has been identified in cells infected with 5′ end DI virions. The inability to see a leader-like RNA might be expected since leader RNA has not been observed in VSV infected cells.[32] Attempts to demonstrate expression of 5′ end DI genomes by genetic complementation of temperature sensitive (*ts*) VSV mutants of Group I(L), II(NS), II(M), and IV(N) have also failed.[33] These data indicate that there is no expression of such DI particle genomes in infected cells.

III. RNA SYNTHESIS BY 3′ END DI PARTICLES

Early reports dealing with the in vitro transcribing abilities of a 3′ end DI virion, DI-LT, also were conflicting. Mori and Howatson[28] reported that this particle synthesized 5 to 15S RNAs in vitro at a rate even faster than that observed with the standard virion. On the other hand, Reichmann et al.[23] found only low levels of transcription from a preparation of 3′ end DI particles. These apparently contradictory reports may be attributable to the presence of two similar-sized DI particles (one transcribing, one not) in the seed stock used by Reichmann et al.[23] Recently Holland[42] has obtained evidence to support this possibility.

Colonno et al.[32] have examined RNA synthesis by the DI-LT particle derived from the heat-resistant VSV_{IND} strain. They first showed that leader RNA transcribed from VSV completely annealed to DI-LT RNA. Thus, if the leader RNA is the initiator of transcription, DI-LT should produce at least a leader RNA in vitro. Examination of the product demonstrated not only the synthesis of leader RNA but also four polyadenylylated, capped, and methylated RNAs that sedimented between 12 and 17S. These mRNAs were of the same size as those produced by transcription of the 3′ half of the VSV genome. Furthermore, this DI particle synthesized RNA at the same rate as VSV.

RNA synthesis by DI-LT in cultured cells has been examined also.[34] Addition of cycloheximide to cells just prior to infection limits virion expression to primary transcription,[35] i.e., synthesis of mRNAs from the input virion. Under these conditions purified DI-LT synthesized RNA at the same rate as VSV. Analysis of the product on SDS-sucrose gradients showed that DI-LT made 12 to 17S RNAs. These transcripts were complementary to virion RNA and were polyadenylylated, suggesting that they were mRNAs.

Following infection without cycloheximide VSV synthesized RNA at a faster rate than DI-LT. Interestingly, DI-LT also made more RNA without cycloheximide than it did in its presence. For the DI-LT infection this slightly greater synthesis appeared to be due to the presence of a new species of RNA which migrated at 28S. The 12 to 17S mRNAs described above also were observed in the absence of cycloheximide. The facts that the 28S RNA did not anneal to 5′ end DI particle RNA which contains part of the L gene and did not contain poly(A), together with the sensitivity of its synthesis to cycloheximide, indicated that it was not mRNA for the L protein. Its 28S size was maintained on denaturing dimethylsulfoxide gradients and thus was not due to adventitious association of labeled transcripts with the DI-LT template. the DI-LT genome is itself 28S and hence this newly synthesized RNA species could represent a replication product. The cycloheximide sensitivity is consistent with this possibility. The labeled 28S RNA contained species that annealed to DI-LT genomic RNA and 12 to 17S mRNAs. Thus both (+) and (−) strand RNAs were made, indicating that both steps of replication had occurred.

The synthesis of messenger and replicative RNA species by DI-LT has been studied as a function of time. Shortly after infection the 12 to 17S mRNAs are the predominant viral RNA made (Figure 1). With time replicative 28S RNAs make up a larger percent of the RNA. The reason for this change has not been determined but perhaps is a result of viral protein synthesis by DI-LT (see below). Specifically N protein may be required for replication since both (+) and (−) strand replicative RNAs sediment as ribonucleoprotein particles containing N protein.[36]

Analysis in the same fashion of species present in the replicative RNA fraction demonstrates that 1 hr after infection there is considerably more (+) strand than (−) strand (Figure 2). After 2, 3, and 4 hr of synthesis the relative amounts of the two strands are equivalent. During this time there is an overall increase in total replicative RNA (see above). This is consistent with a replication scheme for VSV in which a full-length (+) strand RNA is copied to form a new full-length (−) strand RNA.

Primary transcription by a 3′ end DI particle in the absence of helper virus was expected since the DI virion carries into the cell the (−) strand RNP, NS, and L proteins necessary for this process. However, replication was unexpected since VSV replication depends on synthesis of proteins and since DI-LT does not code for all the viral proteins.[37] A similar dependence of replication on protein synthesis is seen in DI-LT infected cells. However, no L protein mRNA or protein (see below) are made during this infection. These data suggest that replication requires the synthesis of one or several of the proteins made by DI-LT, i.e., N, NS, M, or G, but not the synthesis of the L protein. In this infection, the only L protein potentially involved in replication is that which enters with the DI virion.

Proof that the 12 to 17S RNAs are really mRNAs comes from the synthesis of viral proteins in cells infected with DI-LT. DI-LT infected cells were labeled with ^{35}S-methionine and the cytoplasmic proteins examined on SDS-polyacrylamide gels. Although DI-LT is not efficient at shutting off host protein synthesis, labeled M protein is demonstrated easily (Figure 3, lane 3). The proteins N and NS migrate close to host proteins and could not be distinguished from them when total cytoplasmic extracts were analyzed. However, DI RNPs containing newly made (+) and (−) strand RNAs were isolated and found to have ^{35}S-methionine N protein (Figure 3, lane 5). The recent demonstration that DI-LT is an effective cell-killing particle is evidence for the production of both N and NS since cell killing by VSV depends upon the synthesis of these two proteins.[38,39] Notably absent from the gels are labeled L and G proteins. The L protein is not coded for by the DI-LT particle and its absence is expected. The absence of G protein on these gels has not been explained as yet.

A second means of detecting protein synthesis specified by DI-LT is genetic complementation of *ts* VSV mutants with DI-LT. Single infections with either the *ts*-mutant or DI-LT should produce little virus at the nonpermissive temperature. However, if in a mixed infection at the restrictive temperature DI-LT makes an appropriate gene product, larger yields of infectious *ts* virus should be produced. Results with three *ts*-mutants are shown in Table 1. Both NS and G protein mutants are complemented by DI-LT indicating synthesis of these proteins. Similar positive results have been obtained with N and M protein mutants. As expected, no complementation is observed with the L protein mutant since this protein is not coded for by the DI-LT.

Lastly, synthesis of G and M proteins has been demonstrated by exploiting the dif-

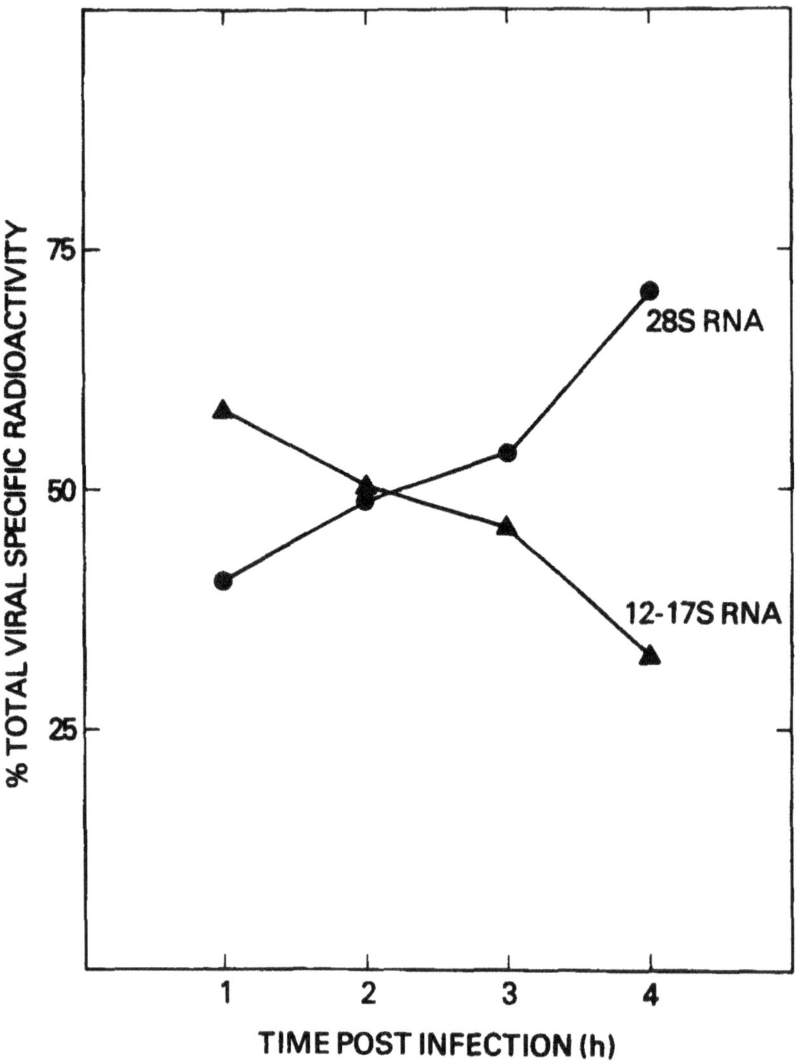

FIGURE 1. Time course for appearance of mRNAs and replicative RNAs in DI-LT infected cells. BHK-21 cells were treated with actinomycin D (10 μg/mℓ) for 30′ at 37°C. DI-LT was added at 200 physical particles per cell. At 20′ post infections, ^{3}H-uridine was added (500 μC/mℓ final) and the sample diluted 1:1 with medium containing actinomycin D. Samples were removed at the indicated times and cytoplasmic extracts prepared.[34] These were adjusted to 1% SDS, heated and the RNAs separated by centrifugation on 10 to 30% sucrose in 1 m*M* EDTA, 1 m*M* Tris-HCl, pH 7.5 and 0.1% SDS (SW41, 20,000 rpm, 16 hr, 20°C). Samples from uninfected cells were handled similarly. The gradients were fractionated and radioactivity of an aliquot determined after trichloroacetic acid (TCA) precipitation. To determine viral specific radioactivity, values for uninfected control cells were substracted from the corresponding DI-LT infected fraction. Radioactivity of 28S RNA (●——●) and 12 to 17S RNA (▲——▲) were tallied and expressed as percent of total viral specific radioactivity.

ferences in electrophoretic mobilities of these proteins for New Jersey (NJ) and Indiana (IND) serotypes of VSV. Chow et al.[40] coinfected cells with DI-LT which was derived from VSV_{IND} and helper VSV_{NJ}. The progeny DI particles contained proteins of both serotypes. Thus, DI-LT expressed genes for G and M proteins when helper virus of another serotype was present.

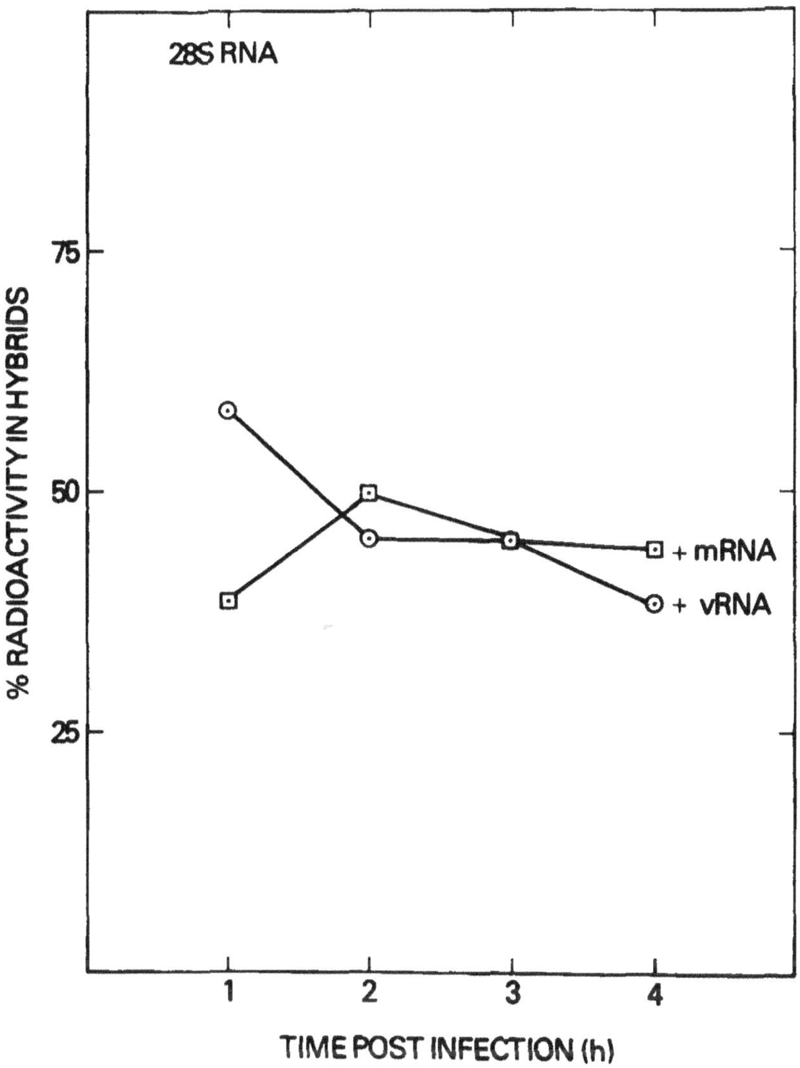

FIGURE 2. Time course for appearance of + and − strand replicative RNAs in cells infected with DI-LT. The 28S RNA fractions from each time point remaining after the analysis in Figure 1 were salt-ethanol precipitated, resuspended in 1 m*M* EDTA, 1 m*M* Tris-HCl, pH 7.5, and heated at 100°C for 3 min. Aliquots were annealed in 0.4 *M* NaCl, 0.01 *M* NaPi, pH 6.8, to excess unlabeled DI-LT virion RNA to determine the percent of + strand 28S material (⊙——⊙) or to excess unlabeled mRNAs to quantitate − strand 28S material (⊡——⊡). The values are expressed as percent RNase resistant radioactivity after 24 hr of annealing.

IV. CONCLUSIONS

RNA synthesis by VSV DI particles falls into two classes as might be predicted from the area of the VSV genome conserved in them. DI particles retaining a portion of the 5′ half are capable of only very limited RNA synthesis while a DI particle with the 3′ half of the VSV genome is capable of full expression of its RNA.

The implications of the RNA products from 5′ end DI particles remain to be explored. The in vitro production of a small RNA made by the 5′ end DI particles may

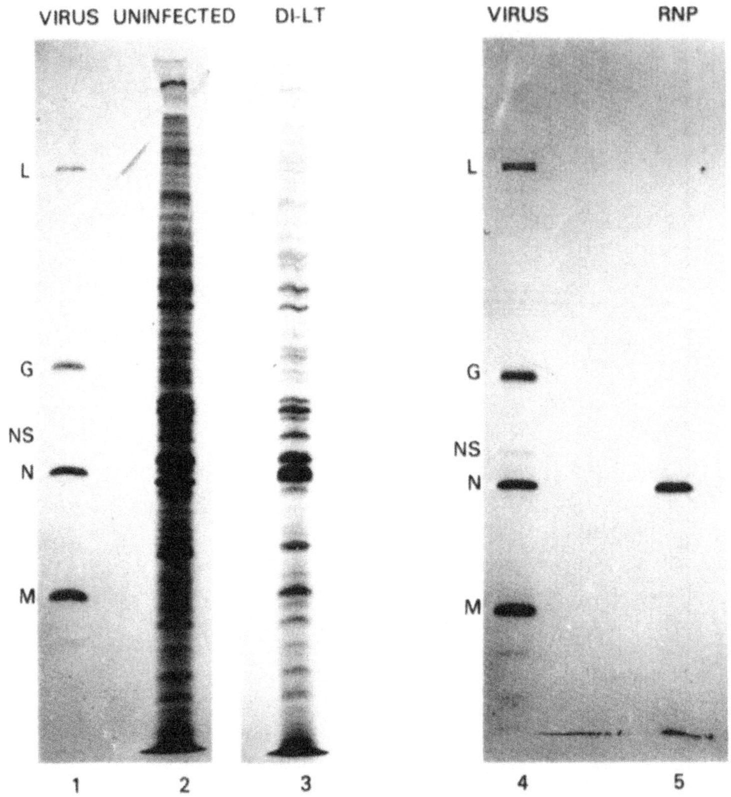

FIGURE 3. Radioautograms of SDS-polyacrylamide gels of ^{35}S-methionine labeled proteins made in DI-LT infected cells. BHK-21 cells were infected as outlined in Figure 1. ^{35}S-methionine was added at 3 hr post-infection; cells were harvested at 4 hr and cytoplasmic extracts were made. Lanes 1 and 4 illustrate the separation of VSV proteins that were also labeled with ^{35}S-methionine. The pattern of uninfected cells is shown in Lane 2. DI-LT infected cells are presented in Lane 3. For Lane 5, DI-LT RNPs were prepared by centrifugation of the cytoplasmic extract on 5 to 20% sucrose gradients in 1 *M* NH_4Cl, 10 m*M* NaCl, 10 m*M* Tris-HCl, pH 7.5, 20 m*M* EDTA (SW41, 40,000 rpm, 1.25 hr).[35] The DI-LT RNP was located by TCA precipitation and pelleted onto a glycerol cushion (SW41, 40,000 rpm, 3 hr). Aliquots were electrophoresed in a Laemmli gel system.[41]

result from an abortive replication or transcription process. The short transcript produced is similar to the B particle leader RNA in that they are both rich in adenosine. However, oligonucleotide composition studies indicate that they are not the same species.[21] Since such small RNAs have not been detected in cultured cells infected with VSV, their relationship to in vivo replication of the virus is not known.

The in vitro synthesis of a normal leader RNA and mRNAs by the 3′ end DI particle presages an extended role for this DI virion in vivo. DI-LT can contribute to protein synthesis during mixed infections with wild-type VSV[40] and can rescue mutants temperature sensitive for N, NS, M, and G proteins. The ability of this DI particle to direct the synthesis of viral proteins may potentiate an immune response in animals infected with helper and this DI particle. Consequently this DI may interfere with the growth of the helper in animals — interference with viral replication and potentiation of the immune response.

TABLE 1

Complementation Studies with VSV Temperature Sensitive Mutants and DI-LT[a]

ts-Mutant (Viral Protein)	DI-LT	PFU/mℓ 32°	PFU/mℓ 39.5°	CI[b]
—	+	$<10^1$	$<10^1$	
G-11(L)	−	1.0×10^3	$<10^1$	
	+	5.5×10^2	$<10^1$	0.55
G-22(NS)	−	1.0×10^3	$<10^2$	
	+	2.8×10^4	4.0×10^1	28
0/45(G)	−	5.0×10^1	n.d.	
	+	2.5×10^4	8.5×10^2	480

[a] BHK-21 monolayers were infected with *ts*-mutants (3 PFU/cell) with or without DI-LT (3 physical particles per cell) at 39.5°. After 7 hr, the supernatant was harvested and plaqued at 32° and 39.5°.

[b] CI = complementation index = $\frac{(ts \times DI\text{-}LT)^{32°} - (ts \times DI\text{-}LT)^{39.5°}}{(ts)^{32°} + (DI\text{-}LT)^{32°}}$

ACKNOWLEDGMENTS

We would like to thank Drs. Howard A. Nash and Ernst Freese for critical reading of this manuscript.

REFERENCES

1. **Huang, A. S. and Baltimore, D.**, Defective interfering animal viruses, in *Comprehensive Virology*, Vol. 10, Fraenkel-Conrat, H. and Wagner, R. R., Eds., Plenum Press, New York, 1977, 73.
2. **Kang, C. Y. and Prevec, L.**, Proteins of vesicular stomatitis virus. I. Polyacrylamide gel analysis of viral antigens, *J. Virol.*, 3, 404, 1969.
3. **Wagner, R. R., Schnaitman, T. C., Snyder, R. M., and Schnaitman, C. A.**, Protein composition of the structural components of vesicular stomatitis virus, *J. Virol.*, 3, 611, 1969.
4. **Perrault, J. and Holland, J. J.**, Absence of transcriptase activity of transcription-inhibiting ability in defective interfering particles of vesicular stomatitis virus, *Virology*, 50, 159, 1972.
5. **Schnitzlein, W. M. and Reichmann, M. E.**, The size and cistronic origin of defective vesicular stomatitis virus particle RNAs in relation to homotypic and heterotypic interference, *J. Mol. Biol.*, 101, 307, 1976.
6. **Ball, L. A. and White, C. N.**, Order of transcription of genes of vesicular stomatitis virus, *Proc. Natl. Acad. Sci., U.S.A.*, 73, 442, 1976.
7. **Abraham, G. and Banerjee, A. K.**, Sequential transcription of the genes of vesicular stomatitis virus, *Proc. Natl. Acad. Sci., U.S.A.*, 73, 1504, 1976.
8. **Colonno, R. J. and Banerjee, A. K.**, Mapping and initiation studies on the leader RNA of vesicular stomatitis virus, *Virology*, 77, 260, 1977.
9. **Moyer, S. A. and Banerjee, A. K.**, Messenger RNA species synthesized in vitro by the virion-associated RNA polymerase of vesicular stomatitis virus, *Cell*, 4, 37, 1975.
10. **Rose, J. K. and Knipe, D.**, Nucleotide sequence complexities, molecular weights and poly(A) content of vesicular stomatitis virus mRNA species, *J. Virol.*, 15, 994, 1975.
11. **Both, G. W., Moyer, S. A., and Banerjee, A. K.**, Translation and identification of the viral mRNA species isolated from subcellular fractions of vesicular stomatitis virus infected cells, *J. Virol.*, 15, 1012, 1975.

12. **Morrison, T., Stampfer, M., Baltimore, D., and Lodish, H. F.**, In vitro translation of vesicular stomatitis virus messenger RNAs and the existence of a 40S "plus" strand, in *Negative Strand Viruses*, Mahy, B. W. H. and Barry, R. D., Eds., Academic Press, New York, 1975, 293.
13. **Stampfer, M. and Baltimore, D.**, Identification of the vesicular stomatitis virus large protein as a unique viral protein, *J. Virol.*, 11, 520, 1973.
14. **Knipe, D., Rose, J. K., and Lodish, H. F.**, Translation of individual species of vesicular stomatitis viral mRNA, *J. Virol.*, 15, 1004, 1975.
15. **Stamminger, G. and Lazzarini, R. A.**, Analysis of RNA of defective VSV particles, *Cell*, 3, 85, 1974.
16. **Leamnson, R. N. and Reichmann, M. E.**, The RNA of defective interfering vesicular stomatitis virus particles in relation to viral cistrons, *J. Mol. Biol.*, 85, 551, 1974.
17. **Lazzarini, R. A., Weber, G. H., Johnson, L. D., and Stamminger, G. M.**, Covalently linked message and anti-message (genomic) RNA from a defective interfering vesicular stomatitis virus particle, *J. Mol. Biol.*, 99, 289, 1975.
18. **Perrault, J.**, Cross-linked double-stranded RNA from a defective vesicular stomatitis virus particle, *Virology*, 70, 360, 1976.
19. **Perrault, J. and Leavitt, R. W.**, Inverted complementary terminal sequences in single-stranded RNAs and snap-back RNAs from vesicular stomatitis defective interfering particles, *J. Gen. Virol.*, 38, 35, 1978.
20. **Petric, M. and Prevec, L.**, Vesicular stomatitis virus — a new interfering particle, intracellular structures and virus-specific RNA, *Virology*, 41, 615, 1970.
21. **Emerson, S. U., Dierks, P. M., and Parsons, J. T.**, In vitro synthesis of a unique RNA species by a T particle of vesicular stomatitis virus, *J. Virol.*, 23, 708, 1977.
22. **Keene, J. D., Shubert, M. F., Lazzarini, R. A., and Rosenberg, M.**, Nucleotide sequence homology at the 3′ termini of RNA from vesicular stomatitis virus and its defective interfering particles, *Proc. Natl. Acad. Sci. U.S.A.*, 75, 3225, 1978.
23. **Reichmann, M. E., Villarreal, L. P., Kohne, D., Lesnaw, J., and Holland, J. J.**, RNA polymerase activity and poly(A) synthesizing activity in defective T particles of vesicular stomatitis virus, *Virology*, 58, 240, 1974.
24. **Bishop, D. H. L. and Roy, P.**, Genome homology of vesicular stomatitis virus and defective T particles and evidence for sequential transcription of the virion ribonucleic acid, *J. Virol.*, 9, 946, 1972.
25. **Emerson, S. U. and Wagner, R. R.**, Dissociation and reconstitution of the transcriptase and template activities of vesicular stomatitis B and T virions, *J. Virol.*, 10, 297, 1972.
26. **Leppart, M., Kort, L., and Kolakofsky, D.**, Further characterization of Sendai virus DI-RNAs: a model for their generation, *Cell*, 12, 539, 1977.
27. **Huang, A. S.**, Viral pathogenesis and molecular biology, *Bacteriol. Rev.*, 41, 811, 1977.
28. **Mori, H. and Howatson, A. F.**, In vitro transcriptase activity of vesicular stomatitis virus B and T particles: analysis of product, *Intervirology*, 1, 168, 1973.
29. **Kang, C. Y., Glimp, T., Clewley, J. P., and Bishop, D. H. L.**, Studies on the generation of vesicular stomatitis virus (Indiana serotype) defective interfering particles, *Virology*, 84, 142, 1978.
30. **Stampfer, M., Baltimore, D., and Huang, A. S.**, Ribonucleic acid synthesis of vesicular stomatitis virus. I. Species of ribonucleic acid found in Chinese hamster ovary cells infected with plaque-forming and defective particles, *J. Virol.*, 4, 154, 1969.
31. **Huang, A. S. and Manders, E. K.**, Ribonucleic acid synthesis of vesicular stomatitis virus. IV. Transcription by standard virus in the presence of defective interfering particles, *J. Virol.*, 9, 909, 1972.
32. **Colonno, R. J., Lazzarini, R. A., Keene, J. D., and Banerjee, A. K.**, In vitro synthesis of messenger RNA by a defective interfering particle of vesicular stomatitis virus, *Proc. Natl. Acad. Sci., U.S.A.*, 74, 1884, 1977.
33. **Reichmann, M. E., Pringle, C. R., and Follet, E. A. L.**, Defective particles in BHK cells infected with temperature-sensitive mutants of vesicular stomatitis virus, *J. Virol.*, 8, 154, 1971.
34. **Johnson, L. D. and Lazzarini, R. A.**, Replication of viral RNA by a defective interfering vesicular stomatitis virus particle in the absence of helper virus, *Proc. Natl. Acad. Sci., U.S.A.*, 74, 4387, 1977.
35. **Marcus, P. I., Engelhardt, D. L., Hunt, J. M., and Sekellik, M. J.**, Interferon action: inhibition of vesicular stomatitis virus RNA synthesis induced by virion-bound polymerase, *Science*, 174, 593, 1971.
36. **Soria, M., Little, S. P., and Huang, A. S.**, Characterization of vesicular stomatitis virus nucleocapsids. I. Complementary 40S RNA molecules in nucleocapsids, *Virology*, 61, 270, 1974.
37. **Perlman, S. M. and Huang, A. S.**, RNA synthesis of vesicular stomatitis virus. V. Interactions between transcription and replication, *J. Virol.*, 12, 1395, 1973.

38. **Marcus, P. I., Sekellick, M. J., Johnson, L. D., and Lazzarini, R. A.,** Cell killing by viruses. V. Transcribing defective interfering particles of vesicular stomatitis virus function as cell killing particles, *Virology*, 82, 242, 1977.
39. **Marvaldi, J. L., Lucas-Lenard, J., Sekellick, M. J., and Marcus, P. I.,** Cell killing by viruses. IV. Cell killing and the inhibition of cell protein synthesis require the same gene functions of vesicular stomatitis virus, *Virology*, 79, 267, 1977.
40. **Chow, J. M.,** Schnitzlein, W. M., and **Reichmann, M. E.,** Expression of genetic information contained in the RNA of a defective interfering particle of vesicular stomatitis virus, *Virology*, 77, 579, 1977.
41. **Laemmli, U. K.,** Cleavage of structural proteins during the assembly of the head of bacteriophage T4, *Nature (London)*, 227, 680, 1970.
42. **Holland, J. J.,** personal communication, 1978.

Chapter 11

RHABDOVIRUS DEFECTIVE PARTICLES: ORIGIN AND GENOME ASSIGNMENTS

M. E. Reichmann and W. M. Schnitzlein

TABLE OF CONTENTS

I. CHARACTERIZATION OF DEFECTIVE INTERFERING (DI) PARTICLES

A. General Properties of DI Particles

Undiluted passages of animal viruses in tissue cultures result in the generation of truncated particles which interfere with viral infection. The phenomenon is also termed autointerference and was first observed by von Magnus in influenza-infected chick embryos.[1] Shortly afterwards, Cooper and Bellett[2] extended this observation to vesicular stomatitis virus (VSV) in chick embryo cell monolayers. The authors suggested that the inhibition was due to a transmissible interfering agent (T) which could have been either an incomplete form of VSV or an interferon-like substance.[3] Incomplete forms of VSV were indeed observed in electron micrographs by Reczko,[4] but the first positive correlation between autointerference and the presence of shorter VSV particles was reported by Hackett.[5] Finally, with the application of rate zonal centrifugation in the purification of viruses, it became possible to separate the short VSV particles free of complete virions.[6-8] In the purified preparations, infectivity was associated with the complete particles while autointerfering ability resided in the shorter particles.

Meanwhile, defective particles in other viral systems were being reported with increasing frequency. The ubiquity of this phenomenon suggested that it may have an important role in viral infections. In view of this, Huang and Baltimore[9] formulated the generalized properties of autointerfering incomplete particles and renamed them defective interfering (DI) particles. These characteristics were largely based on physical and chemical investigations of the VSV DI particles which were as follows:

1. They contained all viral proteins.[10,11]
2. The sedimentation coefficient of the RNA contained in these particles was only 19S compared to a 40S sedimentation coefficient of virion RNA. This corresponded approximately to one third of the viral genome[12] and also reflected the relative sizes of two particles.
3. The interference by these particles did not involve interferon production, was intracellular and sensitive to ultraviolet (UV) light. It was concluded that a functional RNA was required.[13]
4. The autointerference was specific to the viral serotype from which the short particles were derived.[13,14]

B. Variability of DI-Particle RNAs

Various laboratories reported the generation of a particle one third the size of the infectious virion and, in the jargon of the day, this was termed the wild-type T particle.[15-17] However, Petric and Prevec,[18] while investigating a heat-resistant (HR) mutant of VSV, discovered a new T particle which corresponded approximately to one half of the size of the virus and which contained one half of the viral genome. This long T particle also differed from the wild-type short T by being capable of interfering heterotypically with New Jersey serotype viral infections as well as homotypically with its own Indiana serotype virus.[19] At that time, Pringle[20] had isolated chemically induced temperature-sensitive *(ts)* mutants of VSV Indiana serotype which fell into four complementation groups. In an attempt to utilize the mutants in elucidating the important steps for autointerference in the infectious cycle, it was observed that T particles of a great variety of sizes were generated by these mutants.[21] Thus, T-particle RNAs ranging from 12 to 50% of the viral genome became available in a purified form. A reinvestigation of the "wild-type" DI-particle RNAs indicated that different isolates of the wild-type virus also generated DI particles with different genomic sizes. This is illustrated in Figures 1 and 2 where gel and sucrose gradient profiles of RNA molecules from various particles are shown. It was also observed that the DI particle generated by Glasgow wild-type Indiana virions exhibited abnormal sedimentation properties compared to its mobility in gels (Figure 1F and 1C, respectively). While the low sedimentation coefficient suggested a low molecular weight for this RNA, its slow migration in gels indicated a high molecular weight. This discrepancy was explainable by assuming a high degree of secondary structure (high frictional coefficient). This feature was later observed in other DI-particle RNAs[22-25]and will be discussed later in connection with the mechanism of their generation.

The isolation of different DI particles from infections by HR, *ts* mutant and wild-type virions of different origins suggested a possible genetic predisposition to generate a particular type of DI particle.[20] But Holland et al.[26] have shown that extensive clonal purification of a wild-type or the *ts* G31 mutant virion, from which all seeds of DI particles were removed, rendered inocula capable of generating a variety of different sized DI particles. However, even in this case, some selection occurred during prolonged passages. Moreover, the nature of the generated DI particles or their selection was also found to be a function of the host cell.

While all observations previously discussed dealt with the Indiana serotype of VSV and its mutants, similar conclusions were reached with respect to DI particles of the New Jersey serotype of VSV and some other rhabdoviruses. Viral isolates of the New Jersey serotype seem to generate a greater variety of DI particles than its Indiana counterpart. The DI particles can often be very similar in size. Thus, the RNA of particles isolated from a single sucrose gradient zone may be quite heterogeneous.[27,28] A similar heterogeneity was recently observed with some DI particles of Chandipura, Piry, and Mokola.[29]

II. CISTRONIC CONTENT OF DI-PARTICLE RNAS

A. Polarity of DI-Particle RNAs

The negative strand nature of VSV and DI-particle RNAs was demonstrated by annealing the latter to cytoplasmic extracts of virus-infected cells. Simultaneously, it was shown that DI-particle RNA was not complementary to the viral RNA and was, therefore, of the same sense.[30,31] This confirmed earlier suggestions that the DI particles are deletion mutants of the parental virion.[6] Viral- and DI-particle RNAs were also annealed to transcription products obtained in vitro.[32,33] Kinetic in vitro experiments demonstrated that RNA sequences contained in wild-type DI particles were absent in the early stages of viral RNA transcription but appeared at a later stage of the reaction.[33] On the other hand, early reports of in vivo experiments indicated that the DI-particle RNA sequences were complementary in various degrees to all mRNA species.[34]

B. Mapping of DI-Particle RNAs

1. Indiana Serotype-VSV

The first attempt to define the genomic content of DI-particle RNA was of a hypothetical nature. It tried to explain the difference in heterotypic interfering ability of the long and short DI particles of Prevec and their inability to replicate independently of complete virions.[19,35] The requirement for complete virions was assumed to be a result of the deletion of the RNA polymerase cistrons (proteins L and NS).[35] Thus, the inability of DI-particle RNA to synthesize the replicating enzymes explained the dependence of its replication on the virion. Furthermore, it was assumed that the differences between the Indiana and New Jersey serotype virions were mainly reflected in the glyco- (G) and matrix (M) proteins and their respective cistrons. It was therefore

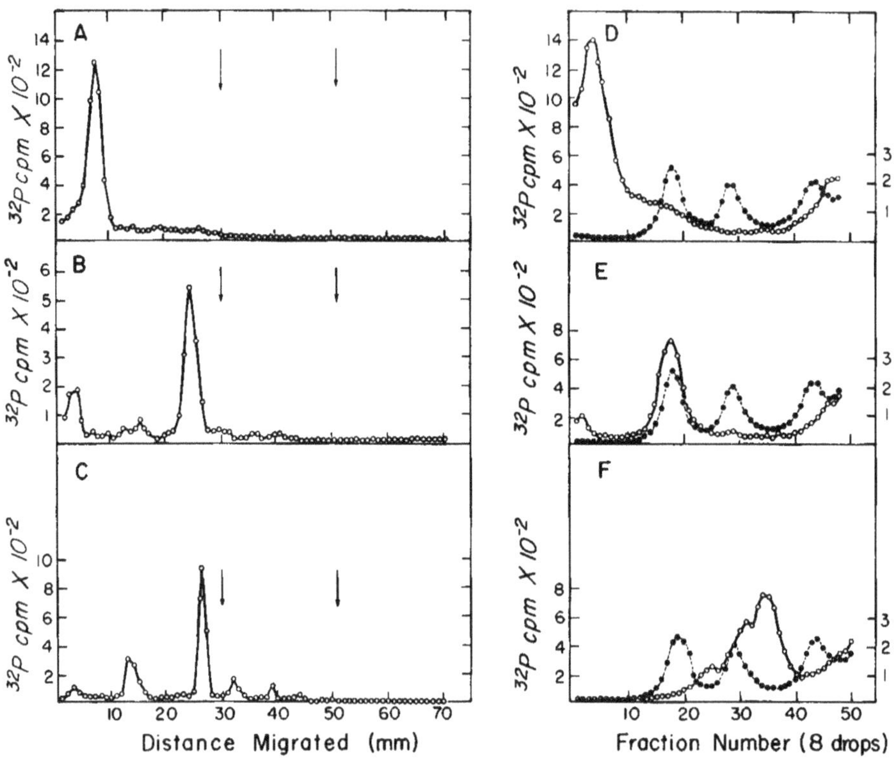

FIGURE 1. Characterization of virion RNA and various DI particle RNA species of Indiana serotype VSV.[68] (A, B, and C) Polyacrylamide gel electrophoresis RNA profiles of Gla virion, HR DI particles, and Gla DI particles, respectively, labeled with ^{32}P. Arrows indicate positions of ^{14}C-labeled BHK ribosomal RNA markers. (D, E, and F) Sucrose gradient profiles of ^{32}P-labeled RNA of Gla virion, HR DI particles, and Gla DI particles, respectively (O—O—O). The positions of ^{3}H-labeled BHK ribosomal and tRNA markers are also shown (•—•—•). Isolation procedures, details of the polyacrylamide gel electrophoresis and sucrose gradient techniques, and the adopted nomenclature are as published before.[22,23]

postulated that these two cistrons composed the short DI-particle RNA, which could not be recognized by the nonhomologous New Jersey viral replicase. On the other hand, the long, heterotypically interfering DI-particle RNA was postulated to also contain the cistron coding for N protein which was supposedly similar in the two viral serotypes. Therefore, this nucleotide sequence could be recognized by the polymerase of both viruses (see Figure 3A).

The availability of DI-particle RNAs of various sizes and improvements in the techniques of viral mRNA fractionations made it finally possible to determine the cistronic content of DI-particle RNAs experimentally. The rationale of these experiments was based on the finding that the individual mRNA fractions had unique nucleotide sequences and did not compete with each other in annealing experiments with virion RNA. They could therefore be used to determine the presence or absence of specific nucleotide sequences in DI-particle RNAs and answer the question whether interference is correlated with a single defined genetic locus.

In the early work, mRNAs were fractionated by sucrose gradient centrifugation and polyacrylamide gel electrophoresis into 13S, 18S, and 30S size classes. These were annealed to three wild-type, one *ts* mutant, and the HR mutant-generated DI-particle RNAs.[22] The results of these experiments suggested the absence of a nucleotide se-

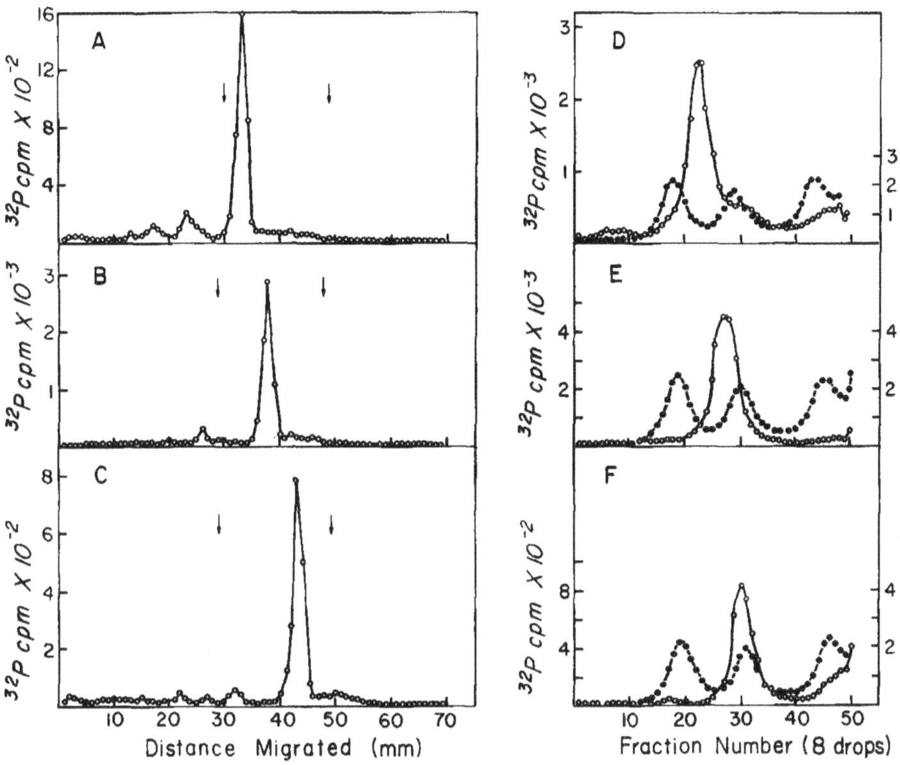

FIGURE 2. Characterization of various DI particle RNA species of Indiana serotype VSV.[68] (A, B, and C) Polyacrylamide gel electrophoresis RNA profiles of How, Hu, and MS DI particles, respectively, labeled with ^{32}P. Arrows indicate positions of ^{14}C-labeled BHK ribosomal RNA markers. (D, E, and F) Sucrose gradient profiles of ^{32}P-labeled RNA of How, Hu, and MS DI particles, respectively (O—O—O). The positions of BHK ribosomal and tRNA markers labeled with ^{3}H are also shown (•—•—•). For further details see Figure 1.

quence linked with autointerference. This was reflected in the ability of the HR DI-particle RNA to anneal to the 13S and 18S mRNAs, while all other particle RNAs annealed to the 30S mRNA. One of the wild-type DI-particle RNAs (How) was complementary to both groups of mRNA species. However, competition experiments between the DI-particle RNAs annealing to the 30S mRNA revealed that all sequences of the short particle RNAs were contained in the long particle RNAs. Thus, at least these particles appeared to have been derived from a common genetic locus. Based on these findings, a map was proposed in which the HR DI-particle RNA was located in the 13S and 18S cistrons while the other particle RNAs originated in a fixed internal part of the 30S and extended towards the 13S and 18S mRNA cistrons (Figure 3B). An alternate map, consistent with the experimental data, but postulating a short common nucleotide sequence for all DI-particle RNAs, was also proposed[36] (Figure 3C).

Similar results were obtained by Stamminger and Lazzarini,[37] whose annealing experiments were carried out with one wild-type and the HR DI-particle RNAs. Their wild-type DI-particle RNA behaved like the How DI-particle RNA insofar that it contained a short sequence complementary to the 13S mRNA species while most of it annealed to the 30S mRNA species. Competition experiments between wild-type and HR DI-particle RNAs failed to reveal any common sequences between the two and led to a map shown in Figure 3D.

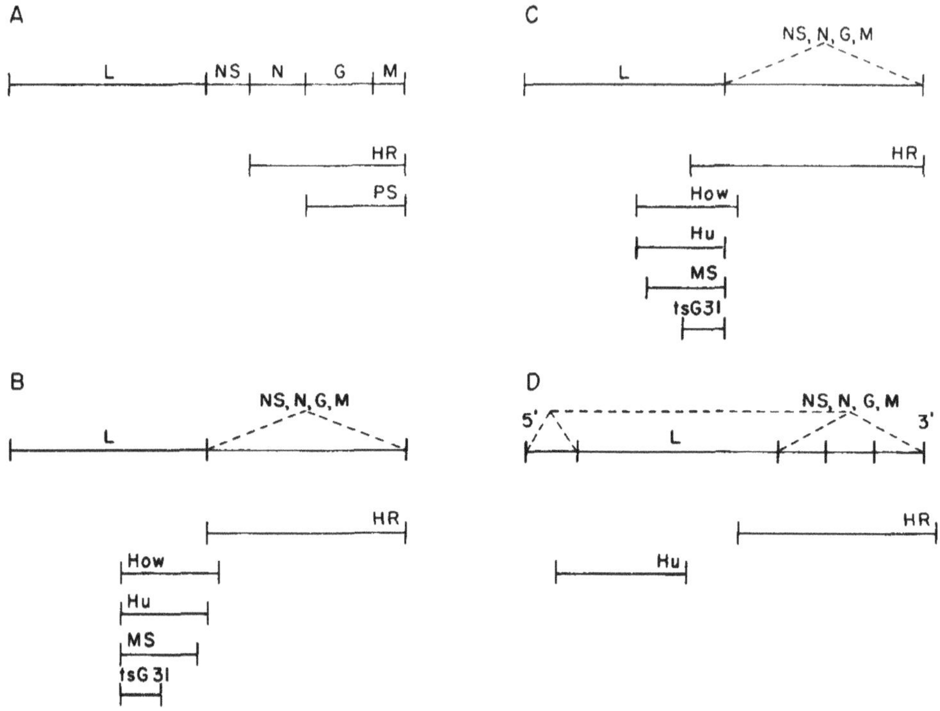

FIGURE 3. Proposed maps showing the location of various DI particle RNAs relative to the viral genome. The position of the viral cistrons are indicated by the proteins they code for. The nomenclature for the various DI particles follows that of Schnitzlein and Reichmann.[23] The maps were proposed by Prevec (A),[35] Leamnson and Reichmann (B) and (C),[22,36] and Stamminger and Lazzarini.[37]

Subsequent work revealed the need for a more rigorous purification procedure of both DI-particle RNAs and mRNA species.[23] This was accomplished by elution from polyacrylamide gels, following electrophoresis. RNAs purified in this manner demonstrated the absence of any nucleotide sequences of DI-particle RNAs, except those of HR, complementary to the 13S and 18S mRNA species. Moreover, the investigation was extended to a greater number of DI-particle RNAs, some of which corresponded in size to that of the HR DI-particle RNA. The unique nature of the later DI-particle RNA was thus established.[23] A new map was then proposed with essentially the same features as the map shown in Figure 4. At that time, the map was based on the following observations:

1. Roy and Bishop's demonstration[33] that only late (180 min) viral in vitro RNA transcripts contained sequences complementary to wild-type DI-particle RNA suggested that these sequences were located distally from the 3′ end of the virion. This taken together with the annealing data dictated that the 30S mRNA cistron should comprise the 5′ half of the viral genome.
2. Stamminger and Lazzarini reported that the wild-type DI-particle RNA annealed to the 3′ terminal fragments of the 30S mRNA.[37] Based on the previous conclusion, this end would have to be equivalent to the 5′ terminus of the viral RNA.
3. Since the nucleotide sequences of the shorter DI-particle RNAs were always contained in the larger DI-particle RNAs,[22,23] it follows that the additional sequences extend from the 5′ end to the 3′ end of this viral cistron, as illustrated in Figure 4. Originally, the 5′ termini of DI-particle RNAs were not completely aligned with

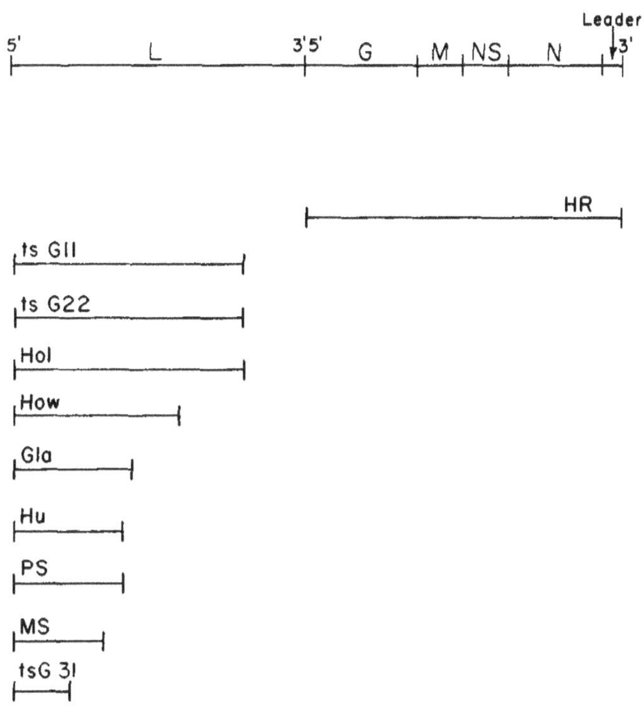

FIGURE 4. Map showing the position of various DI particle RNAs in relation to the cistrons and termini of the viral genome. The locations of the viral cistrons are indicated by the proteins they code for. The gene order and the position of the leader sequence are based on experimental findings.[43-45,48] The nomenclature of the various DI particles conforms with that of Schnitzlein and Reichmann.[23]

the 5′ termini of the virion RNA in order to allow for the possibility of nonequivalence at this end. Since the conserved end was also likely to be required for interference, the different termini of DI and virion RNAs would allow for a greater affinity of the replicative enzymes toward the DI-particle RNA template. This increased affinity was frequently postulated to be the basis of interference.[38,39] Recently, Perrault et al.[40] presented evidence that the 5′ end of virion RNA can be protected from endonuclease digestion by annealing with a short nucleotide sequence complementary to the 5′ end of DI-particle RNA. Based on this evidence, the map in Figure 4 is presented with the 5′ termini aligned.

As seen in Figure 4, the HR DI-particle RNA has a unique location on the map. The particle is also unique in its biological properties insofar that it interferes heterotypically with the New Jersey Concan subgroup virions[19] although it does not interfere with the Hazelhurst subgroup virions.[41,42] No other DI-particle RNA like the HR has been described in the literature. However, in our laboratory, the mutant *ts* O 45 generated a particle with an RNA of a similar cistronic content but with a much higher degree of secondary structure than the HR DI-particle RNA.

The precise location of the individual cistrons, as shown in Figure 4, has also been determined.[43-46] In addition to nucleotide sequences corresponding to these cistrons, it was shown that the 3′ terminus of the virus contains an extracistronic leader sequence of approximately 70 nucleotides.[47,48] A transcript of this sequence has been obtained in vitro. Annealing experiments with this transcript have shown that the HR DI-particle RNA also contains the 3′ terminal leader sequence. However, no annealing between the leader sequence and the RNAs of other DI particles could so far be demonstrated.[49] Recent sequencing of the 3′ termini of wild-type DI-particle and viral RNAs showed extensive homologies in the first 17 nucleotides with only a 3-nucleotide mismatch between nucleotides 8 to 13 from the 3′ end.[50] Annealing experiments between the leader sequence transcript and this region of DI-particle RNA would result after ribonuclease digestion in duplexes of 8 and 5 nucleotides long, respectively. Such structures have been shown to not be stable at room temperature.[51,52]

2. New Jersey Serotype-VSV

DI particles generated by New Jersey serotype of VSV have also been isolated and mapped.[28] All of these DI particles contained RNAs with nucleotide sequences corresponding to part of the L protein cistron, and the sequences of the shorter RNAs were always contained in larger RNAs. By analogy to the Indiana serotype, these DI-particle RNAs also probably originated at the 5′ end of the viral genome.

Several isolates of the New Jersey serotype virion exhibited very limited homology among their RNAs,[28,53,54] as discovered during the mapping of their respective DI-particle RNAs. In view of this, the New Jersey serotype was recently divided into two subroups.[41] In spite of this limited homology, DI particles generated by one subgroup interfered equally well with infections by virions of both subgroups.[28] The shortest of these RNAs contained a maximum sequence of 200 nucleotides homologous to the virion RNA of the other subgroup. These results suggested that, if a conserved nucleotide sequence is required for interference, such a sequence would not exceed 200 nucleotides.

III. ORIGIN OF DI PARTICLES

A. Generation of DI-Particle RNAs

In all cases, DI-particle RNAs are located at one of the termini of the viral genome. This location suggests that these RNAs are generated by an interrupted transcription of a complete negative (HR DI-particle RNA) or positive (all other DI-particle RNAs) viral RNA strand. The specific sizes of DI-particle RNAs require either that the termination take place at specific sites on the viral genome or, in case of a random termination, that a subsequent selection of more efficiently replicating species occur. However, that the process is more complex than just random termination and selection is indicated by the failure to generate DI particles from UV irradiated virions.[26] An early useful clue in the elucidation of the mechanism by which DI particles might be generated was the observation of secondary structure in DI-particle RNAs[22] and the discovery of a complete hairpin-like structure of one DI-particle RNA.[24] Various degrees of "snap-back" structures were reported subsequently.[25] These RNAs clearly contained covalently linked nucleotide sequences corresponding to the negative as well as to the positive strand of viral RNA. The self-annealing parts of the molecule always contained the two ends.[55,56] These inverted complementary terminal sequences resulted in self-annealing, panhandle-like structures visible in electron micrographs. Also, at higher concentrations, multimeric structures were formed by intermolecular end to end annealing.[55] Virion RNA did not exhibit self-annealing properties of this kind.[56] A similar structure of DI-particle RNA was also reported for Sendai and alpha viruses.[57-59] In the latter case, however, inverted terminal sequences were also present in the viral RNA.[60]

The secondary structure of DI-particle RNAs suggested a mechanism by which these particles could be generated. During normal transcription of the complete positive or negative viral RNA strand, the transcribing enzyme could switch templates by turning around on its own nascent polynucleotide chain. Termination would result when the 5′ terminus of the nascent chain was reached. The resulting RNA would contain complementary 5′ and 3′ sequences of a length dependent on the point where the switch in templates took place.[61,62] Alternately, the switch in templates could involve adjacent nascent molecules. This mechanism would explain why generation of DI particles is favored at high multiplicities of infection since the intracellular concentration of nascent chains would be presumably greater than at low multiplicities of infection. Template jumping as a basis for generation of DI-particle RNAs of alpha viruses, which constitute internal deletions, has also been suggested. The deletion is supposedly facilitated by the interaction of the two sticky ends of the viral genome resulting in the looping out of this template.[58,59]

B. Selective Advantage of DI-Particle RNAs

Incomplete transcripts of the viral genome are probably generated at comparable frequencies by mechanisms other than the template switching discussed above. Perhaps the reason these are not detected may be due to a lack of any replicative advantages which would lead to their enrichment and subsequent selection. Mechanistically, the conservation of an identical replicase recognition site on the negative and positive RNA strands, as afforded by the DI-particle RNA structures discussed above, is equivalent to the generation of multiple initiation sites in DI particles of some DNA viruses.[63-65] In the case of defective RNA viruses, a replicative advantage of a kinetic nature assuming a faster turnover of shorter RNA molecules has also been suggested.[66] This selective advantage seems to be operative in the generation of alpha DI particles in addition to the complementary termini of their RNAs.[67] Since the virion RNAs themselves contain complementary ends, this feature alone would not impart by itself a replicative advantage to the DI particles. However, in the case of VSV DI particles, the shorter size of their RNA does not seem to constitute a replicative advantage. This conclusion is based on the following observations:

1. It has been shown that the specific interfering ability per DI particle remains approximately the same even though their RNAs differ in size by as much as a factor of four.[53]
2. In infections containing mixed populations of DI particles, the selection process did not always operate in favor of the shortest DI-particle RNA.[26] Therefore, it appears that in the case of VSV DI particles, the predominant selective pressure is based on the conservation of a 3′ terminal sequence of nucleotides complementary to the 5′ terminal sequence of the viral genome and of the DI-particle RNAs. This is also probably true in DI particles which do not exhibit a high degree of secondary structure in their RNA, since the sequence may be too short to be detectable by the techniques commonly applied for their measurements.

It should be noted that the above mentioned selection process can not be applied to the HR DI particle, since its RNA did not originate from the 5′ end of the viral genome. Instead, a similar template switch resulting in 3′ and 5′ complementary termini could have generated this DI-particle RNA during the transcription of the negative viral RNA strand. In this case, the HR DI-particle RNA would have a 3′ terminal sequence homologous with that of the virion RNA and not with the other DI-particle RNAs. Therefore, VSV DI-particle RNAs may be selected on the basis of having complementary ends and one terminus which is homologous to one of the virion genome ends. While the hypotheses may seem plausible in the light of up-to-date investigations, ultimately the experimental verification of the mode of action of DI particles rests on a successful isolation of a replicative complex which would facilitate in vitro studies of replicative phenomena.

ACKNOWLEDGMENT

The work done in the authors' laboratory, which is quoted in the article, was supported by grants USPH AI 12070, NSF GB 34171, and American Cancer Society VC 92 to M. E. Reichmann.

REFERENCES

1. **von Magnus, P.**, Incomplete forms of influenza virus, *Adv. Virus Res.*, 2, 59, 1954.
2. **Cooper, P. D. and Bellett, A. J. D.**, A transmissible interfering component of vesicular stomatitis virus preparations, *J. Gen. Microbiol.*, 21, 485, 1959.
3. **Bellett, A. J. D. and Cooper, P. D.**, Some properties of the transmissible interfering component of VSV preparations, *J. Gen. Microbiol.*, 21, 498, 1959.
4. **Reczko, E.**, Electronenmikroskopische Untersuchungen am Virus der Stomatitis Vesicularis, *Arch. Gesamte Virusforsch.*, 10, 588, 1960.
5. **Hackett, A. J.**, A possible morphological basis for the autointerference phenomenon in vesicular stomatitis virus, *Virology*, 24, 51, 1964.
6. **Huang, A. S., Greenawalt, J. W., and Wagner, R. R.**, Defective T particles of vesicular stomatitis virus. I. Preparation, morphology, and some biologic properties, *Virology*, 30, 161, 1966.
7. **Crick, J., Cartwright, B., and Brown, F.**, Interfering components of vesicular stomatitis virus, *Nature (London)*, 211, 1204, 1966.
8. **Hackett, A. J., Schaffer, F. L., and Madin, S. H.**, The separation of infectious and autointerfering particles in vesicular stomatitis virus preparations, *Virology*, 31, 114, 1967.
9. **Huang, A. S. and Baltimore, D.**, Defective viral particles and viral disease processes, *Nature (London)*, 226, 325, 1970.
10. **Wagner, R. R., Schnaitman, T. A., and Snyder, R. M.**, Structural proteins of vesicular stomatitis viruses, *J. Virol.*, 3, 395, 1969.
11. **Kang, C. Y. and Prevec, L.**, Proteins of vesicular stomatitis virus. I. Polyacrylamide gel analysis of viral antigens, *J. Virol.*, 3, 404, 1969.
12. **Huang, A. S. and Wagner, R. R.**, Comparative sedimentation coefficients of RNA extracted from plaque-forming and defective particles of vesicular stomatitis virus, *J. Mol. Biol.*, 22, 381, 1966.
13. **Huang, A. S. and Wagner, R. R.**, Defective T particles of vesicular stomatitis virus. II. Biologic role in homologous interference, *Virology*, 30, 173, 1966.
14. **Crick, J. and Brown, F.**, Interference as a measure of cross-relationship in the vesicular stomatitis group of rhabdoviruses, *J. Gen. Virol.*, 18, 79, 1973.
15. **Brown, F., Martin, S. J., Cartwright, B., and Crick, J.**, The ribonucleic acids of the infective and interfering components of vesicular stomatitis virus, *J. Gen. Virol.*, 1, 479, 1967.
16. **Nakai, T. and Howatson, A. F.**, The fine structure of vesicular stomatitis virus, *Virology*, 35, 268, 1968.
17. **Wild, T. F.**, Replication of vesicular stomatitis virus: characterization of the virus-induced RNA, *J. Gen. Virol.*, 13, 295, 1971.
18. **Petric, M. and Prevec, L.**, Vesicular stomatitis virus — a new interfering particle, intracellular structures, and virus-specific RNA, *Virology*, 41, 615, 1970.
19. **Prevec, L. and Kang, C. Y.**, Homotypic and heterotypic interference by defective particles of vesicular stomatitis virus, *Nature (London)*, 228, 25, 1970.
20. **Pringle, C. R.**, Genetic characteristics of conditional lethal mutants of vesicular stomatitis virus induced by 5-florouracil,5-azacytidine and ethyl methane sulfonate, *J. Virol.*, 5, 559, 1970.
21. **Reichmann, M. E., Pringle, C. R., and Follett, E. A. C.**, Defective particles in BHK cells infected with temperature-sensitive mutants of vesicular stomatitis virus, *J. Virol.*, 8, 154, 1971.

22. **Leamnson, R. N. and Reichmann, M. E.**, The RNA of defective vesicular stomatitis virus particles in relation to viral cistrons, *J. Mol. Biol.*, 85, 551, 1974.
23. **Schnitzlein, W. M. and Reichmann, M. E.**, The size and the cistronic origin of defective vesicular stomatitis virus particle RNAs in relation to homotypic and heterotypic interference, *J. Mol. Biol.*, 101, 307, 1976.
24. **Lazzarini, R. A., Weber, G. H., Johnson, L. D., and Stamminger, G. M.**, Covalently linked message and antimessage (genomic) RNA from a defective vesicular stomatitis virus particle, *J. Mol. Biol.*, 97, 289, 1975.
25. **Perrault, J.**, Cross-linked double stranded RNA from a defective vesicular stomatitis virus particle, *Virology*, 70, 360, 1976.
26. **Holland, J. J., Villarreal, L. P., and Breindl, M.**, Factors involved in the generation and replication of rhabdovirus defective T particles, *J. Virol.*, 17, 805, 1976.
27. **Perrault, J. and Holland, J. J.**, Variability of vesicular stomatitis virus autointerference with different host cells and virus serotypes, *Virology*, 50, 148, 1972.
28. **Schnitzlein, W. M. and Reichmann, M. E.**, Interference and RNA homologies of New Jersey serotype isolates of vesicular stomatitis virus and their defective particles, *Virology*, 77, 490, 1977.
29. **Repik, P. and Reichmann, M. E.**, Characterization of the mRNA species and DI particles of four animal rhabdoviruses, in *International Virology IV*, Centre for Agricultural Publishing and Documentation, Wageningen, 1978, 239.
30. **Schaffer, F. L., Hackett, A. J., and Soergel, M. E.**, Vesicular stomatitis virus RNA: complementarity between infected cell RNA and RNAs from infectious and autointerfering virus fractions, *Biochem. Biophys. Res. Commun.*, 31, 685, 1968.
31. **Schaffer, F. L., Hackett, A. J., and Soergel, M. E.**, Vesicular stomatitis virus autointerference, in *Fed. Proc. Fed. Am. Soc. Exp. Biol.*, 28, 1867, 1969.
32. **Bishop, D. H. L. and Roy, P.**, Properties of the product synthesized by vesicular stomatitis virus particles, *J. Mol. Biol.*, 58, 799, 1971.
33. **Roy, P. and Bishop, D. H. L.**, The genome homology of vesicular stomatitis virus and defective T particles and evidence for the sequential transcription of the virion RNA, *J. Virol.*, 9, 946, 1972.
34. **Schincariol, A. L. and Howatson, A. F.**, Replication of vesicular stomatitis virus. II. Separation and characterization of virus-specific RNA species, *Virology*, 49, 766, 1972.
35. **Prevec, L.**, Physiological properties of vesicular stomatitis virus and some related rhabdoviruses, in *Viruses, Evolution and Cancer*, Kurstak, E. and Mamamorosch, K., Eds., Academic Press, New York, 1973, 677.
36. **Reichmann, M. E. and Leamnson, R. N.**, Temperature sensitive mutants of vesicular stomatitis virus and interference, in *Mechanisms of Viral Disease*, Vol. 1, Robinson, W. S. and Fox, C. F., Eds., W. A. Benjamin, Menlo Park, 1974, 101.
37. **Stamminger, G. and Lazzarini, R. A.**, Analysis of the RNA of defective VSV particles, *Cell*, 3, 85, 1974.
38. **Stampfer, M., Baltimore, D., and Huang, A. S.**, Ribonucleic acid synthesis of vesicular stomatitis virus. I. Species of ribonucleic acid found in Chinese hamster ovary cells infected with plaque-forming and defective particles, *J. Virol.*, 4, 154, 1969.
39. **Perrault, J. and Holland, J. J.**, Absence of transcriptase activity and transcription-inhibiting ability in defective interfering particles of vesicular stomatitis virus, *Virology*, 50, 159, 1972.
40. **Perrault, J., Semler, B. L., Leavitt, R. W., and Holland, J. J.**, Inverted complementary terminal sequences in defective/interfering particle RNAs of vesicular stomatitis virus and their possible role in autointerference, in *Negative Strand Viruses and the Host Cell*, Mahy, B. W. J. and Barry, R. D., Eds., Academic Press, New York, 1978, 527.
41. **Reichmann, M. E., Schnitzlein, W. M., Bishop, D. H. L., Lazzarini, R. A., Beatrice, S. T., and Wagner, R. R.**, Classification of the New Jersey serotype of vesicular stomatitis virus into two subtypes, *J. Virol.*, 25, 446, 1978.
42. **Metzel, P. S., Schnitzlein, W. M., and Reichmann, M. E.**, Characterization of distinct vesicular stomatitis virus, New Jersey serotype, isolates with respect to nucleic acid homologies, interference by DI particles and protein structure, in *Negative Strand Viruses and the Host Cell*, Mahy, B. W. J. and Barry, R. D., Eds., Academic Press, New York, 1978, 515.
43. **Ball, L. A. and White, C. N.**, Order of the transcription of genes of vesicular stomatitis virus, *Proc. Natl. Acad. Sci. U.S.A.*, 73, 442, 1976.
44. **Abraham, G. and Banerjee, A. K.**, Sequential transcription of the genes of vesicular stomatitis virus, *Proc. Natl. Acad. Sci. U.S.A.*, 73, 1504, 1976.
45. **Ball, L. A.**, Transcriptional mapping of vesicular stomatitis virus *in vivo*, *J. Virol.*, 21, 411, 1977.

46. **Reddy, D. V. R., Schnitzlein, W. M., and Reichmann, M. E.**, The use of defective interfering particle RNA probes in the determination of the order of *in vitro* transcription of vesicular stomatitis virus genes, *J. Virol.*, 21, 432, 1977.
47. **Colonno, R. J. and Banerjee, A. K.**, A unique RNA species involved in initiation of vesicular stomatitis virus RNA transcription *in vitro*, *Cell*, 8, 197, 1976.
48. **Colonno, R. J. and Banerjee, A. K.**, Mapping and initiation studies on the leader RNA of vesicular stomatitis virus, *Virology*, 77, 260, 1977.
49. **Colonno, R. J., Lazzarini, R. A., Keene, J. D., and Banerjee, A. K.**, *In vitro* synthesis of messenger RNA by a defective interfering particle of vesicular stomatitis virus, *Proc. Natl. Acad. Sci. U.S.A.*, 74, 1884, 1977.
50. **Keene, J. D., Schubert, M., Lazzarini, R. A., and Rosenberg, M.**, Nucleotide sequence homology at the 3′ termini of RNA from vesicular stomatitis virus and its defective interfering particles, *Proc. Natl. Acad. Sci. U.S.A.*, 75, 3225, 1978.
51. **Niyogi, S. K. and Thomas, C. A., Jr.**, The specific association of ribooligonucleotides of known chain length with denatured DNA, *Biochem. Biophys. Res. Commun.*, 26, 51, 1967.
52. **Niyogi, S. K.**, The influence of chain length and base composition on the specific association of oligoribonucleotides with denatured deoxyribonucleic acid, *J. Biol. Chem.*, 244, 1576, 1969.
53. **Reichmann, M. E. and Schnitzlein, W. M.**, Defective interfering particles of vesicular stomatitis virus, in *Microbiology — 1977*, Schlessinger, D., Eds., American Society for Microbiology, Washington, D.C., 1977, 439.
54. **Clewley, J. P., Bishop, D. H. L., Kang, C.-Y., Coffin, J., Schnitzlein, W. M., Reichmann, M. E., and Shope, R. E.**, Oligonucleotide fingerprints of RNA species obtained from rhabdoviruses belonging to the vesicular stomatitis virus subgroup, *J. Virol.*, 23, 152, 1977.
55. **Perrault, J. and Leavitt, R. W.**, Characterization of snap-back RNAs in vesicular stomatitis defective interfering virus particles, *J. Gen. Virol.*, 38, 21, 1978.
56. **Perrault, J. and Leavitt, R. W.**, Inverted complementary terminal sequences in single-stranded RNAs and snap-back RNAs from vesicular stomatitis defective interfering particles, *J. Gen. Virol.*, 38, 35, 1978.
57. **Kolakofsky, D.**, Isolation and characterization of Sendai virus DI-RNAs, *Cell*, 8, 547, 1976.
58. **Kennedy, S. I. T.**, Sequence relationships between the genome and the intracellular RNA species of standard and defective-interfering Semliki forest virus, *J. Mol. Biol.*, 108, 491, 1976.
59. **Kennedy, S. I. T., Bruton, C. J., Weiss, B., and Schlessinger, S.**, Defective interfering passages of Sindbis virus: nature of the defective virion RNA, *J. Virol.*, 19, 1034, 1976.
60. **Hsu, M. T., Kung, H.-J., and Davidson, N.**, An electron microscopic study of Sindbis virus RNA, *Cold Spring Harbor Symp. Quant. Biol.*, 38, 943, 1973.
61. **Leppert, M., Kort, L., and Kolakofsky, D.**, Further characterization of Sendai virus DI-RNAs: a model for their generation, *Cell*, 12, 539, 1977.
62. **Huang, A. S.**, Viral pathogenesis and molecular biology, *Bacteriol. Rev.*, 41, 811, 1977.
63. **Brockman, W. W., Gutai, M. W., and Nathans, D.**, Evolutionary variants of simian virus 40: characterization of cloned complementing variants, *Virology*, 66, 36, 1975.
64. **Lee, T. N. M., Brockman, W. W., and Nathans, D.**, Evolutionary variants of simian virus 40: cloned substituted variants containing multiple initiation sites for DNA replication, *Virology*, 66, 53, 1975.
65. **Griffin, B. E. and Fried, M.**, Amplification of a specific region of the polyoma virus genome, *Nature (London)*, 256, 175, 1975.
66. **Mills, D. R., Peterson, R. L., and Spiegelman, S.**, An extracellular Darwinian experiment with a self-duplicating nucleic acid molecule, *Proc. Natl. Acad. Sci. U.S.A.*, 58, 217, 1967.
67. **Stark, C. and Kennedy, S. I. T.**, The generation and propagation of defective-interfering particles of Semliki forest virus in different cell types, *Virology*, 89, 285, 1978.
68. **Lesnaw, J. A., Leamnson, R. N., Unger, J. T., and Reichmann, M. E.**, unpublished data, 1974.

Chapter 12

INTERFERENCE INDUCED BY DEFECTIVE INTERFERING PARTICLES

C. Yong Kang

TABLE OF CONTENTS

I. INTRODUCTION

The phenomenon of defective-interfering (DI) particle-mediated viral interference has been extensively reviewed previously[1,2] and Chapters 9, 10, and 11 in this volume deal with the origin and genome assignments of DI particles and the possible mechanisms of homotypic and heterotypic viral interference.

This chapter will cover two major aspects regarding defective interfering particles of rhabdovirus. First, factors of the host cell and virus which affect the genesis of DI particles will be discussed, and second, what is known about genetic sequence in a variety of DI particles analyzed by oligonucleotide fingerprinting techniques will be described. Finally, models for DI particle generation will be presented.

II. SEARCH FOR FACTORS WHICH ARE RESPONSIBLE FOR THE GENERATION OF DI PARTICLES

It had previously been thought that the viral determinants were the only important factors affecting the genesis of DI particles. For example, DI particles could be generated by virus-specified premature termination during RNA replication, improper cleavage of messenger RNA transcripts, or other mutational events of the virus during replication. However, Holland et al.,[3] in a direct attempt to test such mechanisms, tried to generate DI particles by mechanical shearing of ribonucleoprotein or chemical and UV mutagenesis, but without success. There is evidence which suggests that not only do clones of virus influence the types of DI particle produced, but also that host cells play important roles in the generation of DI particles.[1] One of the earliest studies indicated that the amount of incomplete influenza virus formation is significantly dependent on the type of host cells.[4] Influenza virus grown in Madin-Darby bovine kidney (MDBK) cells gives a high yield of virus in both high and low multiplicities of infection (MOI), whereas the virus grown in embryonated eggs produces a low yield of virus with high MOI. Choppin and Pons[5] demonstrated the MDBK cells produce influenza virus which contain at least five distinct RNA species in the virion (at that time RNA segments in the influenza virion were not characterized completely), whereas under the same conditions HeLa cells produced noninfectious virus is lacking the longest piece of viral RNA. These results indicated that MDBK cells are either incapable of generating DI particles or may take many more serial passages of the virus to generate a detectable level of DI particles. In contrast, DI particles arise quickly when the virus is passed in HeLa cells. These results suggest partial host-cell dependence on DI particle formation. Similar observations were made in the Sindbis virus system. DI particles of Sindbis virus are readily generated in baby hamster kidney cells (BHK-21) and chick embryo fibroblasts during serial passages of the virus, while no DI particles are generated in *Aedes albopictus* cell cultures.[6] Kingsbury and Portner[7] demonstrated that not all tissues of the same animal generate DI particles of Sendai virus with the same efficiency. Embryonated eggs readily generate DI particles, while chick embryo cells did not generate detectable levels of DI particles of Sendai virus. A similar phenomenon was observed with Newcastle disease virus (NDV).[8] NDV serially passaged in chick embryo fibroblasts in culture gives rise to DI particles, while the same virus passaged in embryonated eggs did not. These data suggest that not only do different species of host cell generate DI particles with varying efficiencies, but also that different stages of differentiation of the same species may respond differently. These observations extend further the possibility that cells from different organs in the same species may generate nonidentical DI particles.

The search for factors which may be responsible for the generation of DI particles continues. Most investigation is carried out using vesicular stomatitis virus (VSV) as a model system because the DI particles of VSV are readily separable from infectious particles and the latent period of the virus is short.

Cooper and Bellett[9] first recognized the autointerference phenomenon in the VSV system. Reczko[10] demonstrated varying lengths of bullet-shaped particles of the VSV, and Hackett[11] suggested that these particles might be the interfering component. Huang and Wagner[12] demonstrated that the truncated particles are the DI particles of VSV. Not until very recently, when Holland and co-workers[3] demonstrated that DI particle generation was found to be a rather random process, have we placed emphasis on the generation of DI particles by cloned virus stocks. The generation of particular sizes of DI particles by the infectious bullet-shaped particle (B virion) pool is due to the random generation of DI particles during preparation of first passage pools of cloned B virions.[3] These DI particles breed true during the additional passages needed to produce detectable quantities of DI particles. These investigators have also suggested that different host-cell lines selectively amplify different DI particles indicating a strong selective pressure by host cells for DI particle replication. It now appears that the species of DI particles of VSV generated during serial, undiluted, high multiplicity passages is determined not only by the clone of infectious virus, but also by host cells.

A. Effects of Clonal Isolates of Virus in Generation of VSV DI Particles

It is apparent that the genetic constituent of the infectious viruses, in part, determines the type of DI particles generated during the replications of virus. Reichmann et al.[13] showed that each different clone of temperature-sensitive mutants regularly generated a unique size of DI particles upon serial undiluted high multiplicity passages. This observation was confirmed by Holland and co-workers.[3] The heat resistant variant of VSV (VSV Indiana serotype) also produced long DI particles consistantly.[14,15] In contrast, different clonal isolates of New Jersey serotype VSV (VSV_{NJ}) generated different sizes of DI particles during continuous passage.[16] A clone of VSV_{IND}, passed independently, generated as many as six different patterns of DI particles which included different numbers and different size classes of DI particles.[3] These results are consistant with previous observations showing the generation of different sizes of DI particles by independent clonal isolates of virus propagated in one cell type.[17] Very recently Kang et al.[18] have shown that a single cell-type infected with two different plaque isolates of VSV, generated either the same pattern of DI particles or different patterns depending on the cell types used. BHK-21 cells produced the same pattern of DI particles with two independent clones of virus, while mouse cells (L_2 cells) and rat cells transformed by B77 strain of avian sarcoma virus (R(B77)) produced different patterns of DI particles with the same clone of virus. It is reasonable to conclude that the genetic make up of any particular clone of virus influences but does not totally control the kind of DI particles generated.

B. Effects in Host Cells in Generation of DI Particles

It is clear that both the clone of virus and the host cell determine the type of DI particle generation. Recently Holland and co-workers[3] showed that six different host-cell lines produced different patterns of DI particles generated during serial passages of a single clone of VSV. In addition, HeLa cells did not generate DI particles in the same experimental conditions. These workers concluded that different host-cell lines selectively amplify different DI particles and this suggested a strong role for host-cell factors in DI particle replication. Lunger and Clark[19] found that the same clone of VSV propagated in three different reptilian cell lines and BHK-21 cells produce different sizes of DI particles.

To determine host effects on the generation of DI particles, Kang et al.[18] have plaque-purified VSV in four different mammalian cell lines — human (RSa), hamster (BHK_{21}), rat (R(B77)), and mouse (L_2) — and then made serial undiluted passages in each cell line. Lysates from each passage were analyzed by sucrose-gradient centrifugation to determine the pattern of DI particles synthesized in these cells (Figure 1). The RSa cells produced rather heterogeneous sizes of DI particles, while the BHK_{21} cells produced two distinct sizes of DI particles. In contrast, R(B77) cells and L_2 cells produced three sizes of DI particles each. The three different DI particles produced in R(B77) cells, however, appear to be different from the three DI particles produced in L_2 cells. It is evident from these data that different size classes of DI particles are generated in various host cells from a single clone of virus, which suggests strongly that a host function largely influences the types of DI particle production. The efficiency of DI particle generation also appears to be determined by the host-cell lines. Huang and Baltimore[20] have demonstrated that MDBK cells show no reduction of VSV titers up to five high multiplicity of infection (MOI) passages, whereas Chinese hamster ovary (CHO) cells produced marked reductions of the virus titer in three passages. Similar observations were made by Potter and Stewart[21] who showed that VSV DI particles are generated in chick embryo cells with greater efficiency than in mouse L-cells. More extensive analyses were made by Kang and co-workers[22] to demonstrate that the number of passages required to produce detectable levels of DI particles is dependent on the type of host-cell line (Figure 2). Independent clones of VSV passaged in the four different cell lines yield varying amounts of infectious virus at different numbers of passages. The human RSa cell lines produced a tenfold reduction of infectious virus in each passage with over a three-log reduction from the original titer of virus by the fourth consecutive passage. In contrast, mouse L_2-cells produced a constant amount of virus up to the sixth passage and then showed rapid decline of titers in the subsequent passages. Four to five passages were required before the yield of virus was reduced by approximately three logs from the initial titer when the virus was passed in the hamster, BHK_{21}, cells and rat, R(B77), cells. The rate of reduction of the infectious-virus production was due to the generation of DI particles. The generation time is determined by the host cell rather than by the virus, since three independent clones of virus showed the same pattern of reduction of virus titer in any given cell type.[22]

C. Requirement of Host Function for the Generation of DI Particles

Darnell and Koprowski[23] presented genetic evidence that the generation of DI particles is controlled in some way other than by the virus itself. The replication of West Nile virus, a flavivirus, in cell cultures derived from inbred mice, which are genetically resistant to flavivirus infection, produce interference readily, but cell cultures from congenic inbred mice, which are genetically susceptible to infection, are incapable of inducing viral interference. Thus, the susceptibility to flavivirus infection in mice appears to be directly related to the ability of host cells to produce DI particles. All the evidence so far presented indicates strongly that the induction of DI particles during serial undiluted passages is dependent on a host-cell function. To substantiate further the host function-dependent induction of DI particles, Kang and Allen investigated the induction of DI particles in cells pretreated with actinomycin D,[24] a known inhibitor for DNA-directed information flow,[25] with no known effects on VSV replication.[26] The stock virus was plaque-purified and prepared in cells pretreated with actinomycin D; serial undiluted passages were carried out in cells with or without actinomycin D pretreatment. Figure 3 demonstrates that if cells were pretreated with actinomycin D to suppress cellular macromolecular synthesis, cells were incapable of DI particle in-

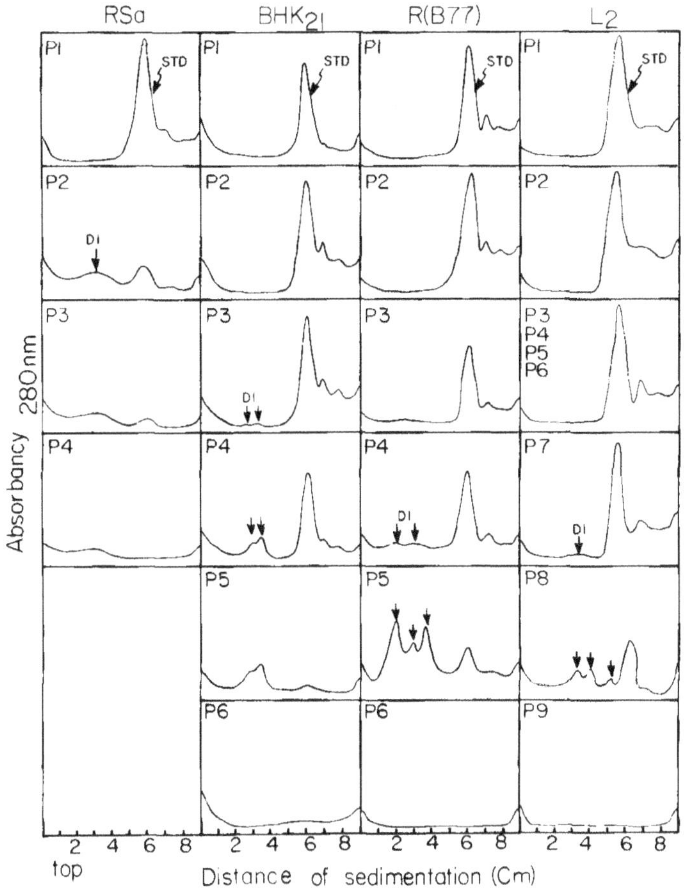

FIGURE 1. Analysis of total virus particles from the continuous undiluted passages of VSV_{IND}. Cloned human (RSa), hamster (BHK_{21}), rat (R(B77)), and mouse (L_2) cells were infected with a multiplicity of infection of 100 PFU/cell of plaque purified prototype strain of VSV_{IND} for the first passage. The subsequent passages were made by infecting 0.5 m*l* of undiluted previous passage virus. Samples of lysate from each culture of each passage were analyzed by sucrose gradient centrifugation.

duction. The replication of DI particles already present in an infectious virus stock, however, were not affected by pretreatment of cells with actinomycin D. Thus, the induction, but not the replication of defective interfering particles, appears to be a host-cell function-dependent process. This observation is confirmed by Dr. C. J. Pfau, Rensselaer Polytechnic Institute, working with an arenavirus. He found that actinomycin D pretreated cells are incapable of generating DI particles of Pichinde virus.[58]

This is the most direct evidence to date to demonstrate that the host function is a crucial factor in generation of DI particles. Identification of such factor(s) will help in finding the molecular mechanism of DI particle generation.

The nucleic acid sequence analyses of DI particle RNAs induced in various cell types will suggest whether or not the DI particle genomes are evolved by random events during the virus replication. The knowledge on the origin of all available DI particle RNAs may lead us to find the molecular mechanism of DI particle generation. The next part of this chapter is concerned with the genetic constituents of different DI particles of VSV.

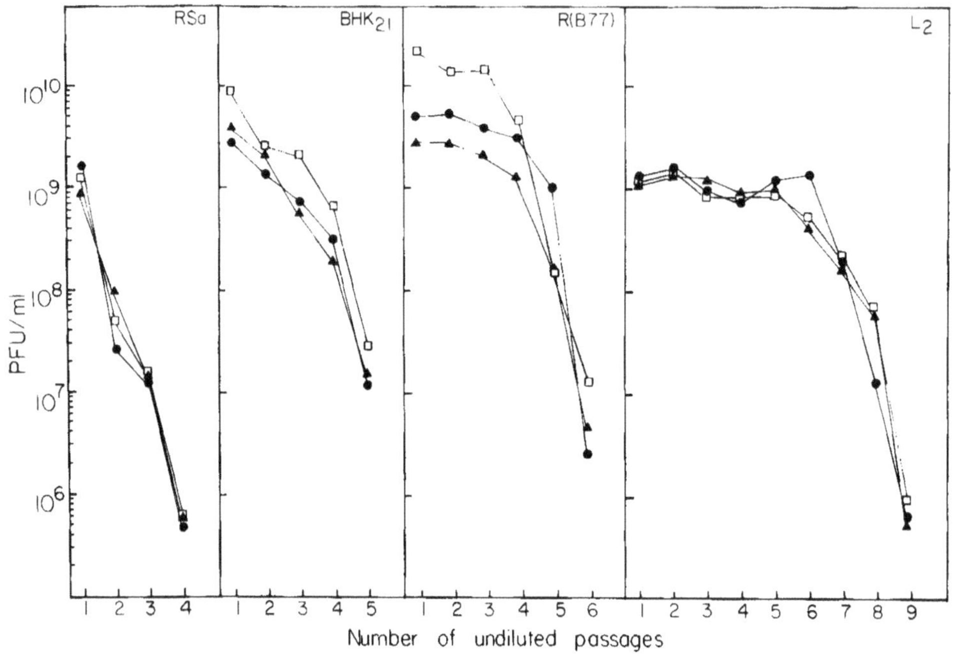

FIGURE 2. Yield of standard B virion during successive undiluted passages of three clonal isolates of VSV_{IND} in four different hosts. Infection of four different cell lines with different clones of VSV_{IND} were carried out with the same technique as described in Figure 1. Cell-free virus fluids were harvested after 7 hr of growth period and the yield of infectious virus was measured by plaque assay (□ Clone No. 1, ▲ Clone No. 2, and ● Clone No. 3).

III. NUCLEOTIDE SEQUENCES IN DI PARTICLE GENOME

DI particles of VSV are truncated forms of infectious bullet (B)-shaped particles, and the size of the DI particles correspond directly to the length of the nucleocapsid. The length of the nucleocapsid is determined by the length of the RNA genome.[27,28] All of the DI particle RNAs are a subgenomic length of infectious B virion RNA, and the nucleotide sequences in the DI particle RNAs represent a part of the infectious B virion RNA.[17,29-32] Thus, DI particles are in a sense deletion mutants of the virus. Most of the VSV DI particle genomes are single-stranded RNA; however, some DI particle RNAs contain varying amounts of double-stranded portions, each representing both parts of the infectious B virion RNA sequence and a complementary sequence (messenger polarity).[33,34] The nucleotide sequences of DI particle genomes are mapped by nucleic acid hybridization using messenger RNAs representing the known portion of infectious B virion genome as has been already discussed in Chapter 9 in this volume.

It is apparent that most of the DI particle RNA represents the 5′ half of the infectious virus genome, since only the 30S messenger RNA, which represents the cistron of L protein, anneals with the DI particle RNAs.[30,31] In contrast, the RNA of the long DI particle which is derived from the heat-resistant strain of the VSV_{IND} (designated DI-LT) anneals with 12 to 18S messenger RNAs which represent the 3′ half of the infectious B virion genome.[29-32] If there are any overlapping sequences between the DI-LT particle RNA and the RNAs from all short DI particles, it would appear to be very limited.[31] The map of the DI-LT RNA sequence has been assigned more definitely by the demonstration of the synthesis of Indiana serotype specific M and G proteins in cells coinfected with DI-LT particles and infectious VSV_{NJ}.[35] In addition, Colonno

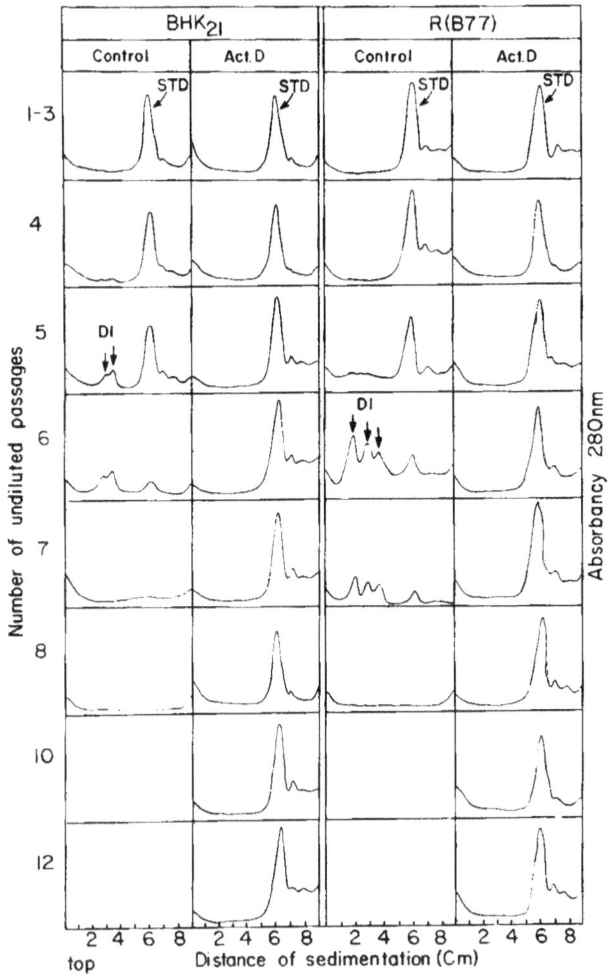

FIGURE 3. Analysis of total virus particles from the continuous undiluted passages made in cells with or without actinomycin D pretreatment. Stock of VSV_{IND} was prepared from consecutive plaque purification in cells pretreated with actinomycin D. The monolayer cultures with (Actinomycin D) or without (control) actinomycin D pretreatment were infected with a multiplicity of infection of 100 PFU/cell for the first passage. For the subsequent passages, 0.5 mℓ/ 100 mm dish of undiluted virus from the previous passages was used. At the end of 7-hr growth period, the total virus particles were analyzed by sucrose gradient centrifugation.

et al.[32] have synthesized in vitro the 12 to 18S messenger RNAs, but not the 30S messenger RNA using DI-LT; and 12 to 17S messenger RNAs are also synthesized in vivo infected with DI-LT. Thus, it is apparent that DI-LT RNA maps on the 3′ half of the genome, which represents cistrons for N, NS, M, and G proteins according to the gene order assigned by Ball and White.[37] All of the other DI particles tested so far map on the different portions of the 5′ half of the genome.

A. Oligonucleotide Fingerprint Analysis of VSV DI Particle RNAs

The technique of oligonucleotide fingerprinting has added a new dimension in deter-

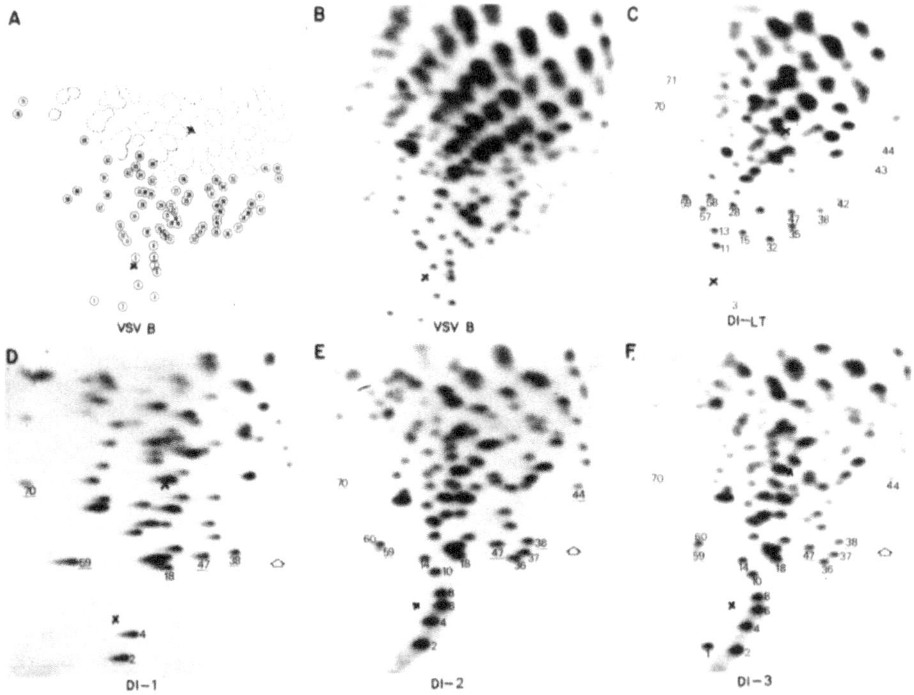

FIGURE 4. The oligonucleotide fingerprints of a B virion and four DI particle RNAs of prototype VSV_{IND}. A nomenclature system of VSV_{IND} large oligonucleotides (A), traced from the T_1 digest of a B virion of VSV_{IND} prototype strain (B) are shown. The fingerprint of the DI-LT from HR strains of VSV_{IND} (C) and fingerprints of the smaller DI particle RNAs—DI-1 (D), DI-2 (E), and DI-3 (F) — are shown. (From Clewley, J. P. and Bishop, D. H. L., in *Negative Strand Viruses and the Host Cell*, Barry, R. D. and Mahy, B. W. J., Eds., Academic Press, London, 1978, 271. With permission.)

mining the RNA sequences of DI particles. This technique allows us to define the RNA sequences in DI particle RNAs generation in the same or different cell lines using the same clone of virus stock. The technique will identify overlapping and/or nonoverlapping sequences between two or more DI particle RNAs and identify oligonucleotides which may not be present in the infectious B virion RNA genome.

Figure 4 illustrates the fingerprints of four different DI particle RNAs together with standard B virion RNA. Each kind of DI particle RNA contains characteristc oligonucleotides and most, if not all, of the infectious B virion RNA. All 14 identifiable oligonucleotide spots of DI-1 (small) RNA are present in DI-2 (medium) RNA, and all 25 oligonucleotide spots present in DI-2 RNA are present in DI-3 (large) RNA (Figure 4). Thus, it would appear from the data that the 29 identifiable oligonucleotide spots of DI-3 RNA contain all the nucleotide sequences of both DI-1 and DI-2 RNAs. This result indicates that all three DI-particle RNAs originated from the same region of the standard B virion genome. In addition, there is only a very weak spot (marked with arrow) at the right side of the fingerprint which is the same size but has different charge as spot 59. The possible origin and significance of this spot will be discussed later in Figures 7 and 8. Figure 4C represents the oligonucleotide fingerprints of the DI-LT RNA originated from the heat-resistant variant of Indiana serotype.[14,15] As with the other DI particle RNAs, all 24 identifiable oligonucleotide spots present in the DI-LT RNA appear to have originated from the infectious B virion RNA. There are no extra spots in DI-LT RNA fingerprint which are not present in the B virion RNA.

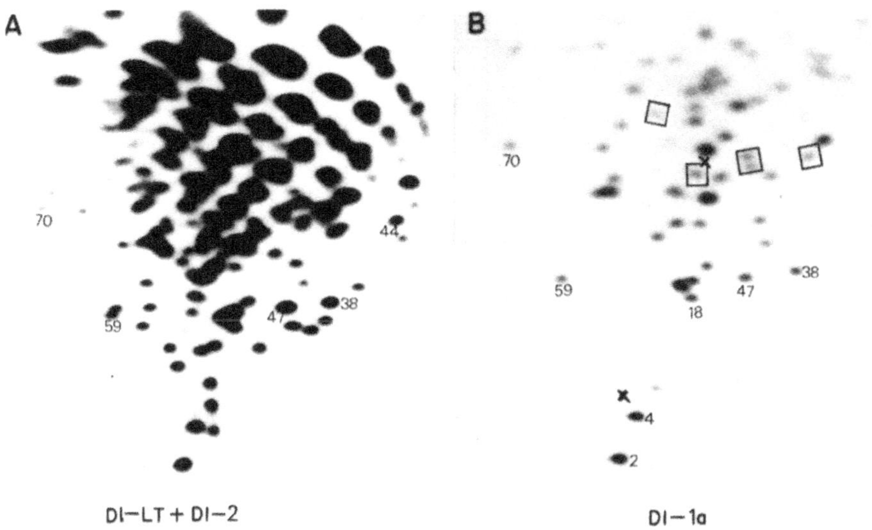

FIGURE 5. The oligonucleotide fingerprints of RNAs from DI-LT plus DI-2 (A) and DI-la (B) are shown. The oligonucleotide spots enclosed by square indicate the spots which are not found in the fingerprints of DI-1.

Four out of 24 oligonucleotide spots in the DI-LT RNA designated as spots 38,47,59, and 70 seem to over lap with DI-1, DI-2, and DI-3 RNAs and one additional oligonucleotide spot, spot 44, in DI-LT RNA overlaps with both DI-2 and DI-3 RNAs. We did not expect any nucleotide sequence homologies between DI-LT and all other DI particle RNAs, since only 12 to 18S messenger RNAs anneal with DI-LT RNA, whereas all other DI particle RNAs anneal with 30S messenger RNA.[29-32] One cannot draw a firm conclusion as yet whether the overlapping spots are indeed identical until the nucleic acid sequences in each of the oligonucleotide spots are determined. However, the pattern of cofingerprints of DI-LT and DI-2 RNAs shown in Figure 5A indicates that the overlapping oligonucleotide spots may well be identical sequences. Figure 5B shows another DI particle RNA, DI-la which has been derived from the original VSV_{IND} (Toronto) by Dr. L. Prevec. This DI particle sedimented slightly ahead of DI-1 in the sucrose gradient. The pattern of fingerprints of DI-1 is very similar to DI-1, except DI-la RNA contains four or five extra spots which are not present in DI-1 RNA. However, all these spots are present in the DI-2, DI-3, and in B virion RNAs. Thus, this DI-la RNA is another size which shares common nucleotide sequences. Very recently, Huang[28] showed a fingerprint of one DI-particle RNA in which the majority of the oligonucleotide spots present in DI-particle RNA are also present in infectious B-virion RNA. She found two extra spots in DI-particle RNA which are not present in infectious B-virion RNA. It is reasonable to assume that the additional sequences she found represent the complementary strand which is derived from a partially double-stranded DI-particle RNA. The DI-011 RNA from Dr. Lazzarini's laboratory has a peculiar pattern of oligonucleotide fingerprints. DI-011 RNA contains not only common nucleotide sequences found in many small DI particle RNAs, but has at least five oligonucleotide spots that are not present in infectious B virion RNA (Figure 6B). These five oligonucleotide sequences may well have been derived from the positive polarity complementary RNA which is present in the partial double-stranded portion of the DI-011 genome as has been reported.[33] In addition, 7 out of 21 DI-011 oligonucleotide spots share common spots with DI-LT RNA. Moreover, there are at least two

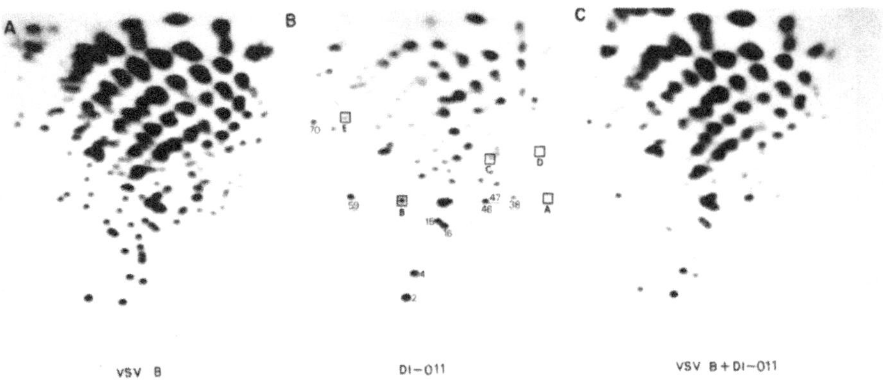

FIGURE 6. The oligonucleotide fingerprints of RNAs from B virion (A), DI-011 (B), and mixture of B virion and DI-011 (C) are shown. The oligonucleotide spots A, B, C, D, and E are not present in the infectious B virion RNA.

oligonucleotide spots (spot 16 and 27) which are not found in either DI-LT or all other smaller DI RNAs. Thus, DI-011 represents rather complicated forms of RNA.

If we assume that any DI-particle RNA must have polymerase recognition sites where RNA synthesis can initiate, it is not unusual to find some nucleotide sequence homologies between all DI-particle RNAs. It is known that the 3′ ends of both the infectious B virion RNA and the DI-particle RNAs contain $_{OH}$UpGpCp and the 5′ ends of both the DI-particle RNAs and infectious B-virion RNAs contain the inverted complementary sequences, GpCpAppp.[38-40,56,57] Most recent data indicate that there are at least eight consecutive nucleotide sequences at the 3′ termini of both DI-particle RNA and the infectious B-virion RNA that are identical. In addition, 14 out of 17 nucleotide sequences at the 3′ termini of DI-particle RNA and the B-virion RNA are the same.[57] The 3′ terminal sequences of the B-virion RNA are $_{OH}$UpGpCpUpUpCpUpGpUpUpUpGpUpUpUpGpGp and the 3′ terminal sequences of DI-particle RNA are $_{OH}$UpGpCpUpUpCpUpGpGpUpGpUpUpUpUpGpGp.[57] If there are about 20 nucleotides of inverted complementary sequences at the 3′ and the 5′ termini, self-annealing could occur when N proteins are removed from the ribonucleoprotein (RNP) molecules. Results of such annealing may create the ribonuclease T_1-resistant partial double-stranded RNA. Fingerprints of such RNA after heat denaturation may reveal extra spots, depending upon sequences in the RNA. When purified B-virion or DI-particle RNAs are self-annealed before the fingerprint analysis, there could be an extra spot which is not present in nonself-annealed RNA preparation. Figure 7 shows the fingerprint analysis of three DI-particle RNAs. As can be seen in Figure 7A, B, and C, all three RNAs showed one extra spot which was not present in normal infectious virus RNA (Figure 4B). This extra spot moved almost to the same position as spot A of DI-011 position (Figure 6). It is also evident that Figure 7D shows a heavy well-isolated spot below spot 2 (with arrow mark) which has a similar charge as spot 59. This spot may well be the incomplete digest of RNA. This spot could be the precursor for spot 59 since spot 59 has a remarkable reduction of its intensity when this extra spot appears. We do not know the origin of the extra spot, shown in Figure 7A, B, and C with arrows, but it is reasonable to assume that the extra spot represents the terminal sequences of the genome, since this spot can be enriched by self-annealing of the purified DI-particle RNAs. One must be aware of such spots in both DI and infectious B-virion RNA in a special condition, since once could create the same spot using infectious B-virion RNA by self-annealing (Figure 8, marked with arrows). When the puri-

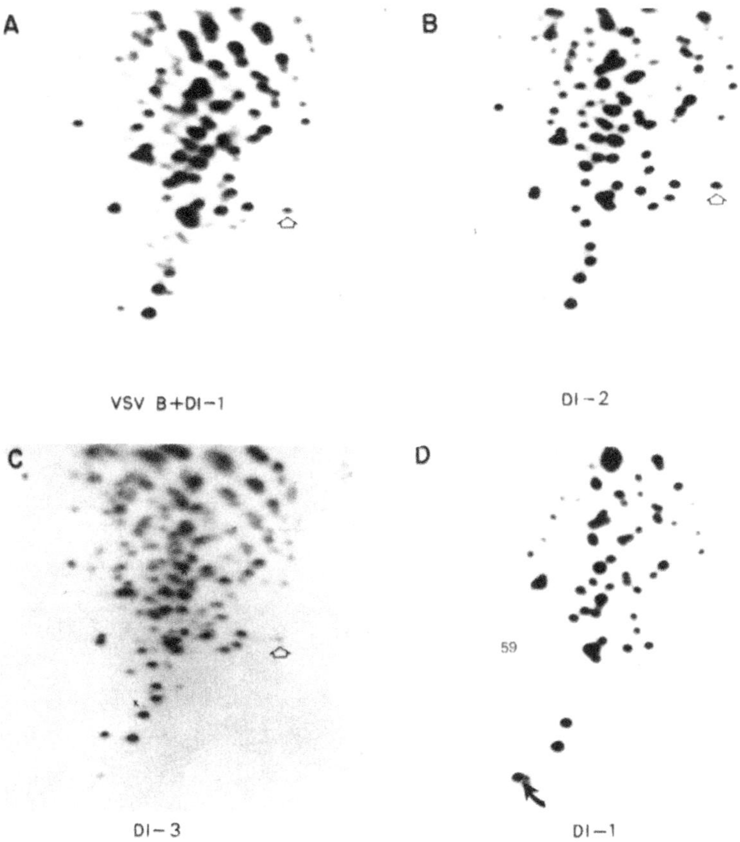

FIGURE 7. The oligonucleotide fingerprints of RNAs from B virion RNA plus DI-1 (A), DI-2 (B), DI-3 (C), and DI-1 (D) are shown. RNAs in panel (A), (B), and (C) were self-annealed at 63°C for 24 hr before digestion with T_1. DI-1 RNA in panel (D) was prepared by the same way as fingerprints shown in Figure 4D but showing incomplete digestion (arrow).

fied B-virion RNA was self-annealed for 24 hr and fingerprinted, a faint extra spot appeared (Figure 8B with arrow). If the same sample were digested with RNase T_1, selected for double-stranded RNA, heated, and fingerprinted, the extra oligonucleotide spot appeared little darker (Figure 8D). This extra spot migrated to the same position as the extra spot found in DI-particle RNAs. The reason that we can enrich this extra spot with smaller DI-particle RNA much more efficiently is because the shorter RNA pieces from small DI-particle RNA can fold back to anneal each end much faster than the full-length B-virion RNA.

In conclusion, all Di-particle RNAs except DI-LT share common nucleotide sequences, most of which are present in infectious B-virion RNA and all of the nucleotide sequences present in DI-LT RNA also come from infectious B-virion RNA. There are some homologies between DI-LT and all other DI-particle RNAs so far examined. It is also apparent that both DI-LT and other DI-particle RNAs do not represent the entire B-virion genome. This result is consistent with the size analysis of DI particles because the combined oligonucleotide of DI-LT and DI-3 RNA (the largest among DI-particle RNAs mapping on the 5′ half of the B-virion genome) do not make up the entire length of the infectious B-virion RNA.

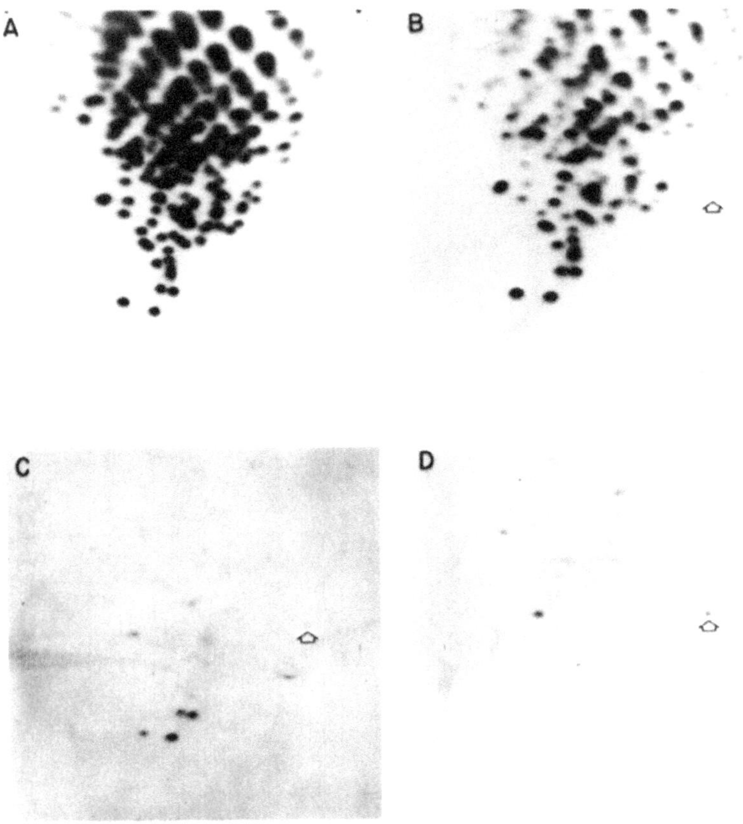

FIGURE 8. The oligonucleotide fingerprints of B virion RNAs before (A) and after self-annealing (B, C, and D).

B. Model for DI-Particle Generation

There are at least four different forms of VSV DI-particle RNAs as shown in Figure 9. These include (1) linear single-stranded RNA, (2) panhandle or circular RNA, (3) partially double-stranded RNA, and (4) completely double-stranded RNA (snap-back RNA).[2,28,41,42] The question is how these different forms of DI-particle RNAs have been generated in infected cells during serial undiluted passages of the virus with a high multiplicity of infection.

Recently, several investigators have proposed models to explain how DI-particle RNA genomes could be generated during the replication of the VSV[28,41] and Sendai virus.[42] All three models are concerned with the generation of panhandle-structured RNA, double-stranded RNA, and circular double-stranded RNAs. Huang,[28] and Kolakofsky and associates[42] each proposed a similar model which predicts that the virion RNA polymerase can copy the nascent RNA strand if the 3′ end of the nascent RNA molecule in some way turns around to form a hairpin. This model predicts that the polymerase starts to copy toward the middle of the genomic RNA but at some point dissociates from the original template, turns around, and copies the nascent RNA chain in the direction of the 5′ end (Figure 10). Huang also predicted that other complementary sequences can readily be generated by introducing a second hairpin loop during the synthesis of DI particle genomic RNA.[28] The resulting RNA genome containing a long stretch of the inverted complementary terminal sequences, may produce a long double-stranded handle on a panhandle structure or double circles if two iden-

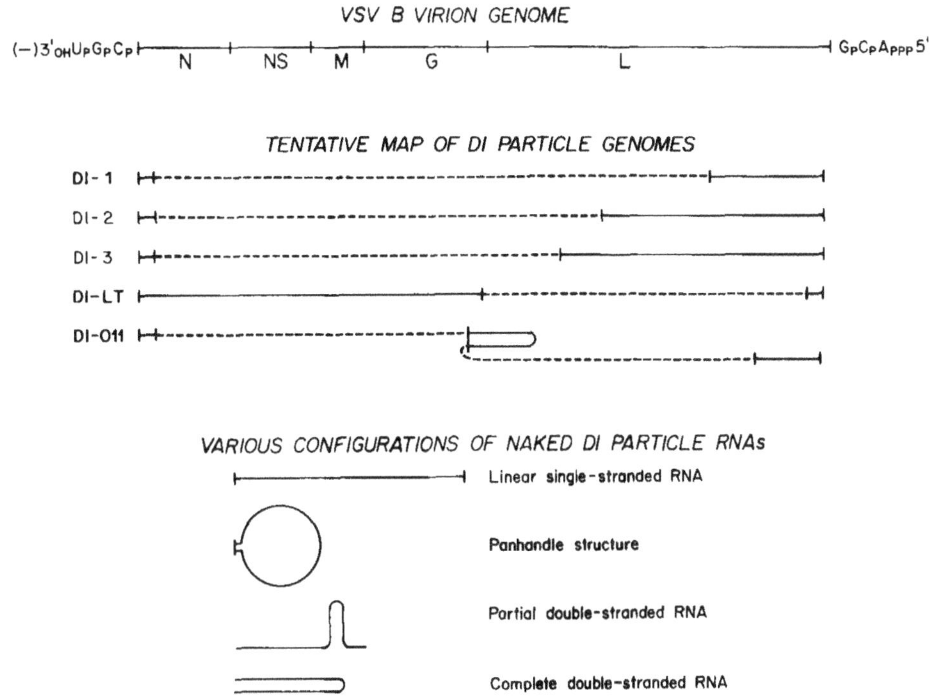

FIGURE 9. Tentative map of DI particle RNA genome based on the oligonucleotide fingerprints and various configurations of DI particle RNAs after removal of proteins.

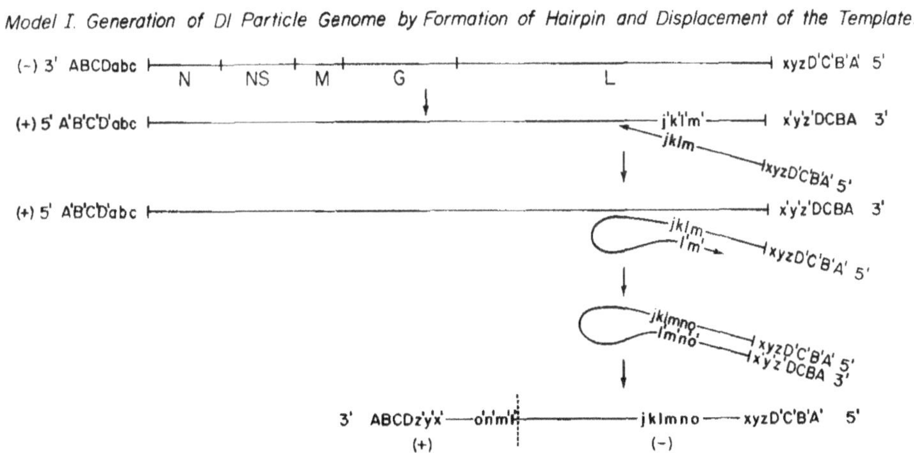

FIGURE 10. Model for generation of DI particle genome by formation of hairpin and displacement of the template.[28,42]

tical molecules annealed each other at low terminal ends.[28,41,42] If the DI-particle RNA contains completely inverted complementary sequences, one can obtain a linear double-stranded RNA molecule which is covalently linked as a hairpin structure. If two identical RNA molecules annealed each other, one can obtain a linear double-stranded RNA with twice the length of the hairpin structure or the circular double-stranded RNA is the termini self-annealed but the rest of the molecules annealed between the two identical molecules.[41] Most of these predictions are based on the structural configuration of the DI particle RNAs examined under the electron microscope after the self-annealing of the purified DI particle RNAs. In addition, Huang proposed part of her model on the basis of her oligonucleotide fingerprints of one DI particle RNA which gives two extra spots which are not found in the standard infectious particle RNA.[28]

The results from Figures 4 through 8 enable us to hypothesize a different model for the generation of all forms of the VSV DI particles.

Replication of the VSV genome starts after the virus-specific protein synthesis is accomplished.[43] The primary replication may initiate by the virion RNA polymerase at the 3′ end of the minus strand genomic RNA using the ribonucleoprotein (RNP) capsid as the template.[44] This primary replication can produce a full-length complementary sequence which has plus polarity, the same as the messenger RNA. There is some evidence that this full-length plus strand RNA rapidly associated with N protein to form a ribonucleoprotein complex.[59] This plus strand ribonucleoprotein complex then acts as a template for the secondary replication to synthesize more negative-stranded genomic RNAs. The possible mechanism of RNA replication has already been discussed by Dr. Wertz in Chapter 5 in this volume. DI-particle RNA could be generated during either primary or secondary replication. Two models presented here are to explain only the generation of the simple forms of the DI-particle RNA which is linear single-stranded RNAs, since these RNAs are most abundant in nature. The model shown in Figure 11 generates DI-particle RNA during the primary replication. This model predicts that the host cell makes a loop at the 3′ half of the RNP which enables the RNA polymerase to read through to copy mainly the 5′ half of the genome without stopping at the loop junction. The DI-1, DI-2, and DI-3 (Figure 11) genomes may have arisen from such a process. The looping of the RNA template and the copy choice of the RNA polymerase could thus generate the linear single-stranded RNA genome of DI particles. As can be noticed, the 3′ terminus of the genome is conserved in this model. This can explain some nucleotide sequence homologies between DI-LT and the other DI particles (Figure 9). This type of mechanism can conserve both 3′ and 5′ termini which are probably crucial regions for the polymerase as recognition sites for the initiation of RNA synthesis. It is also reasonable to assume that the size of the DI particle RNA has inverse relationship to the size of the loop. Since the 3′ and 5′ termini of the genome contain inverted complementary sequences, this type of RNA could also form a panhandle structure or a structure similar to circle if the two ends annealed after the removal of the proteins.[28,41] The model shown in Figure 11 generates a DI particle representing the 3′ half of the genome. A typical example of such a DI particle is the DI-LT RNA. It is virtually the same model except the looping takes place within the 5′ half of the genome instead of the 3′ half. The looping at the 5′ half of the genome could be a rare event since most of the DI-particle RNA represents a part of the 5′ half of the genome. The model II-3 shown in Figure 11 explains how one DI-particle RNA can have sequences from both a part of 5′ half and a small portion of the 3′ half of the genome. This model predicts a double looping of both parts of the 3′ half and 5′ half of the genome. The nucleotide sequences between 2 and 3 are present in neither the DI-LT nor the DI-3 RNAs. An example of such a DI-particle RNA is DI-011 from Dr. Lazzarini's laboratory. DI-011 contains nucleotide sequences present in both DI-LT and DI-3 in addition to some sequences which are present only in infectious B-virions RNA. This DI-011 RNA also contains some sequences which are not present in B-virion RNA and which may represent complementary plus-strand RNA. The formation of such more complex RNA molecules is hard to explain by any single model, but can be explained if we consider all the possibilities presented in Figures 10 and 11.[28,42] All of the models presented in Figures 10 and 11

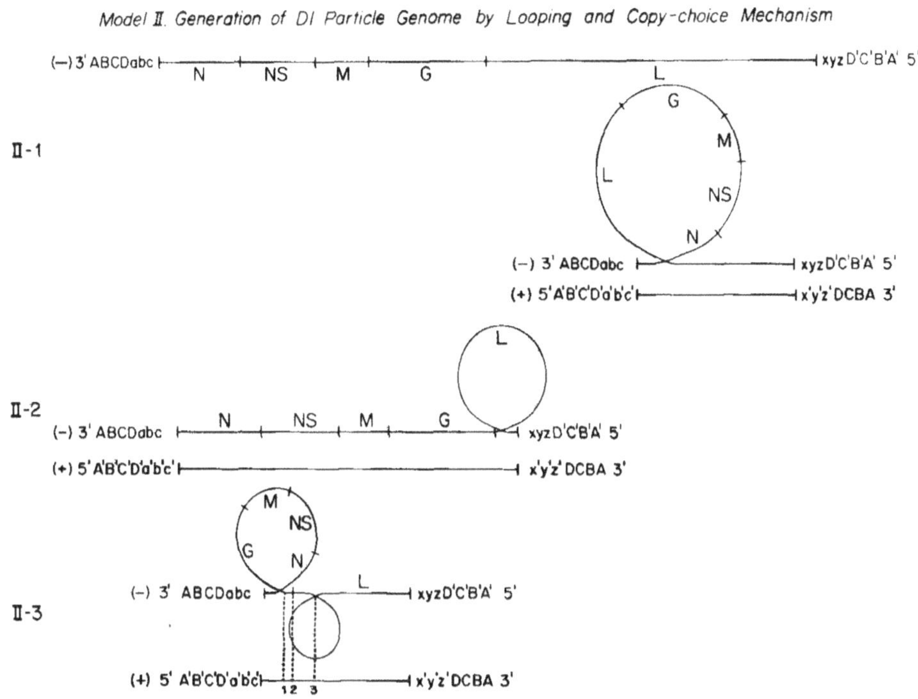

FIGURE 11. Model for generation of DI particle genome by looping and copy-choice mechanism.

are based on the single replication event: the single polymerase on the single template molecule and the DI RNA being generated as a chance event during the replication of B virion genomic RNA. The predicted host function in model II would be the factor(s) responsible for the looping of the genomic RNA.[24] It is also possible that there are many other types of defective particle RNAs that may be generated during replication, but only those defective RNAs containing the polymerase recognition site(s) can be amplified to give rise to a detectable level of DI particles in the later passages.

Another possible mechanism for the generation of DI-particle RNA is by splicing mechanisms. The organization of genes in eukaryotic cells becomes extremely intriguing. Recent developments in molecular biology regarding expression of various kinds of messenger RNAs in eukaryotic systems and in some virus systems enable us to predict numerous possibilities in biological systems. This includes the generation of DI particles. It is now well established that the variable and constant regions of the light chain of mouse immunoglobulin gene is not contiguous,[45] the ovalbumin gene is also split in chicken DNA,[46,47] and the rabbit β-globin gene is also interrupted.[48] In addition, the hexon messenger RNA of adenovirus comes from multiple regions of the adenovirus genome.[49] Recent evidence indicates that adenovirus mRNA,[49] globin mRNA,[50,51] and immunoglobulin light chain mRNA[52] are processed from large nuclear RNA. The molecular mechanism for the messenger RNA processing events is believed to be by a splicing mechanism.[53] Thus, it is reasonable to assume that any eukaryotic cells are able to delete certain portions of RNA by splicing. If we introduce foreign RNA such as virus RNA, it is possible to get the input RNAs being spliced. Model III-1 in Figure 12 generates DI particle RNA by splicing to generate DI-1, DI-2, or DI-3 RNAs. The viral-specific RNA may have certain secondary structures which includes a loop. If the host cell recognizes the specific site on the RNA molcule and removes the loop by excision and ligation, a DI particle RNA can be generated. The

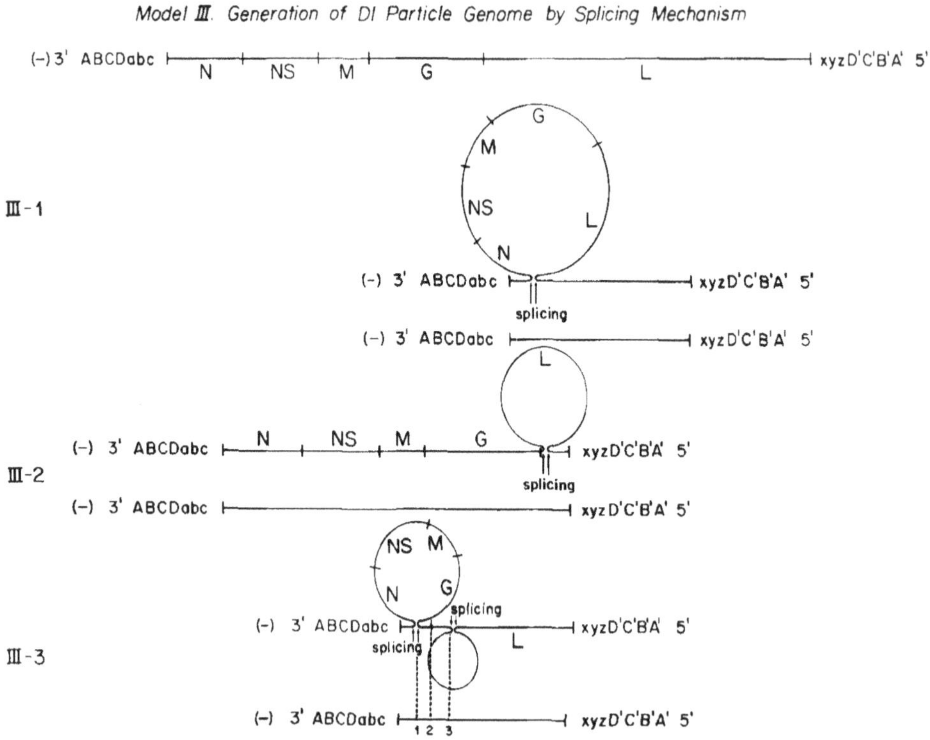

FIGURE 12. Model for generation of DI particle genome by splicing mechanism.

enzymes which are responsible for cutting the RNA may also have a ligase type of activity to join the ends. It is possible that the one molecule may be able to carry two types of enzymatic functions. Such an example exists in other virus systems, for example, the reverse transcriptase activity and the ribonuclease H activity in the RNA tumor virus cannot be separated.[54] Model III-2 in Figure 12 generates DI-LT particle RNA representing the 3′ half of the infectious B-virion genome and Model III-3 generates DI particle containing both a part of the 3′ and a part of the 5′ half of the genome by double splicing mechanism. This splicing model is based on the single template molecule and does not require RNA replication to generate the DI-particle RNAs. It is possible that many more defective RNAs may be generated by this mechanism and the process could be random events, but only those RNAs which contain the polymerase recognition site(s) will be amplified to give rise to a detectable level of DI particles in the later passages. The nucleotide sequences between 1 and 2, and 2 and 3 in Model III-3 are the same region as described in the Model II-3. The host function for this model could be the splicing enzymes. The more complex forms of DI-particle RNAs may have been evolved from the multiple event than a single event. A complete mapping of DI-particle RNA and complete assignments of all the oligonucleotide spots to each cistron will help to explain the mechanism of DI-particle RNA generation. In addition, in vitro attempts to generate DI particles may help us to find the actual mechanism of DI-particle generation.

REFERENCES

1. Huang, A. S., Defective interfering viruses, *Ann. Rev. Microbiol.*, 27, 101, 1973.
2. **Huang, A. S. and Baltimore, D.**, Defective interfering animal viruses, in *Comprehensive Virology*, Frenkel-Conrat, H. and Wagner, R. R., Eds., Vol. 10, Plenum Press, New York, 1977, 73.
3. **Holland, J. J., Villarreal, L. P., and Breindl, M.**, Factors involved in the generation and replication of rhabdovirus defective T particles, *J. Virol.*, 17, 805, 1976.
4. **Choppin, P. W.**, Replication of influenza virus in a continuous cell line: high yield of infected virus from cells inoculated at high multiplicity, *Virology*, 39, 130, 1969.
5. **Choppin, P. W. and Pons, M. W.**, The RNAs of infective incomplete influenza virions grown in MDBK and HeLa cells, *Virology*, 42, 603, 1970.
6. **Igarshi, A. and Stollar, V.**, Failure of defective interfering particles of Sindbis virus produced in BHK or chicken cells to effect viral replication in *Aedes albopitus* cells, *J. Virol.*, 19, 398, 1976.
7. **Kingsbury, D. W. and Portner, A.**, On the genesis of incomplete Sendai virions, *Virology*, 42, 878, 1970.
8. **Roman, J. M. and Simon, E. H.**, Defective interfering particles in monolayer-propagated Newcastle Disease virus, *Virology*, 69, 298, 1976.
9. **Cooper, P. D. and Bellett, A. J. D.**, A transmissible interfering component of vesicular stomatitis virus preparations, *J. Gen. Microbiol.*, 21, 485, 1959.
10. **Reczko, E.**, Electronmikroskopische Untersuchunger am Virus der stomatitis vesicularis, *Arch. Gesamte Virusforsch.*, 10, 588, 1960.
11. **Hackett, A. J.**, A possible morphologic basis for the autointerference phenomenon in vesicular stomatitis virus, *Virology*, 24, 51, 1964.
12. **Huang, A. S. and Wagner, R. R.**, Defective T particles of vesicular stomatitis virus. II. Biologic role in homologous interference, *Virology*, 31, 173, 1966.
13. **Reichmann, M. E., Pringle, C. R., and Follett, E. A. C.**, Defective particles in BHK cells infected with temperature-sensitive mutants of VSV, *J. Virol.*, 8, 154, 1971.
14. **Schincariol, A. L. and Howatson, A. F.**, Replication of Vesicular Stomatitis Virus. I. Viral specific RNA and nucleoprotein in infected L cells, *Virology*, 42, 732, 1970.
15. **Petric, M. and Prevec, L.**, VSV — a new interfering particle, intracellular structures and specific RNA, *Virology*, 41, 615, 1970.
16. **Parrault, J. and Holland, J. J.**, Variability of vesicular stomatitis virus autointerference with different host cells and virus serotypes, *Virology*, 50, 148, 1972.
17. **Roy, P., Hefti, E., and Bishop, D. H. L.**, Complementary RNA species isolated from vesicular stomatitis virus (HR strain) defective virions, *J. Virol.*, 11, 915, 1973.
18. **Kang, C. Y., Glimp, T., Clewley, J. P., and Bishop, D. H. L.**, Studies on the generation of vesicular stomatitis virus (Indiana serotype) defective interfering particles, *Virology*, 84, 142, 1978.
19. **Lunger, P. D. and Clark, H F.**, Host effect on vesicular stomatitis virus morpho genesis and "T" particle formation in reptilian, avian, and mammalian cell lines, *In Vitro*, 11, 239, 1977.
20. **Huang, A. S. and Baltimore, D.**, Defective viral particles and viral disease processes, *Nature New Biol.*, 226, 325, 1970.
21. **Potter, K. N. and Stewart, R. B.**, Comparison of vesicular stomatitis virus defective interfering particle synthesis in chick embryo and L cells, *Can. J. Microbiol.*, 22, 1458, 1976.
22. **Kang, C. Y., Allen, R., and Glimp, T.**, Host cell function dependent induction of defective interfering particles of vesicular stomatitis virus, in *Negative Strand Viruses and the Host Cell*, Barry, R. D. and Mahy, B. W. J., Eds., Academic Press, 1978, 501.
23. **Darnell, M. B. and Koprowski, H.**, Genetically determined resistance to infection with group B arboviruses. II. Increased production of interfering particles in cell cultures from resistant mice, *J. Infect. Dis.*, 129, 248, 1974.
24. **Kang, C. Y. and Allen, R.**, Host function-dependent induction of defective interfering particles of vesicular stomatitis virus, *J. Virol.*, 25, 202, 1978.
25. **Reich, E., Franklin, R. M., Shatkin, A. J., and Tatum, E. L.**, Action of Actinomycin D on animal cells and viruses, *Proc. Natl. Acad. Sci. U.S.A.*, 48, 1238, 1962.
26. **Kang, C. Y. and Prevec, L.**, Proteins of vesicular stomatitis virus. IV. Intracellular synthesis and extracellular appearance of virus-specific proteins, *Virology*, 46, 678, 1971.
27. **Huang, A. S. and Wagner, R. R.**, Comparative sedimentation coefficients of RNA extracted from plaque-forming and defective particles of vesicular stomatitis virus, *J. Mol. Biol.*, 22, 381, 1966.
28. **Huang, A. S.**, Viral pathogenesis and molecular biology, *Bacteriol. Rev.*, 41, 811, 1977.
29. **Leamnson, R. N. and Reichmann, M. E.**, The RNA of defective vesicular stomatitis virus particles in relation to viral cistrons, *J. Mol. Biol.*, 85, 551, 1974.

30. **Schnitzlein, W. M. and Reichmann, M. E.**, The size and cistronic origin of defective vesicular stomatitis virus particle RNAs in relation to homotypic and heterotypic interference, *J. Mol. Biol.*, 101, 307, 1976.
31. **Stamminger, G. and Lazzarini, R. A.**, Analysis of the RNA of defective VSV particles., *Cell*, 3, 85, 1974.
32. **Colonno, R. J., Lazzarini, R. A., Keene, J. D., and Banerjee, A. K.**, In vitro synthesis of messenger RNA by a defective interfering particle of vesicular stomatitis virus, *Proc. Natl. Acad. Sci. U.S.A.*, 74, 1884, 1977.
33. **Lazzarini, R. A., Weber, G. H., Johnson, L. D., and Stamminger, G. M.**, Covalently linked message and anti-message (genomic) RNA from a defective vesicular stomatitis virus particle, *J. Mol. Biol.*, 97, 289, 1975.
34. **Perrault, J.**, Cross-linked double-stranded RNA from a defective vesicular stomatitis virus particle, *Virology*, 70, 360, 1976.
35. **Chow, J. M., Schnitzlein, W. M., and Reichmann, M. E.**, Expression of genetic information contained in the RNA of a defective interfering particle of vesicular stomatitis virus *Virology*, 77, 579, 1977.
36. **Johnson, L. D. and Lazzarini, R. A.**, Replication of viral RNA by a defective interfering vesicular stomatitis virus particle in the absence of helper virus, *Proc. Natl. Acad. Sci. U.S.A.*, 74, 4387, 1977.
37. **Ball, L. A. and White, C. N.**, Order of transcription of genes of vesicular stomatitis virus, *Proc. Natl. Acad. Sci. U.S.A.*, 73, 442, 1976.
38. **Banerjee, A. K. and Rhodes, D. P.**, 3′ terminal sequence of vesicular stomatitis virus genome RNA, *Biochem. Biophys. Res. Commun.*, 68, 1387, 1976.
39. **Colonno, R. J. and Banerjee, A. K.**, A unique RNA species involved in initiation of vesicular stomatitis virus RNA transcription in vitro, *Cell*, 8, 197, 1976.
40. **Hefti, E. and Bishop, D. H. L.**, The 5′ nucleotide sequence of vesicular stomatitis viral RNA, *J. Virol.*, 15, 90, 1975.
41. **Perrault, J. and Leavitt, R. W.**, Inverted complementary terminal sequences in single-stranded RNAs and snap-back RNAs from vesicular stomatitis defective interfering particles, *J. Gen. Virol.*, 38, 35, 1977.
42. **Leppert, M., Kort, L., and Kolakofsky, D.**, Further characterization of Sendai virus DI-RNAs: A model for their generation, *Cell*, 12, 539, 1977.
43. **Huang, A. S. and Manders, E. K.**, Ribonucleic acid synthesis of vesicular stomatitis virus. IV. Transcription by standard virus in the presence of defective interfering particles, *J. Virol.*, 9, 909, 1972.
44. **Emerson, S. U. and Wagner, R. R.**, Dissociation and reconstitution of the transcriptase and template activities of vesicular stomatitis B and T virions, *J. Virol.*, 10, 297, 1972.
45. **Brack, C. and Tonegawa, S.**, Variable and constant parts of the immunoglobulin light chain gene of a mouse myeloma cell are 1250 nontranslated bases apart, *Proc. Natl. Acad. Sci. U.S.A.*, 74, 5662, 1977.
46. **Breathnach, R., Mandel, J. L., and Chambon, P.**, Ovalbumin gene split in chicken DNA, *Nature (London)*, 270, 314, 1977.
47. **Doel, M. T., Houghton, M., Cook, E. A., and Carey, N. H.**, The presence of ovalbumin in RNA coding sequences in multiple restriction fragments of chicken DNA, *Nucleic Acid Res.*, 4, 3701, 1977.
48. **Jeffreys, A. J. and Flavell, R. A.**, The rabbit β-globin gene contains a large insert in the coding sequence, *Cell*, 12, 1097, 1977.
49. **Berget, S. M., Moore, C., and Sharp, P. A.**, Spliced segments at the 5′ terminus of adenovirus 2 late mRNA, *Proc. Natl. Acad. Sci. U.S.A.*, 74, 3171, 1977.
50. **Curtis, P. J. and Weismann, C.**, Purification of globin messenger RNA from dimethylsulfoxide-induced Friend cells and detection of putative globin messenger RNA precursor, *J. Mol. Biol.*, 106, 1061, 1976.
51. **Bastos, R. N. and Aviv, H.**, Globin RNA precursor molecules: biosynthesis and processing in erythroid cells, *Cell*, 11, 641, 1977.
52. **Gilmore-Hebert, M. and Wall, R.**, Immunoglobulin light chain in RNA is processed from large nuclear RNA, *Proc. Natl. Acad. Sci. U.S.A.*, 75, 342, 1978.
53. **Gilbert, W.**, Why genes in pieces?, *Nature (London)*, 271, 501, 1978.
54. **Moelling, K.**, Further characterization of Friend murine leukemia virus reverse transcriptase-RNase H complex, *J. Virol.*, 18, 418, 1976.
55. **Clewley, J. P. and Bishop, D. H. L.**, Evolution of rhabdovirus genomes, in *Negative Strand Viruses and the Host Cell*, Barry, R. D. and Mahy, B. W. J., Eds., Academic Press, London, 1978, 271.
56. **Keene, J. D., Rosenberg, M., and Lazzarini, R. A.**, Characterization of the 3′ terminus of RNA from vesicular stomatitis virus and from its defective interfering particles, *Proc. Natl. Acad. Sci. U.S.A.*, 74, 1353, 1977.

57. **Keene, J. D., Schubert, M., and Lazzarini, R. A.,** Nucleotide sequence homology at the 3′ termini of RNA from vesicular stomatitis virus and its defective interfering particles, *Proc. Natl. Acad. Sci. U.S.A.*, 1978, in press.
58. **Pfau, C. J.,** personal communication.
59. **Lambright, P. and Kang, C. Y.,** unpublished data.

INDEX

A

B

C

E

F

G

H

I

J

K

L

M

N

O

P

R

S

T

U

V

W

X

For Product Safety Concerns and Information please contact our EU representative GPSR@taylorandfrancis.com Taylor & Francis Verlag GmbH, Kaufingerstraße 24, 80331 München, Germany

Batch number: 10731074

Printed by Printforce, the Netherlands